Studies of Brain Function Vol. 19

Leon L. Voronin

Synaptic Modifications and Memory

An Electrophysiological Analysis

With 64 Figures

Springer-Verlag
Berlin Heidelberg New York
London Paris Tokyo
Hong Kong Barcelona
Budapest

Professor Dr. LEON L. VORONIN
Brain Research Institute
Russian Academy of Medical Sciences
Per Obukha 5
Moscow 103064
Russia

ISBN-13: 978-3-642-47619-8 e-ISBN-13: 978-3-642-47615-0
DOI: 10.1007/978-3-642-47615-0

Library of Congress Cataloging-in-Publication Data. Voronin, L. L. Synaptic modifications and memory: an electrophysiological analysis / Leon L. Voronin. p. cm. – (Studies of brain function; vol. 19) Includes bibliographical references and index. ISBN-13: 978-3-642-47619-8
1. Neuroplasticity. 2. Hippocampus (Brain) 3. Memory. I. Title. II. Series: Studies of brain function; v. 19. [DNLM: 1. Electrophysiology. 2. Memory – physiology. 3. Hippocampus – physiology. 4. Receptors, Synaptic – physiology. W1 ST937KF v. 19 1993 / WL 102 V9535s 1993] QP364.V67 1993 153.1'2–dc20 DNLM/DLC for Library of Congress 93-6851

Softcover reprint of the hardcover 1st edition 1993

Media conversion: M. Masson-Scheurer, Kirkel
31/3145-5 4 3 2 1 0 – Printed on acid-free paper

Acknowledgements

Several scientific meetings and several persons are related to the initiation of this monograph. At the end of 1985, following a talk at the Max-Planck-Institute for Biological Cybernetics in Tübingen, Professor Dr. Valentino Braitenberg proposed that I write a book for his series, *Studies of Brain Function*. I accepted this kind proposal, having in mind some initial experience in summarizing earlier results of our group both in a monograph edited by Georgian Academician Tengiz Oniani [1015] and in reviews [1012, 1016, 1017] sponsored by Academician Platon Kostyuk, the Editor-in-Chief of *Neurophysiology* and *Neuroscience*. The cooperation with Dr. Ulrich Kuhnt from Göttingen which started in 1986 following "perestroika" events was of incomparable value both for completing a certain stage in our studies of LTP mechanisms and writing the monograph. Very stimulating were invitations of Drs. Helmut Haas and György Buzsaki to participate in a hippocampal meeting in 1987 [1022], of Hinrich Rahmann to present a review of LTP studies at a meeting in Mainz in 1988 [1023], of William H. Gispen and Steven Rose to participate in meetings on learning and memory in 1989 and 1990, and of Jan Delacour to write a review chapter on hippocampal LTP for his book on memory mechanisms in 1991. These and several other meetings and discussions with many scientists from East and West were very useful and encouraging. Last, but not least, were two invitations of Valentino Braitenberg to Tübingen in 1991 where I essentially completed the writing of the manuscript with his generous support and advice. I initiated the writing of the monograph with the hope that the cooperation with Western scientists and a more precise knowledge of Russian science would serve not only purely scientific purposes but would also further a closer mutual understanding between different nations, thus contributing to peace on this planet.

I would like to thank all those who directly or indirectly contributed to this monograph. It is difficult to list all the names and I confine myself to expressing gratitude to my colleagues who participated in experiments, data processing, discussions and writing of related original publications. Thanks are due to the following for their generosity in giving permission to use the materials on which an essential part of this book is based: Drs. A. Baskys (Chaps. 1, 8, 11, 12), I. E. Kudryashov (Chaps. 1, 3–6, 8, 11), S. L. Buldakova (Chap. 1) and I. N. Sharonova

(Chaps. 1, 3), Professor Dr. V. G. Skrebitsky (Chaps. 1, 3), Drs. L. N. Gusev, V. S. Kesarev, D. R. Shidarev, V. F. Uvarov (Chap. 3), R. G. Kozhedub and S. N. Kozhechkin (Chap. 6), V. I. Derevyagin (Chaps. 8–11), D. B. Logunov (Chaps. 8, 10), G. Hess (Chap. 11, 13, 14), A. G. Gusev (Chaps. 12, 13–15) and V. Roschin (Chap. 13). I am especially obliged to Dr. Ulrich Kuhnt for very stimulating cooperation, resulting in our joint writing of the last experimental Chapters 13–15. I am indebted to Professors O. S. Adrianov, V. G. Skrebitsky and the late Professor M. Ya. Rabinovich for encouragement and advice during the experimental work, as well as in the preparation of the monograph. I am deeply grateful to Valentino Braitenberg and Ulrich Kuhnt who took great trouble in editing the monograph.

Several people gave generous help and advice during the preparation of the manuscript with word processors. I am grateful to Dieter Michael, Ludwig Ehrenreich, Anna Tlustochowski, Victor Istomin, Sergei Molotkov and Nadezhda Nakhutina. I am especially grateful to Margarete Ghasroladashti, Hubert Preissl, Thomas Schillen and Igor Chizhmakov for their generous help at the most difficult final stage of the manuscript preparation. Finally, I cannot find the words to describe my gratitude to my wife, Eleonora Voronina, for her understanding of my interest in science and for entering an essential part of the following text into the computer.

Leon L. Voronin

Contents

Introduction

The "neurosciences" is now a common term to describe several branches of the biology related to studies of the nervous system. In fact, this term also characterizes a multidisciplinary approach to studies of probably the most complicated system of the most complicated living organisms. Investigation of the central nervous system (CNS) is the most quickly developing branch of the neurosciences.

According to Konorski's definition, the CNS "is endowed with two principal properties which are referred to as *reactivity* and *plasticity*. *Reactivity* of the system is its capacity to be activated by stimulation of receptive organs; *plasticity* is the capacity to change its reactive properties as the result of successive activations" ([520], p. 7). According to Kostyuk [535], plastic reorganizations include "any changes in the efficacy or direction of neuronal connectivity which exceed in their duration usual synaptic and spike events" ([535], p. 25).

Neuronal plasticity underlies adaptive modifications of various complexities both in normal CNS, e.g. its development, formation of memory traces and conditioned responses (CRs), as well as in brain pathology, e.g. reorganization and compensation of damaged nervous functions. Clarification of plasticity mechanisms is of great cognitive importance and might be of practical value in the treatment of mental and especially memory disorders. Understanding of these mechanisms should be essential for the development of "artificial intelligence"; the advent of "neurocomputers" is a sign of the influence of the neurosciences on modern technology. The related studies seem to be important from a general, philosophical point of view furthering our understanding of the relationship between physico-chemical processes and the essential components of brain psychic activity, which are learning and memory.

Mechanisms of learning and memory have been of traditional interest in Russian physiology [797]. The aim of my studies and those studies of my colleagues reviewed in this monograph is the attempt to describe mechanisms of memory related CNS plasticity in terms of present-day neurophysiology. The history of this science shows that a proper choice of the experimental paradigm is one of the important conditions for successful progress. As a basic experimental model, we used the phenomenon of hippocampal long-term potentiation (LTP), which we encountered during our joint experiments with Igor Kudryashov in 1973, without knowing the original results of Lømo [622] and Bliss and Lømo [144], which were published as abstracts. Initial detailed descriptions of LTP [145, 143, 1042] and of a related phenomenon, "chronic potentiation" [161, 1000], showed that the problem of LTP mechanisms overstepped the limits of purely neurophysiological problems. At the present time, one can safely say that LTP is attaining the same importance in resolving neurophysiological, biochemical and morphological mechanisms of memory as the action potential of the squid giant axon in understanding the membrane mechanisms of spike generation, or as Pavlovian conditioning in the initial formulations of general laws of animal behaviour. In addition to hippocampal LTP, some more simple phenomena of CNS plasticity were included in our study, especially the phenomenon of low-frequency depression (LFD) of electrical responses. This phenomenon also seems to be of importance in understanding the cellular basis of behavioural modifications. In fact, the LFD is characterized by properties similar to those of behavioural habituation and extinction of CRs [475, 478, 918, 968].

The neurosciences are becoming precise quantitative sciences. This process is aided by the revolutionary development of computer technology and specifically by the appearance of personal computers easily available to any neuroscientists. Therefore, a significant part of this monograph is dedicated to reviews of publications dealing with applications of quantitative methods to studies of the nervous system, and also modifications of quantitative methods for studies of mechanisms of CNS plasticity.

Neurophysiological, biochemical and morphological mechanisms and properties of LTP, habituation and other memory related phenomena of neuronal plasticity are under study in many research groups in various countries. The basic aim of this monograph is the attempt to describe the contribution of those studies in which the author participated directly. The related projects have been carried out since 1973 initially in the Department of Neuronal Mechanisms of Plasticity of the Brain Research Institute, Russian Academy of Medical Sciences and since 1986 also in the laboratory of Dr. Ulrich Kuhnt in the Neurobiology Department of Max-Planck-Institute for Biophysical Chemistry in Goettingen (FRG). Many other research groups have studied related problems and had obtained important results since the pioneering publications of Bliss and Lømo [145] and Bragin and Vinogradova [161]. Extensive reviews of LTP studies are beyond the scope of this monograph, but the interested reader can find references to the principal results of other authors in the appropriate sections and can consult numerous reviews. The reviews [22, 140, 141, 228, 302, 380, 949, 963, 980, 1001, 1016], especially numerous in the past 5 years [23, 24, 93, 146, 199, 236, 395, 396, 503, 562, 585, 642, 643, 917, 964, 1023], are cited in References.

Recording of electrical activity of single neurones by means of intracellular electrodes was the basic technique used. The microelectrode technique [299, 377, 475, 533, 807] is one of the basic methods of the neurosciences up to now, having been employed in the Brain Institute for investigations of intracellular activity of unanaesthetized mammalian cortex since the 1960s [1036]. In addition to intracellular responses, both extracellular single neuronal discharges and field potentials (FPs) were recorded and analyzed in our studies. While writing the text, I have assumed that the basic neurophysiological terminology is familiar to the reader [172, 188, 299, 300, 477, 536, 614]. Special sections are devoted to detailed descriptions of relatively less known fields of neurophysiology such as the analysis of so-called unitary postsynaptic potentials (UPSPs), the quantum hypothesis of synaptic transmission and the model of transmitter depletion. Elementary knowledge of mathematics and statistics is presumed for understanding some parts of Chapters 7 and 9. However, the reader who is not interested in using the related methods in his own work may wish to skip all mathematics and confine himself primarily to the Introductions, Conclusions and Summaries of these chapters, without essential loss in understanding our basic results.

To induce hippocampal LTP, a high-frequency (tetanic) stimulation of a presynaptic pathway is nessesary. Therefore, superficially the LTP looks like a post-tetanic potentiation (PTP), a well known phenomenon of a "relatively prolonged increase in response that occurs after a junctional region has been subjected to repetitive orthodromic activation" ([298], p. 194). PTP was initially discovered at sympathetic ganglia [593] and the spinal cord [617] by recording aggregate electrical responses. The PTP was then found during intracellular recordings of end-plate potentials (EPPs) at neuromuscular junctions and of postsynaptic potentials (PSPs) at various structures of many species (see reviews [102, 441, 1008, 475]). The first detailed studies of post-tetanic modifications of neocortical responses were carried out in Georgia by Drs. Roitbak [836] and Okujava [775], followers of the famous Georgian physiologist, Academician Ivan Beritashvili. The basic mechanism of the short-term PTP is an increase in the number of transmitter portions (quanta) released by the tetanized presynaptic ending per presynaptic spike. This conclusion was based on the data of so-called "quantal analysis" of neuromuscular transmission [260–262, 485, 486] and supported by the absence of changes in subsynaptic receptor sensitivity [451, 563]. PTP has been used by many theorists in an attempt to provide a neurophysiological basis for information storage, memory and conditioning [117, 118, 298, 301, 333, 534]. However, the duration of PTP is at most only several minutes, as had been shown in many studies on various junctions under normal conditions, including earlier studies on neocortical synapses [21, 775, 997, 1007–1009, 1041]. Therefore, only short-term memory might be explained on the basis of such PTP-like "processes related to the transmitter release from the presynaptic ending" [117].

In addition, a prolonged facilitation (scores of minutes and hours) was found in some particular cases. These included changes in monosynaptic spinal cord responses after tetanization of degenerating afferent fibres [304, 534] and after repeated or long-lasting (minutes or scores of minutes) afferent tetanic stimulations [121, 597, 796, 921, 923]. Similar long-lasting stimulations had been shown to produce comparable prolonged facilitation in other preparations: crayfish neuromuscular junction [50, 54] and some central synapses of the marine mollusc *Aplysia* [793]. There are also several special cases of prolonged facilitation (up to scores of minutes) of neocortical (mainly polysynaptic) responses: single unit discharges after combinations of direct cortical stimuli [147] or after local cortical stimulation by steady currents [126], field potentials (FPs) after stimulation of pain afferents [556], direct cortical responses after surface tetanic stimulation under deep Nembutal anaesthesia [837] and pyramidal tract responses (PTRs) after antidromic stimulation with blocking of cortical synaptic transmission [127]. Discovery of hippocampal LTP [143, 145] came to the attention of many scientists due to at least two reasons: its unusual duration (days and weeks) and its good reproducibility even after very short tetanic stimulation (seconds or less) with parameters comparable to the natural activity of hippocampal neurones. The interest in this phenomenon was strengthened by the presumably important role of the hippocampus in memory and learning [195, 234, 279, 436, 456, 457, 668, 776, 780, 774, 799, 853, 868, 926, 927, 945, 969, 1000, 1142].

1 Hippocampal Long-Term Potentiation and Its Characteristics

1.1 Brief Overview of the Structure of the Hippocampus

I shall consider major morphological features of the hippocampus before going into further considerations. Only a basic scheme of the hippocampal morphology will be given. More detailed descriptions can be found in the original writings of Cajal [202], Filimonov [337] and Lorente de No [625], as well as in more recent publications [19, 20, 164, 385, 456, 457, 623, 781, 943, 948, 1000]. Most of these publications contain information on the electrophysiological responses recorded in the hippocampus.

The major cytoarchitectonic areas of the hippocampal formation are the following: *regio superior* or CA1, *regio inferior* or CA2–CA3, which will be called CA3 for short, taking into account that the transitional region CA2 is usually not clearly distinguishable, the dentate gyrus (DG) and hilus of the DG or CA4. Area CA1 contains relatively small (soma diameters of about 20 μm) pyramidal neurones. Areas CA3 and CA4 consist of the largest cells (about 30 μm). The DG contains small (less than 18 μm) granular neurones. The remarkable feature of the hippocampus is its laminated structure [29, 1000] (but see [19]). The orientation of the lamella is approximately transverse to the longitudinal axis of the hippocampus, and each lamella contains a trisynaptic chain consisting of the following links: (1) the entorhinal cortex (EC)–DG connected by axons of the perforant pathway; (2) DG–CA3 connected by the mossy fibres; (3) CA3–CA1 connected by the Schaffer collaterals. Stimulation of the DG (Fig.1.1, S1) with recording from area CA3 (Fig.1.1, R) and also activation of the CA3–CA1 pathway (see below, Sect. 1.5) was used in our studies. In addition to the input from the EC, the hippocampus has another powerful input (mainly to area CA3 and the DG) from the medial nucleus of the septum: see stimulating electrode S2 (Fig.1.1) in the septo-fimbrial area (SFA). There are also commissural fibres from pyramidal neurones connecting hippocampal formations of the two hemispheres, association fibres from CA3 to the DG, and also, apparently, association connections within each cytoarchitectonic area especially powerful in area CA3 [718]. The basic direction of outgoing axons is towards the EC. In addition, an essential target of axons of CA3 neurones is the lateral nucleus of the septum.

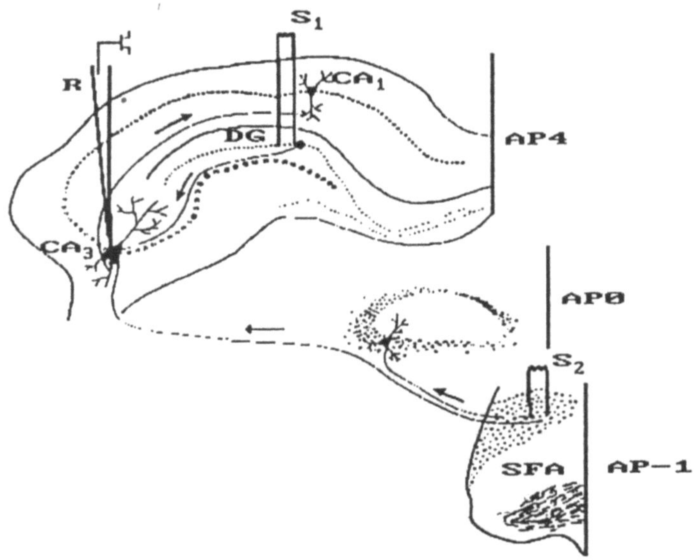

Fig. 1.1. Schematic representation of the major hippocampal connections investigated. Three levels of hippocampal sections are shown from the stereotaxic atlas of the rabbit brain [336]. The sections correspond approximately to the most usual recording (AP4, AP0) and stimulation (AP-1) sites. *R* recording microelectrode; S_1 and S_2 stimulating electrodes. See text for other notations

1.2 Electrical Responses of the Hippocampus

Occurrence of extremely high-amplitude (up to 10–20 mV with microelectrode recording) extracellular field potentials (FPs) is a particular feature of hippocampal electrophysiology, especially in experiments with intracerebral stimulation. The unusually large FP amplitude is explained by a compact package of the basic neuronal elements of the hippocampal structures and regular lamellar organization of the afferent connections. Three major primary components of the hippocampal FPs can be distinguished. The biphasic or negative component 1 with the shortest latency is recorded occasionally (triangles in Fig. 1.2A1, B1). It reflects an activation of presynaptic fibres, antidromic activation of neurones located within the recording area (Fig. 1.3B2) or both, depending on the recording and stimulation conditions. A slow wave which we shall term component 2 (Fig. 1.2B, cross) reflects postsynaptic currents. It is positive at the cell body level (Fig. 1.5) and reverses slightly below (Fig. 1.3A1). Note that in all our illustrations upward deflections indicate an increase in positive potential of the recording electrode relative to the reference electrode. The maximal amplitude of the negative wave is usually observed at the level which corresponds to the location of activated presynaptic terminals. Component 2 is frequently termed "extracellular excitatory

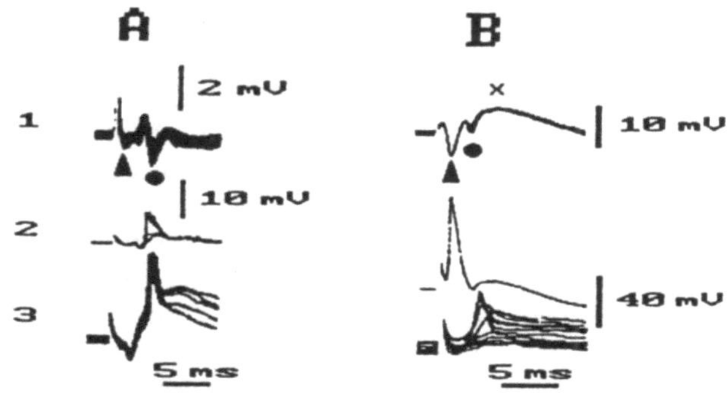

Fig. 1.2A,B. Examples of FPs induced by the septo-fimbrial stimulation in area CA3 to demonstrate FP components (*1*) and their relations with single neurone responses (*2* and *3*) in two experiments (**A** and **B**). *Triangles, dots* and *crosses* mark first, second and third FP components, respectively. The second component in **A1** corresponds to orthodromic spikes of two neurones (**A2** and **A3**) recorded in the same electrode track. The first and second components in **B1** correspond, respectively, to antidromic (**B2**) and orthodromic (**B3**) neuronal spikes recorded intracellularly later in the same experiment

postsynaptic potential" (extracellular EPSP), "population EPSP," "synaptic potential," or simply "EPSP". However, we shall not use these terms, firstly, in order to avoid confusion, since the term "EPSP" was introduced for responses recorded intracellularly, and secondly, because other processes, including inhibitory synaptic currents and asynchronous spike discharges, can participate in the formation of this component. In principle, FP amplitude can greatly depend on factors other than the postsynaptic current, for example on the resistance of the external medium [42]. The third and the most typical component of the hippocampal FP is a fast wave which is negative near the level of the pyramidal cell bodies. This component (Figs. 1.2A,B, 1.3A, dots) is usually called "population spike". It was shown [28] that the orthodromic population spike corresponds in time to action potentials of a large number of neurones of a corresponding hippocampal region and reflects synchronous discharges of these neurones. The nature of component 3 has the clearest interpretation, and it was characteristics (amplitude and latency) of this component that we measured in most of our experiments with FP recordings. When no pronounced orthodromic population spike was recorded (e.g. with the use of near-threshold currents or with recording from the dendritic level), the amplitude of component 2 (or total amplitude of components 2 and 3) was measured from zero line to the maximum. The appearance of FPs of sufficiently high amplitude and of typical wave form is a good electrophysiological control to locate the recording electrode in the hippocampal structures. The optimum position of a stimulating electrode can also be chosen according to the FP characteristics.

Extracellular spike discharges of single neurones were recorded from the hippocampus in many studies. The appearance of a large number of extracellular discharges creating a typical noise can serve as an additional criterion to detect the pyramidal layer during the electrode insertion toward the dorsal hippocampus through the neocortex and subcortical white matter.

The first detailed studies of intracellular potentials of hippocampal neurones were reported in the early 1960s [26, 480, 922]. Well-marked inhibitory postsynaptic potentials (IPSPs) and the infrequent occurrence of distinct primary EPSPs [26, 43, 300, 480] were found to be characteristic features of intracellularly recorded responses of hippocampal neurones to intracerebral stimulation in vivo (see reviews [193, 784]). For example, according to Andersen and Lømo [25], only four primary EPSPs were found in several hundred recorded neurones. The neurones with the primary EPSPs were situated mainly outside the pyramidal layer. Intracellular recording from DG neurones in vivo was not as successful [27] as from the pyramidal neurones. Stable intracellular recordings from small DG neurones were obtained only with the development of the in vitro slice technique [252, 253, 287].

1.3 Initial Experiments on Long-Term Potentiation

In the early 1970s, Bliss and co-workers [142, 144] presented the first preliminary descriptions of LTP with recording of FPs evoked in the DG after stimulation of the input from the EC. Subsequently, this phenomenon was described in more detail for both anaesthetized [145] and unanaesthetized [143] rabbits. The LTP was defined as a prolonged (from 30 min to 3 days) change in three parameters: the amplitude of the slow wave (component 2), the amplitude of the population spike (component 3) and the latency of the latter. LTP developed after one or two tetanizations of 10–20 s duration with 10–100 Hz frequencies. One, two or all three of these parameters could be potentiated. Independently of the initial communications of Bliss and Lømo [144, 145], Bragin and Vinogradova [161] described a long-lasting enhancement (up to several days) of single neurone discharges in area CA3 after repeated ("chronic") stimulation of the DG. One can assume that this "chronic" potentiation has mechanisms similar to those of LTP evoked by one or a small number of tetanizations.

Shortly after the beginning of our experiments on the magnocellular CA3 hippocampal region, two attempts to produce potentiation in another pyramidal cell hippocampal region (CA1) were reported in preliminary forms as abstracts. In one of these communications [30], LTP of FPs was briefly described after single tetanization of the Schaffer collaterals. In the other work [504], no "chronic" potentiation of CA1 neuronal discharges was found after repeated tetanizations of the same pathway.

1.4 Working Hypothesis and Early Results

As was mentioned in the Introduction, the phenomenon of PTP attracted the attention of many scientists interested in neuronal mechanisms of memory and conditioning. Already the first detailed descriptions of the hippocampal LTP [143, 145] contained the suggestion that this phenomenon might be closely related to memory processes. More strongly, this hypothesis might be reformulated as a suggestion that neuronal mechanisms which underlie LTP are common to those underlying memory and conditioning [1042]. The common mechanisms might be based on hypothetical "modifiable" synapses, a concept used in many theoretical models (e.g. [168, 179, 294, 386, 387, 516, 668, 791, 882, 903, 1010, 1019, 1021]). In this case, LTP studies are of great interest not only as a problem related to general neurophysiology, but also for understanding basic mechanisms of memory, and specifically of behavioural conditioning [797].

In order to study the related cellular mechanisms, it was of importance to demonstrate reproducible LTP in the hippocampal pyramidal layer because previous intracellular recordings from the small DG cells were very unstable [27]. The "chronic" potentiation, as described for hippocampal pyramidal neurones [161], was induced after repeated stimulations over a comparatively long period and was demonstrated with successive recordings from several neurones. This procedure seemed not to be practical for a precise analysis with intracellular recordings. In this connection, our first step was an attempt to produce LTP after a single tetanization with recordings of FPs from the CA3 region which contains large pyramidal neurones suitable for intracellular recording from awake animals. Before proceeding to a more detailed analysis of LTP by means of known neurophysiological methods, it seems reasonable to check some corollaries from the formulated hypothesis. In fact, if LTP and CR share common neuronal mechanisms, the former should possess some characteristic properties of the latter, namely, extinction, self-restoration after extinction, specificity for activated inputs, and involvement of "reinforcing" intracerebral systems [241, 360, 367, 386, 387, 450, 688, 777, 937, 1107].

1.5 Methods

Experiments with recording of hippocampal responses described in this and in the following chapters were performed mainly on unanaesthetized rabbits. The animal was placed in a hammock, the head being fixed to a metal frame by means of special blocks previously attached to the skull. A preliminary operation (1–3 days before the first experiment) was carried out under sodium pentobarbitone (Nembutal) anaesthesia (30–40 mg/kg). Stimulating electrodes (Fig. 1.1) were inserted into the SFA and one of the following structures: the DG, the Schaffer collaterals, or a symmetrical point contralateral to the recording site. The electrodes were placed according to stereotaxic coordinates [336] and adjusted according to electrophysiological criteria (amplitude and wave form of the FPs). Testing stimuli (up to 1 mA at 0.1–0.5 ms stimulus duration) were delivered usually with intervals of 2 to 10 s. Every testing series consisted of 10 to 200 stimuli (usually 10 or 20). Intervals between series were from 5 to 30 min. In most cases, the current strength was continually controlled by the registration of a voltage drop across a resistance in the stimulation circuit. Those experiments in which the strength of the testing current changed after the tetanic stimulation were excluded from the analysis. In many experiments, separate adjacent electrodes (usually bipolar) were used for testing and tetanization to exclude a possible influence of local changes under the testing electrode which might result from the tetanization.

Recording microelectrodes filled with K^+ acetate were inserted into the anterior segments of the dorsal hippocampus corresponding to the CA3 or CA1 area, usually as deep as the pyramidal layer judged by FP characteristics. In some experiments, the position of the stimulating and recording electrodes was checked by means of electrolytic destruction followed by a morphological control. Single or averaged responses were recorded. On-line averaging of 10 to 200 responses (usually 10 or 20) was performed with ART-1000 analyzer (Saip, France). Measurements, computations and off-line averaging, when necessary, were performed with equipment TAS (Leitz, FRG) and computer P6060 (Olivetti, Italy) in our earlier experiments. Various types of laboratory computers were used in more recent experiments.

Here and in the following chapters, the evaluation of the statistical significance of differences between groups of measurements were performed using mainly nonparametric statistics. "Statistically significant" difference means a difference at $P<0.05$ if it is not specified otherwise.

One experimental series presented in this chapter was conducted with in vitro preparations. Hippocampal slices were prepared from 1–2 month old mice, essentially as described by Skrebitsky and Vorobyev [910]. FPs were recorded from the CA1 region after the stimulation of the Schaffer collateral-commissural system. A pair of glass micropipettes filled with Yamamoto solution [1127] were used for the stimulation. Otherwise, the methods of testing and tetanization were similar to those used in the experiments conducted with in vivo preparations. The data obtained in the in vitro preparations will be specified below; in all other cases, our experiments were carried out under in vivo conditions.

Additional explanations will be given below and can be found in original publications from our group [1005, 1042, 1044, 1045, 1046, 1031, 1032, 1036].

1.6 Long-Term Potentiation of Field Potentials in the CA3 and CA1 Regions

In accordance with the cited data of other authors, a slow positive wave with the amplitude of several mV was recorded at the level of the pyramidal layer in areas CA3 and CA1 after testing stimulation of the SFA (Fig. 1.2B) or Schaffer collaterals (Fig. 1.5), respectively. The positive wave inverted and was below the depth of the maximum amplitude (Fig. 1.3A1). A high-amplitude negative population spike (up to 10–15 mV, component 3) was superimposed on the slow wave, provided that sufficiently strong or frequent stimuli (more than 2–3 Hz) were delivered.

The correspondence in time of the population spikes with neuronal discharges was confirmed under our conditions with single neurone recordings (Fig. 1.2A). More than 100 units were recorded from the CA3 area and more than 50 cells were recorded from CA1 area in our initial experiments. The correspondence of the orthodromic discharges of single neurones with the population spike (component 3) was evident in the experiment illustrated in Fig. 1.2A, where FP recorded during tetanic stimulation (Fig. 1.A1) were compared with spikes of two different neurones (Fig. 1.2A2 and A3) recorded later in the same experiment. In another experiment, illustrated in Fig. 1.2B, the intracellularly recorded antidromic discharge (Fig. 1.2B2), evoked by a low-frequency testing stimulation, corresponded to the first population spike (component 1 according to our designation). In response to 10 Hz tetanic stimulation (Fig. 1.2B3), the antidromic discharge was blocked and a partial orthodromic spike appeared. The latter corresponded in time to the orthodromic population spike of the FP (component 3, Fig. 1.2B1, dot). Neuronal discharges recorded both at the pyramidal layer and outside this layer (presumed "nonpyramidal cells", see Section 5.3) coincided in time with the population spikes in CA1 and CA3 areas.

A prolonged potentiation of the FP amplitude was induced in most experiments by one or two conditioning trains (10–20 Hz, 7–20 s) delivered through the same or adjacent stimulating electrodes. The potentiation persisted, as a rule, throughout the whole post-tetanic period in experiments with post-tetanic recordings for more than 1 h and sometimes up to 6 h (Fig. 1.3). The amplitude of the slow wave (component 2) and especially of the orthodromic population spike (component 3) was increased. In many cases, the population spike was small (Figs. 1.4, 1.5A) or absent in control responses before the tetanic stimulation (Fig. 1.3A1) and it appeared after the tetanization (Fig. 1.3A2–10).

The amplitude of the small early wave (component 1) increased in only one experiment (Fig. 1.2B1) where this component was of unusually large amplitude (see [1042] and Sect. 5.8 for an additional discussion related to this experiment). Typically, the "presynaptic" component 1 was not changed significantly, sometimes even slightly decreasing (Fig. 1.3B, triangles).

In the illustrated experiment (Fig. 1.3B), no significant change was found after the first tetanization. The second tetanization induced both an increase in the rise time of the slow wave and the appearance of a population spike (Fig. 1.3A2). The increased response persisted throughout the experiment (Fig. 1.3A, B). Figure 1.5 illustrates an increase in the rise time and amplitude of the positive slow wave recorded from CA1 area after tetanization of the Schaffer collaterals. The population spike was very small before tetanization (Fig. 1.5A). Larger population spikes with reduced latencies were evoked after the tetanus.

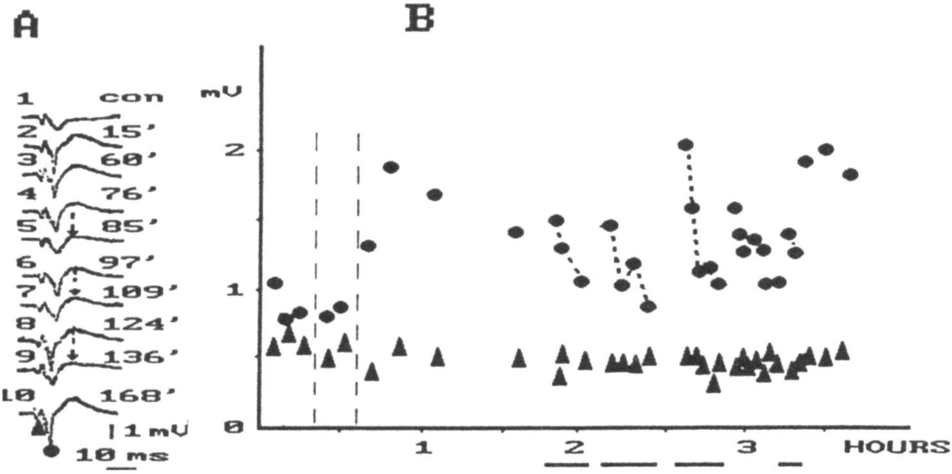

Fig. 1.3A,B. Long-term potentiation and low-frequency depression of CA3 responses to the septo-fimbrial stimulation. **A** Averaged responses (from ten consecutive stimulus presentations at 0.2/s) before (**A1**) and after (**A2–A10**) tetanization. *Right hand numbers* denote time (in min) after tetanization. *Broken lines with arrow heads* connect responses recorded at the beginning (**A4**, **A6** and **A8**) and at the end (**A5**, **A7** and **A9**) of three series of continuous low-frequency stimulation. **B** Amplitude time course in the same experiment. *Triangles* and *dots* are amplitudes of presumed presynaptic and postsynaptic components, respectively, measured from the average FPs (see **A10**). First and second tetanic trains (10 and 20 Hz, respectively) are marked by *vertical broken lines*. *Oblique broken lines* correspond to the periods of continuous 0.2/s stimulation. The *last three dots and triangles* correspond to testing stimulation with 0.5–1-min intervals

In general, a well-pronounced LTP of more than 30 min duration after tetanization of the SFA was observed in 21 out of 26 initial experiments with FP recordings from CA3. LTP at the CA1 was observed after 15 out of 39 tetanizations of the Schaffer collaterals (19 experiments on 9 rabbits). There were no noticeable long-lasting effects after 17 out of the 39 tetanizations. A long-lasting (dozens of minutes) amplitude decrease, i.e. post-tetanic depression (PTD) was induced in seven cases. The comparison of the post-tetanic effects in these two hippocampal areas shows that LTP in the CA1 area in our preparation is less reproducible than in the CA3 area.

1.7 Low-Frequency Depression and Self-Restoration of the Potentiated Response

The extinction of a CR if no reinforcement is applied as well as CR self-restoration after extinction are known characteristic features of behavioural conditioning [797]. If mechanisms which underlie LTP and behavioural conditioning are similar, one can expect that the potentiated response should not simply decrease with time but should rather be extinguished as a result of a low-frequency testing without tetanic stimulation. One might expect a recovery of the "extinguished" response after an interruption during the testing procedure. Such properties were indeed found in most of our experiments with FP recordings from both CA3 and CA1 regions. It should be noted that a decrease in the amplitude of the population spike or slow wave of the FP ("habituation", see Sect. 2.5) with their recovery during intervals between series was sometimes observed even with control testing stimuli before tetanic stimulation [413, 961]. However, similar effects were found to be more common and more prominent after the tetanization. There was no significant decay of the post-tetanic response at initial stages of the experiments when responses were collected with long (5 to 30 min) intervals between testing series as illustrated in Fig. 1.3B where each series consisted of ten stimuli delivered with 5 to 15 s interstimulus intervals. In contrast, the response amplitude was usually decreased significantly after several minutes of continuous presentation of the low-frequency testing stimuli. The oscillograms in Fig. 1.3A2, 3 show that the averaged post-tetanic response decreased only slightly during 1 h after tetanic stimulation when separated low-frequency trains were used for the testing (Fig. 1.3B, four initial dots after the second tetanization). Figure 1.3A4 and 5 shows a low-frequency depression (LFD) of the potentiated response when the testing procedure was switched to a continuous stimulation at 0.2 Hz for 9 min. Partial recovery of the response amplitude was observed during the following 12 min period without testing stimulation (Fig. 1.3A6). The second period of the continuous low-frequency stimulation (Fig. 1.3A7) and the second period without stimulation (Fig. 1.3A8) resulted in repeated LFD and a more complete recovery, respectively. Corresponding series of LFD are shown in Fig. 1.3B as dots connected with interrupted lines. Testing stimulation with 0.5–1 min intervals (Fig. 1.3A10 and the last three dots in Fig. 1.3B) shows that the potentiated response was not actually diminished in comparison with the response recorded shortly after tetanization (Fig. 1.3A2). The first, "presynaptic", component typically showed neither LTP nor LFD (Fig. 1.3A, B, triangles). This suggests the absence of local changes at the stimulation site and the absence of significant changes in the presynaptic afferent volley (see Sect. 1.10 for discussion). Similar LFD of the FPs, especially of the population spikes, was found in our recording from the CA1 region [891].

To check how general and reproducible the LFD is, we performed a series of experiments on hippocampal slices in collaboration with Drs. A.Baskys and S.Buldakova. We measured FPs in area CA1 after stimulation of the Schaffer collaterals-associative fibres in the *stratum radiatum*. A prominent LFD of the potentiated population spike was found in eight out of ten slices during a testing stimulation with frequencies from 0.1 to 0.3 Hz. In the experiment illustrated (Fig. 1.4), an approximately twofold decrease in the population spike amplitude was evident after the third series of 50–70 testing stimuli delivered at 0.3 Hz before the tetanic stimulation (Fig. 1.4, left part of the graph). A prominent manifold depression of the population spike amplitude was observed during similar low-frequency

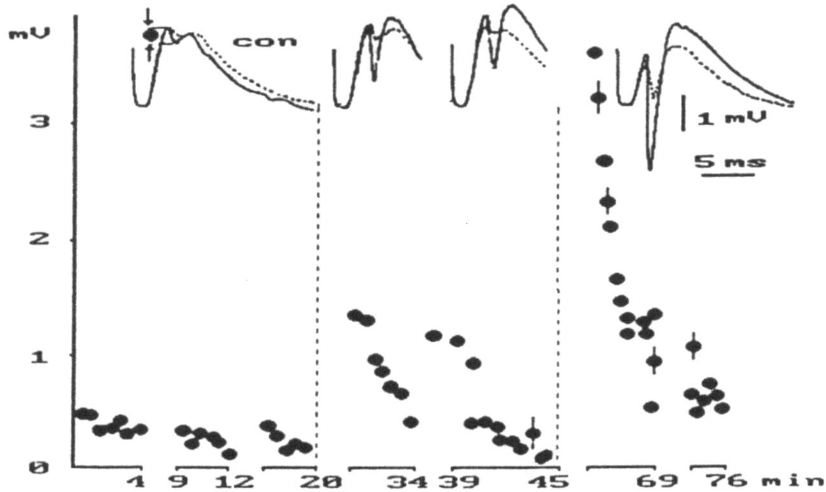

Fig. 1.4. Long-term potentiation; and low-frequency depression of CA1 responses recorded in in vitro hippocampal slice preparation; after Schaffer collateral stimulation. Oscillograms show examples of first (*continuous curves*) and last (*dotted curves*) FPs from each stimulus series. *Ordinate in the graph* the mean amplitude of the population spike; (N=10 before tetanic stimulation and after the first tetanization; N=5 after the second tetanization). *Error bars* (±SEM) are shown where these are larger than the symbols. *Abscissa* time in minutes from the beginning of the experiment. Stimuli within each series were delivered with intervals of about 3 s. No stimulation was given during intervals shown as *breaks* at abscissa (10 and 20 min after the first and second tetanization, respectively). [1018]

testing after the first tetanization (Fig. 1.4, first vertical dotted line) and especially after the second tetanization (Fig. 1.4, second vertical dotted line). Similar to in vivo observations, the depression in the slice preparation was not due to a simple decay of the potentiation with time. This becomes clear from the restoration of the response amplitude after periods of 5–6 min without testing stimulation (Fig. 1.4)

1.8 Specificity of Long-Term Potentiation for the Tetanized Input

In addition to responses evoked by the stimulation of the Schaffer collaterals, FPs evoked by the stimulation of the symmetrical contralateral point of the hippocampus were recorded in three experiments from area CA1. No changes in responses to the stimulation of the contralateral homotopic point (Fig. 1.5B) were found in these experiments during LTP of the responses evoked by the Schaffer collateral stimulation (Fig. 1.5A). Therefore, the LTP was homosynaptic. In contrast, when PTD instead of LTP took place in the same experiment, the responses evoked through the control non-tetanized input were also severely depressed

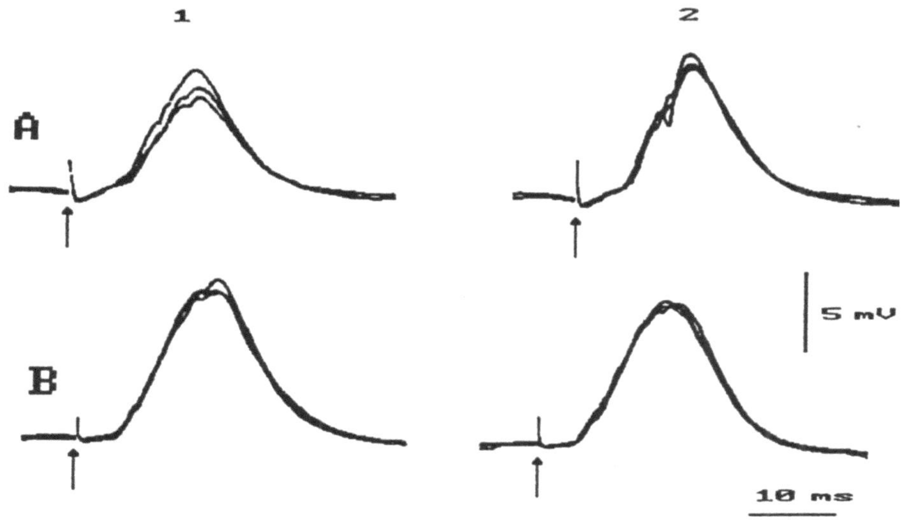

Fig. 1.5A,B. Specificity of long-term potentiation. Responses recorded after stimulation of the Schaffer collaterals (**A**) and contralateral field CA1 (**B**) before (*1*) and after (*2*) tetanization of the Schaffer collaterals. Stimulus artifacts are marked by *arrows*

[891]. Tetanic stimulation of the contralateral homotopic point (five trials) never induced LTP

In our more recent experiments [546], FPs were recorded from area CA3 after stimulation of the DG and medial septum. Tetanic stimulation was applied through one of these inputs with the other serving as control. Only a small increase in the amplitude of the control FP induced by stimulation of the medial septum was evident in the illustrated experiment from this series (see Fig. 6.4B2) during a prominent LTP of the FP induced by the DG stimulation (Fig. 6.4A2). Moreover, during the LTP of the FP evoked by the septal stimulation, the dento-hippocampal response was depressed below the pretetanic level (Fig. 6.4A3). Similar heterosynaptic depression of the control responses was found after 10 out of 26 tetanizations delivered in this series. No significant depression was observed in other cases, as illustrated in Fig. 6.4B2.

The above recordings from area CA1 provided the first evidence of LTP specificity for the tetanized input [891]. The specificity of LTP was confirmed by other authors in various systems (see section: Discussion).

1.9 Involvement of Reinforcing Systems During Tetanization

It was suggested that mechanisms turning into action during tetanization might be essentially similar to those that take part in the formation of behavioural CRs.

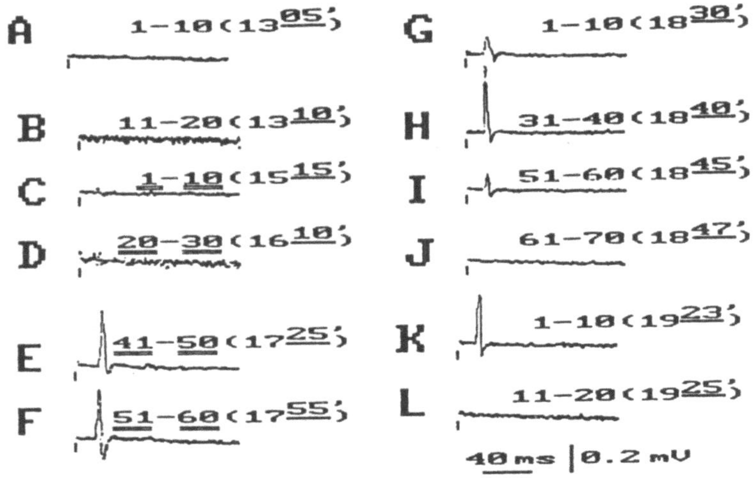

Fig. 1.6A–L. One-session "local conditioned startle response" with tetanic stimulation of the septo-fimbrial area as an additional reinforcement. Averaged (N=10) EMG to sound click recorded from foreleg muscles are shown in **A**, **B** before pairings, in **C–F** during four series of pairings and in **G–L** during extinction series. **K** Demonstrates a spontaneous recovery after a pause without stimulation of about 30 min, with subsequent faster extinction in **L**. *Digits above the traces* are ordinal numbers of control trials or numbers of extinction trials; *underlined digits* are ordinal numbers of pairings. *Digits in parentheses* denote time (h, min). Click presentations are indicated by *small vertical bars*

This suggestion implies an involvement of "reinforcing" factors (see Sect. 1.10) which should imitate a "biologically significant" unconditioned stimulus (UCS) necessary for the elaboration of CRs [797]. Therefore, the tetanizations which induce LTP presumably activate intracerebral "reinforcement" systems in addition to the activation of direct afferent pathways under study. In fact, the tetanized structures (EC, DG, hippocampus proper) contain both "positive" points supporting the self-stimulation and "negative" points inducing escape responses [180, 777]. The results of the following preliminary experimental series confirmed the involvement of the reinforcing systems during tetanization which induced LTP. It turned out that septal tetanization with stimulus strength adequate to induce LTP could serve as an artificial "additional reinforcement" for the elaboration of a CR of a simple type – so-called "local conditioned startle response" (LCSR). It was previously shown that to elaborate the LCSR, click conditioned stimulus (CS) had to be combined with a complex stimulation consisting of the direct stimulation of the foreleg representation in the rabbit motor cortex and an "additional reinforcement" [1043]. In these previous studies, a stimulation of the lateral hypothalamus was used as the additional reinforcement. In the present experiments, performed in collaboration with Drs. I.Kudryashov and S.Ioffe, we used not hypothalamic but septo-fimbrial stimulation with the same parameters which were suitable for LTP induction in the same animal but with somewhat shorter train duration (from 3 to

5 s instead of 7 to 20 s). Otherwise, the basic procedure to elaborate the LCSR was similar to that in the studies cited above.

It is seen from Fig. 1.6A,B that before pairings, future CS (click) does not produce any foreleg electromyographic response (EMG). The same rabbit was used in this experiment as previously tested in experiments with LTP induction (see Fig. 5.1). A short-latency EMG to the click was elaborated after several series of pairings (Fig. 1.6C–F). The response exhibited typical features of a CR. These included an extinction (Fig. 1.6G–J), a spontaneous recovery after the extinction (Fig. 1.6K), and so-called "training of extinction", which is a faster extinction in the subsequent series (Fig. 1.6K,L) in comparison with the initial ones (Fig. 1.6G–J). We performed experiments on 4 rabbits. LTP was observed in three animals. In one case, LTP was preceded by a prolonged PTD of about 50 min duration. One rabbit served as a control: no significant LTP was found in this animal after several tetanic stimulations of the SFA. The septo-fimbrial stimulation with the same parameters (frequency, strength and pulse duration) which induced LTP without a depression proved to be effective for the elaboration of the LCSR in response to the CS after 30–40 pairings in two animals. One of these experiments and the experiment on the control animal without LTP were "blind", so that the investigator who carried out the experiment with the elaboration of the CR did not know whether LTP was induced or not. In the control experiment on the animal which did not show LTP, no CR was elaborated despite of more than 100 pairings with SFA stimulations.

1.10 Discussion

Experiments with FP recordings from the in vivo preparation described in the present chapter allowed us to reveal LTP in the hippocampus proper, i.e. in areas CA3 and CA1. Especially well reproducible was LTP in area CA3 after tetanic stimulation of the SFA or DG. The LTP appeared to be similar to that originally found in the DG [145]. The demonstration of the possibility to induce LTP in the magnocellular pyramidal layer of the CA3 region was encouraging for our further work with intracellular techniques aimed at the elucidation of cellular mechanisms which underlie both short- and long-term plasticities.

The occurrence of LTP in the Schaffer collaterals-CA1 system described briefly by Andersen at al. [30] was confirmed in our in vivo experiments. A well pronounced LTP in this system was reported later for the hippocampal slices [31, 505, 506, 639, 871, 962] and anaesthetized animals [962] by other authors (see reviews cited in the Introduction for more recent references related to LTP in various preparations). According to data from Vinogradova's laboratory obtained with recordings of single neurones, long-lasting post-tetanic effects were much more reproducible in the Schaffer collaterals-CA1 system of the hippocampal slices than in the corresponding system of in vivo preparations [505, 506]. The same was our impression: LTP in area CA1 of the in vivo preparation was induced in only about half of the cases. The reason for the difference in LTP reproducibility between the in vivo and in vitro experiments is not quite clear. An involvement of inhibitory systems was suggested which might prevent the LTP induction in the in vivo preparation [506, 507]. These inhibitory systems might be partially damaged as a result of the slice preparation. More recent in vitro experiments indicated that pharmacological blockade of postsynaptic inhibition might, in fact, significantly facilitate LTP induction [78, 1094]. Similarly, another type of experiments [595] showed a reproducible LTP after a "patterned" stimulation in the theta-rhythm frequency which seemed to suppress postsynaptic inhibition [594, 512, 788].

More recently, a distinct LTP was demonstrated for all three major hippocampal regions – the DG, CA3 and CA1 in in vitro preparations [10, 31, 160, 162, 291, 912, 910, 1004, 1127, 1128] (see reviews cited in the Introduction for more recent references). Similar studies using in vivo preparations were not as numerous as those using the slice techniques until recently (see e.g. [594] and below).

LTP in the septum-DG pathway was not discovered for the anaesthetized rats [832], but was found in unanaesthetized [702] or slightly anaesthetized and immobilized rats [557, 559, 560]. In contrast to the latter studies and to our data on the DG and CA3 regions of unanaesthetized intact animals, only a short-term PTP of the septo-hippocampal responses was found in area CA1 of anaesthetized rabbits with partially removed neocortex [929].

It was shown that the pathway from the rat contralateral EC to the DG did not display LTP [603, 1105]. The failure was explained by the small efficacy of the input to the DG from the contralateral EC [603, 1105]. This explanation was supported by two sets of observations: (1) conjoint tetanization of the contralateral and ipsilateral EC–DG pathway induced LTP [603]; (2) it was possible to induce LTP when the crossed projection increased its input to the DG as a result of a postlesion "sprouting" [1105]. The absence of LTP in the commissural projection to the DG is in accordance with our data showing that the stimulation of the contralateral side which evoked FPs without a population spike did not induce LTP in area CA1. It was found later [148, 192] that a more effective activation of the commissural projections to the CA1 area from the contralateral CA3 region induced LTP comparable to that induced by the Schaffer collateral stimulation. In general, LTP seems to be a universal phenomenon for the hippocampal synapses. It is well reproducible in various pathways in both the in vivo and in vitro preparations of various species.

Longevity is the most remarkable characteristic of LTP which gave the name to the phenomenon. As already noted, the potentiated response was demonstrated to persist for up to 3 days in the initial study on chronic rabbits [143]. In a special study on freely moving rats [281], a FP potentiation induced in the DG by several trains to the EC was shown to persist for at least 2 weeks. Post-tetanic effects were found even 2 months after the last tetanization. The above mentioned "patterned" tetanic stimulation was recently shown to induce especially persistent LTP in the CA1 area [934]. However, some other publications suggest that, at least following more traditional tetanic patterns (trains of several seconds), LTP might decay gradually [72, 257, 809]. Somewhat surprisingly, no evidence of saving was found when LTP was reinduced [257].

Repetitions of conditioning trains has been shown to induce a step-wise response increment [145] and to increase LTP longevity [72, 73]. According to the latter authors, LTP mean decay constant is equal to 4, 20 and 37 days after 1, 3 and 12 conditioning trains, respectively. Although De Jonge and Racine [257] found no lasting effects of repeated inductions of LTP on the decay rates, a great number of conditioning trains used in this study might "saturate" the underlying mechanisms and in addition might evoke some changes related to the PTD. As is evident from our data, a gradual decrease in the response amplitude during one experimental daily session observed by some authors (e.g. [143, 145, 699]) may be at least partially explained by the LFD of the potentiated response. In fact, according to our data, the response amplitude might not actually decrease throughout the experiment in either the in vivo and in vitro preparations when low testing frequencies and long intervals between the testing trains are used. On the other hand, one can expect different LTP time-courses depending on structures, pathways and conditions of LTP induction [809]. More recent data strongly suggest also the existence of several (at least two or three) LTP phases with different time courses and different mechanisms (e.g. [348, 826]).

The existence of the LFD of the potentiated response was confirmed more recently in anaesthetized rats [79]. It should be noted that these authors used too high a testing frequency (1 Hz), which, according to our observations, is intermediate between that favourable for the production of a frequency potentiation and LFD. The LFD of the potentiated responses and their spontaneous recovery are very similar to analogous properties of behavioural conditioning, and support the involvement of common mechanisms.

Another important property of LTP encountered in our studies is its specificity for the tetanized pathway. The specificity of LTP in the Schaffer collateral-CA1 pathway was confirmed in more detailed studies on hippocampal slices [31, 295, 639] and was described later for other pathways [2, 559, 602–604, 696, 940]. Moreover, some of these authors found a depression of the response to the control stimulation of the non-tetanized control pathway. According to our pilot experiments with recordings from area CA1, the PTD was heterosynaptic [3, 891]. On the other hand, more recent data indicate that PTD might be also homosynaptic and might be induced in the hippocampus and other structures under certain conditions distinct from those necessary to induce LTP [45, 159, 220, 459, 932]. Additional studies are necessary to understand the relationship between PTD, which, according to our data, is heterosynaptic and of shorter duration than LTP, and the new type of neuronal plasticity which was termed long-term depression (LTD).

On the other hand, a heterosynaptic, i.e. unspecific, LTP was described later in in vitro studies [723, 1127]. In these publications, an "unspecific" increase in FP, EPSP and IPSP was found in area CA3 when tetanic and control stimuli were applied to different parts of the DG. In a more recent publication [159], even more complicated post-tetanic changes including heterosynaptic LTP and PTD (or LTD, see Sect. 1.8) were found in a non-tetanized input after LTP was induced in one of three inputs to CA3 neurones (mossy fibres, Schaffer collaterals and fimbria). The heterosynaptic potentiation might be explained, in principle, on the basis of either activation of common afferent fibres by the control and tetanic stimuli or existence of interneurones which are common in control and tetanized pathways. The latter explanation seems especially probable, taking into account known collateral connections between CA3 neurones [718]. A polysynaptic nature of the potentiated responses was not completely ruled out in the above-cited publications with "heterosynaptic" LTP: no exact latency measurements were presented and no other data to support monosynaptic nature of the recorded responses were given. The lack of a paired-pulse facilitation (PPF, see Sect. 7.10.2 for definition) in the control pretetanic measurements with the testing heterosynaptic stimulation does not seem to be completely convincing evidence for the activation of separated pathways because common pathways might not be effective before the tetanization [719]. It should be stressed that the unspecific LTP was more evident in late components in some examples, and may be accompanied by epileptiform afterdischarges [1127]. In fact, Higashima and Yamamoto [429] retested the specificity in the DG–CA3 pathway in the absence of seizure discharges. In agreement with the above stated suggestion (see also [1016]), they found an input specific potentiation of short-latency FPs and heterosynaptic enhancement of long-latency FPs and EPSPs.

Nevertheless, the problem of the unspecific LTP does not seem to be resolved, especially in view of the complicated heterosynaptic effects shown recently [159]. It should be noted that LTP in area CA3 is distinct from the LTP in both area CA1 and the DG, as judged by some other properties [412, 497, 220] (see also section General Conclusions). On the other hand, it should be stressed that LTP specificity (homosynapticity) was convincingly demonstrated for both area CA1 and the DG and therefore it can be taken as one of characteristic features of the hippocampal LTP.

The next LTP feature which was not specially investigated in our studies is the existence of a threshold intensity to induce LTP (see also the data of Chap. 3, indicating the lack of LTP induction after a weak stimulation). This property was noted in some initial publications on LTP [143, 871] and was later termed "cooperativity" [699, 599]. According to our own and literary data, such a threshold stimulus strength is usually near the population spike threshold. A similar property, which could be termed "associativity", became apparent in experiments with concurrent activation of ipsi- and contralateral EC [603, 604] or medial and lateral perforant paths [696, 699]. The concurrent activation of these inputs to the DG induced LTP at intensities where identical but independent stimulation of the two pathways either failed or was less effective. LTP appearance after sprouting of fibres from collateral EC to DG, which I referred to above [1105], can be understood on the basis of cooperativity. The existence of cooperativity and associativity [1093] was confirmed in more recent publications and extended for other hippocampal pathways (see reviews cited in the Introduction). The failure to induce LTP after a suppression of synaptic transmission in slice preparations [296,

297, 557, 1096] might be also related to the same properties. Attempts to estimate EPSP amplitude necessary to induce LTP in the Schaffer collaterals-CA1 connection showed that it might be rather small: about 1 mV [424], which might correspond to the activation of only a few fibres [1054]. However, it should be noted that the depolarization during the tetanus itself might be larger due to frequency facilitation (FF) and involvement of "silent" fibres [1054]. The nature of the "cooperativity" became more clear in recent years (see section General Conclusions).

Our data, obtained with the elaboration of the behavioural CR of a simple type, supported the suggestion that some of the properties of LTP might be explained by the necessity of not only a strong activation of postsynaptic elements but also of activation of hypothetical "reinforcing" (or "facilitating") systems [241, 360, 367, 386, 387, 450, 777, 937, 1107]. Principles of LTP formation were previously schematized [1016] using an analogy with a "three-factor" principle of the CR formation. This principle postulates a temporary conjunction of three necessary components [386, 1019]. The first component of this scheme is the activation of afferent fibres with "plastic" or "modifiable" synapses [168]. The second and the third components are a sufficiently strong activation of the target postsynaptic neurones and a "reinforcing" system, respectively. The activation of the "reinforcing" system by itself may or may not evoke changes in the activity of the target neurones. A peculiarity of LTP formation within this scheme, in comparison with the formation of behavioural CRs, is the involvement of the three factors (presynaptic, postsynaptic and reinforcing ones) due to a tetanic stimulation of the same point of the brain. In other words, it was suggested [1016] that the "three component" modifiable synapses, which might be termed "Griffith synapses" [386, 387], underlie LTP formation. More recent data in this line, which may be interpreted as a discovery of so-called Hebbian-like synapses in the hippocampus, will be considered in the section General Conclusions.

In addition to the above-mentioned properties of LTP, it is worth stressing that LTP is induced by a stimulation within a broad frequency range – from 3 to 400 Hz, which is in the same range as natural cellular activity [295, 871] (see also [197]). Several (from 8 to 20) short (from 4 to 10 pulses) high-frequency trains of 100–400 Hz were found to evoke LTP most reproducibly [280]. These trains are similar to the natural activity of hippocampal neurones. Although direct comparisons are lacking, more recent data indicated that the already mentioned "patterned" stimulation consisted of either a single pulse followed by a short burst [841] with a delay of about 200 ms, or several short trains with intervals of about 200 ms ("theta" stimulation) is even more effective [594, 595, 934]. The "patterned" stimulation might be even more similar to the natural activity of hippocampal neurones.

In one publication [906], a long-lasting increase in the population spike amplitude was reported in the DG of freely moving rats after one series of 30 to 42 stimuli of increasing current intensities at 0.2 Hz (but not at 0.04 Hz). The authors [906] attributed the difference with the literature data to the high current intensities used in their study. It might also be suggested that these data [906] are related to behavioural conditioning, especially to so-called CR to time or "trace assimilation" of the rhythm of a previous stimulation demonstrated at the neuronal level [522, 523, 732, 1006, 1009, 1010, 1021, 1040, 1041]. Such a trace assimilation of the rhythm of the previous stimulation is especially well pronounced in hippocampal neurones [521, 1002]. To check the involvement of the time-related mechanisms, it would be useful to compare the aftereffects of periodical stimulation with those of a random stimulation at a similar mean frequency.

Some additional LTP properties, such as the erasure of LTP by electroconvulsive seizure [426], might be explained by superposition of the above-mentioned heterosynaptic PTD or LTD (see also [11, 295]) and await further studies. A somewhat more complex "depotentiation" phenomenon was described later [603]. It was found that the response potentiated by a concurrent stimulation of ipsi- and contralateral EC could be diminished by ipsilateral tetanization. The authors interpreted these data as an indication of the existence of special "modifiable" synapses distinct from those mediating "classical" conditioning. However, this phenomenon bears a resemblance to the PTD in control (non-tetanized) pathway (Fig. 6.4). Persistence in time of the "depotentiation" was not convincingly demonstrated. Some recent observations already mentioned might indicate the existence of a more specific LTD. However, the relationship between "depotentiation", heterosynaptic PTD and specific LTD and LTP are awaiting further studies.

Some of the above-mentioned LTP properties are not well established and need future experimental support, but most of them had been demonstrated convincingly and independently by several research groups. It seems important to stress that all these properties are not only distinct from those of short-term PTP but, additionally, are very similar to analogous properties of memory and behavioural conditioning. This analogy between LTP and behavioural conditioning is supported by more recent correlative studies of LTP and CRs (see General Conclusions). Studies of cellular mechanisms of LTP seem to be especially attractive in view of these striking similarities, and encouraged our further experimental work in this direction.

Some preliminary considerations of possible neuronal mechanisms of LTP might be done on the basis of the data of this chapter obtained with FP recordings. Two types of "local" effects under stimulating electrodes could be considered. The first one is of artefact nature and includes changes in the resistance and location of the electrode. As noted in Section 1.5, these changes were controlled at least partially by routine measurements of current strength, by using constant current stimulator, biphasic stimuli and especially separate stimulating electrodes for testing and tetanization. The second type is a possible post-tetanic change in fibre excitability, e.g. resulting from long-lasting ionic concentration shifts. In principle, postactivation increases in excitability are known for various types of afferents including intrahippocampal fibres [630, 1092], but they are of too short duration to explain LTP. More prolonged post-tetanic changes were described in the Schaffer collateral terminals [858], but it was a reduction in the excitability, so that they could not explain LTP as a result of an increase in the number of activated fibres activated. Generally, the following observations testify against local effects: (1)

the lack of long-lasting post-tetanic increase in presynaptic fibre excitability; (2) the lack of long-lasting post-tetanic changes in K^+ concentration [10, 353, 406, 555]; (3) the lack of essential threshold changes in stimulus-response curves of FPs ([145]; see also Chap. 3); (4) the lack of LTP in one target structure with LTP occurrence in another [22, 603, 1105]; (5) LTP revealed by testing through electrodes different from those which were used for the conditioning tetanization [145]. The occurrence of LTP in the CA3 region in the in vitro experiments after removal of the granular and molecular layers of the DG testifies against an involvement of changes in soma-dendritic membrane excitability of presynaptic neurones [1127]. Finally, in some experiments, an early component was distinct in FP recordings which reflected a presynaptic volley (Fig. 1.3, triangles). There were no reproducible long-lasting increases in this component during LTP according to our and literary data [31, 145]. Summarizing, it can be stated that changes in the afferent volley are not a probable candidate to explain LTP mechanisms. Therefore, the post-tetanic increase in FP amplitudes should reflect changes either in synaptic or postsynaptic (membrane) loci.

The synaptic location of LTP mechanisms is supported by an increase in the amplitude and a decrease in the rise time of the earlier FP component which presumably reflects monosynaptic activation according to latency measurements. However, these observations require more direct verification if account is taken for the complicated nature of the "aggregate" FPs induced by strong activation. Moreover, some authors [145, 32, 1106] considered the discrepancy between changes in amplitude of the population spike and the slow component of FPs (so-called E–S potentiation [32]) as a support of an involvement of "membrane" (postsynaptic) mechanisms in LTP induction. However, the results of our experiments and the above cited literary data, which showed LTP specificity for the tetanized pathway, are in accordance with a synaptic localization of the underlying mechanisms, and argue against a general increase in postsynaptic excitability. Generally speaking, an interpretation of changes in aggregate FPs is ambiguous: shifts in different FP components may arise due to different contributions from elements situated at different distances from the recording electrode. In fact, the method of current source density analysis indicated shifts of the main current sink during LTP [62, 704, 955]. The E–S potentiation may in principle be explained not by the increase in the postsynaptic excitability but by other mechanisms, for example by a different LTP in excitatory and inhibitory pathways [4, 395] or by a lasting (several minutes) post-tetanic hyperpolarization (see Sect. 5.8). Therefore, methods more precise than the recording of the aggregate FPs are required to specify neuronal mechanisms underlying LTP.

1.11 Summary

A brief description of the structure of the hippocampus and of field potentials (FPs) recorded from various hippocampal areas was presented. Initial literary data showing long-term (hours and days) potentiation (LTP) of FPs in the dentate gyrus (DG) were cited. It is suggested that LTP directly reflects mechanisms underlying the elaboration of behavioural conditioned reflexes (CRs). LTP of FPs recorded from unanaesthetized rabbits was demonstrated in area CA3 after one or two tetanizations of the septo-fimbrial area (or the DG) and in area CA1 after tetanization of the Schaffer collaterals. The following properties of the LTP were found: (1) low-frequency depression ("extinction") of the potentiated FP during repeated presentations of testing stimuli with intervals of 1 to 10 s; (2) spontaneous recovery of the depressed responses after an interval of several minutes without testing stimulation; (3) "specificity" of the potentiation for the tetanized pathway (judged by the lack of changes in commissural FPs during LTP in the Schaffer collaterals-CA1 system); (4) a probable involvement of "reinforcing" brain systems during conditioning tetanization (judged by the possibility of elaborating a CR of a simple type reinforced by the septo-fimbrial tetanization adequate to induce LTP). These properties of hippocampal potentiation, together with properties known from the literature (longevity, "associativity", stepwise increment after repeated tetanizations, electro-shock erasure, correlation between LTP and behavioural conditioning) are regarded as support for the hypothesis of mechanisms common to LTP and CRs. The available data testify against either local effects in the stimulated region or changes in presynaptic volley as an essential mechanism of LTP formation and support the hypothesis of its synaptic localization. However, this hypothesis requires more direct support with the aid of more precise methods than the recording of aggregate FPs. The demonstration of LTP in the hippocampus proper at magnocellular pyramidal layers provides the possibility for further analysis of the neuronal mechanisms of this phenomena using the microelectrode technique.

2 Unitary and Minimal Postsynaptic Potentials (Literature Review)

2.1 Introduction

The natural continuation of our study at this stage seems to be the application of the microelectrode technique to analysis of LTP mechanisms. Various variants of this technique had been successfully used for studies of the CNS by many workers (see references in the Introduction). Up to recent times, mainly stimulations of peripheral receptors, nerve trunks or structures which produced activity of a large number of afferent fibres were used while recording from CNS neurones. However, it is sometimes more useful to analyze postsynaptic potentials (PSPs) generated in a target (postsynaptic) neurone by activation of a small number of afferent (presynaptic) fibres. I shall term such PSPs "minimal" PSPs or MPSPs. Even a more precise method is a recording of PSPs generated by activation of a single afferent fibre. These PSPs are termed "elementary, "individual" or "unitary" PSPs (UPSPs). Initial detailed studies of UPSPs at neuro-neuronal synapses have been done with intracellular stimulation of giant axons in both invertebrates [305, 409, 951, 1073] and in some non-mammalian vertebrates [56, 673, 674, 844] though first related publications with recordings from spinal motoneurones are dated back to the early 1960s [488, 570, 571]. Due to more recent improvements in methods of stimulation, recording and analysis, the studies in this field have attracted considerably more attention. I shall mention here the following improvements which make it easier to collect data and interpret results related to the analysis of MPSP and UPSP at the level of vertebrate CNSs: elaboration of the methods of stimulation of single afferent fibres [462, 570, 708]; intracerebral microstimulation [48, 47, 60, 393, 468, 540, 820, 893]; improvement of methods of intracellular recording due to bevelling of the microelectrodes [77, 724, 377]; intracellular staining [77, 354, 484, 750]; elaboration of new isolated preparations including brain slices, isolated spinal cord and cell culture. More recently, recordings of postsynaptic currents (PSCs) instead of PSPs became available due to development of the whole-cell patch technique ([139, 307, 677]; see also Sect. 2.2.6). The last but not least improvement is the growing use of computers, specifically, the averaging technique which permits low-amplitude signals to be distinguished from technical and synaptic noises.

Thus, before going into experimental analysis of LTP mechanisms, I shall try to bring together the methods used to record MPSPs and UPSPs, and to analyze basic data obtained by these methods. Special attention will be given to the analysis of plasticity mechanisms.

2.2 Methods Related to Unitary and Minimal PSP Recording

2.2.1 Intracellular Stimulation of the Afferent Unit

By definition, the most direct method of UPSP detection is intracellular stimulation of an afferent (presynaptic) neurone (or its axon) with a simultaneous intracellular recording from the related target neurone. This method was used in a number of initial studies on various invertebrates, especially on the sea snail *Aplysia* (e.g. [475, 481]). Logunov and Balaban [620] identified a neurone in the snail CNS which they termed LPa7. Intracellular stimulation of this neurone produced monosynaptic EPSPs in giant identified neurones of the parietal ganglion (see [855] for the nomenclature of snail CNS neurones). This finding provided a convenient object for UPSP recording at the invertebrate CNS, and will be used in our studies (Chaps. 8–10).

Intracellular recordings from presynaptic structures at the vertebrates were until recently used in practice only in some special cases such as in experiments on neurones of the chick ciliary ganglion [673, 674] as well as on giant neurones of the *medulla oblongata* and the spinal cord of the fish and lamprey [56, 92, 268, 528, 675, 845]. When we began our studies of LTP mechanisms, there were only a few publications dealing with intracellular stimulation of afferent neurones at the mammalian brain. For example, one research group [601, 1101] gave a preliminary account on results of intracellular recordings from motoneurones of the cat spinal cord with intracellular stimulation of neurones of the dorsal roots ganglion. More recent developments have allowed the expantion of the application of simultaneous intracellular recordings from two neuronal elements and the analysis of UPSPs at both the spinal cord (see below Sect. 2.4.3) and hippocampal levels (Sects. 7.8, 7.9). Various variants of in vitro techniques (see previous section) are especially convenient for the double intracellular recordings at the vertebrate CNS. The relevant publications will be cited in Section 2.4.1 and considered later in more detail (Sects. 7.8 and 7.9). An interesting variant of the cell culture technique is the growing of single identified neurones taken from invertebrate CNSs (see [761]). Such neurones, plated together with an appropriate partner, form effective synaptic connections which can be studied with double intracellular recordings. Nevertheless, the number of publications related to the UPSP recordings at higher levels of the mammalian brain (mainly in vitro) are very limited up to the present time.

2.2.2 Macrostimulation of Afferent Fibres

Activation of a single afferent fibre with macroelectrodes can be easily performed at various neuromuscular preparations innervated, as a rule, by a single axon. A similar situation takes place in some autonomic ganglia which are innervated either by a single fibre [134, 265, 673, 767, 954]) or by a relatively small number of fibres with different thresholds [135, 692, 850]. However, multiple innervation is typical for CNS neurones, so that pathways suitable for the stimulation of a single fibre can be found only in exceptional cases. The connection climbing fibre – Purkinje cell of the cerebellum – is an example of such an exception [407, 616].

The method of splitting afferent nerves [570, 805] or spinal cord dorsal roots [453, 462, 570–576, 708] into thin bundles (filaments) consisting of a small number of fibres of group Ia was widely adopted in studies on the cat spinal motoneurones. This procedure enables only one fibre or a limited number of fibres to be excited by electrical stimulation or muscle stretch. In some cases, it was possible to isolate rhythmic EPSPs from background synaptic activity. These EPSPs were taken as UPSPs arising due to a rhythmic discharge of one afferent fibre [190, 191, 311].

2.2.3 Intracerebral Microstimulation

The methods of a local extracellular microstimulation of afferent units by current pulses [47, 540] or by microiontophoresis [468] might substitute a more laborious method of dual intracellular penetration. However, in distinction from the method of intracellular stimulation, there is usually the risk of exciting several presynaptic units. Several methods can be used to reduce this risk or to control the contribution of additional presynaptic units. These methods include a simultaneous recording of the presynaptic discharge, selection of the threshold current, and application of appropriate control procedures such as minimally subthreshold stimulation with recording of the presynaptic unit or above-threshold stimulation after the presynaptic unit is lost. A variant of this method is the averaging of the target intracellular activity triggered by spontaneous spikes of the extracellularly recorded presynaptic neurone [47, 392, 1027]. In practice, this method diminishes the risk of the activation of additional presynaptic neurones but creates additional complications due to the possible occurrence of postsynaptic activity arising from common inputs to the presynaptic and target neurones [368].

2.2.4 Recording of "Minimal" PSPs

When the number of stimulated units is not known, but a slight reduction in the stimulus strength has led to the disappearance of the response, MPSPs are said to be recorded [462, 488, 577, 578, 805]. It should be noted that not all responses recorded in numerous publications with near threshold stimuli should be considered as MPSPs. This term should be limited to cases when at least a part of the following criteria are fulfilled: occasional response "failures" at a constant stimulus strength and polymodal distributions of amplitudes of recorded PSPs [889, 972, 1045, 1074]; "all-or-none" behaviour of the response with changes in the stimulus strength and appearance of "steps" and "plateaus" in the stimulus-response curve [577]. Usually stimulus currents of not more than several μA were used to record MPSPs [942]. As a rule, the averaging technique has to be used to reveal the MPSPs.

The method of MPSP recording may supplement or even substitute the technically more difficult method of UPSP recordings in many cases. For example, recordings of MPSPs were used to analyze synaptic connections in the in vivo neocortex [47, 883, 884, 1074], where simultaneous intracellular penetrations into two neurones is subject to very great difficulties.

2.2.5 Additional Procedures for the Detection of Single UPSPs

Averaging is the most powerful and universal technique to reveal UPSPs and MPSPs when their amplitudes are comparable to or below the noise amplitudes. However, in some cases, e.g. for the quantal analysis (see Chap. 7), it is necessary to record single (non-averaged) UPSPs and MPSPs. In this situation, averaging may be also useful to choose an appropriate time for the detection and measurement of the postsynaptic response.

The PSPs of neuromuscular junctions (end-plate potentials, EPPs) usually considerably exceed the noise level [485]. However, the EPP amplitude is reduced in solutions with changed Ca^{2+} and Mg^{2+} concentrations used in many studies to facilitate the quantal analysis [157, 260, 485, 495]. In vitro experiments on neuroneuronal synapses were also frequently conducted with similarly changed ionic compositions [212, 213, 475, 700].

On the other hand, acetylcholinesterase (AChE) inhibitors were used to increase "single quantum" EPPs [152, 495]. Such a procedure might increase EPP amplitude by as much as two to three times [331]. To record PSPs of large amplitudes, units (fibres or neurones) with a large R_{in} can be selected [714] since the UPSP amplitude is proportional to the input resistance [495]. Some other procedures may also prove to be useful. They include frequency filtration [138, 212, 570], artificial shifts in the level of RMP [578] and intracellular Cl^- injection [319, 524]. The latter procedure has been used to increase the amplitude of reversed unitary IPSPs.

2.2.6 Measurements of UPSPs and MPSPs Amplitudes and PSCs

In most cases, the PSP amplitude is measured at the time corresponding to the maximum of the averaged response. However, in many cases, more sophisticated measurements are used such as the measurements of the PSP area or the charge transferred through the membrane. An example is an initial study of UPSPs of spinal motoneurones [308]. However, this method was abandoned later in favour of a measurement of an average amplitude over several points near the EPSP maximum [464]. In some cases, PSP rise-time is measured instead of the amplitude [972]. Measurements of PSCs under the voltage-clamp conditions might give a better signal-to-noise ratio than conventional voltage measurements [57]. However, until recently, this technique was applied to detailed analysis of unitary PSCs at central neurones only in a few instances, namely, to the Ia afferent-motoneurone synapses at the spinal cord [339], excitatory connections between cultured mouse spinal neurones [756] and Mauthner cell inhibitory synapses [320]. The development of the whole-cell voltage clamp technique opens new possibilities for analysis of minimal and unitary PSCs in the CNS (see, e.g. [100, 662]). Especially promising for analysis of unitary PSCs seems to be "perforated" whole-cell recording with a patch electrode which contained nystatin [438]. The technique minimizes losses of cytoplasmic components and might be especially valuable for studies of long-term plasticities in the CNS.

2.2.7 *Mono- and Polysynaptic UPSPs*

Detection of monosynaptic PSPs is naturally facilitated after stimulation of one afferent unit, especially at mammalian CNS, where, as a rule, an activation of a large number of presynaptic fibres is required to evoke a spike and to secure a polysynaptic transmission. The exact number of the fibres may vary from about ten to several scores and hundreds. It depends upon the threshold of the target neurone (from 1 to 20 mV) and UPSP amplitude (usually from 0.02 to 0.6 mV, see Table 7.1). More recent measurements with the whole cell clamp technique showed that the amplitude of unitary hippocampal EPSP can reach several mV [34]. Nevertheless, a single unitary EPSP seems to be able to discharge a target neurone in the mammalian CNS only in exceptional cases, when "giant" or multiple synapses are activated such as climbing fibre-cerebellar Purkinje cell [407, 616]. The other possibility favouring the discharge might be a particular state of the corresponding interneurones, for example, their high-frequency activity [1075].

In the invertebrate CNS, such suprathreshold activation by a single UPSP is much more common [475]. It should be noted that when polysynaptic PSPs are produced by stimulation of one presynaptic unit, they need not necessarily be unitary according to the definition, since a secondary multiplication in polysynaptic nets may lead to a synchronous activation of more than one presynaptic neurone. Such cases are presumably infrequent in the vertebrate CNS, but this problem has not yet been sufficiently investigated. Available data obtained at the spinal cord level indicate that amplitudes of polysynaptic UPSPs might be usually not higher than 5–10 μV [1075], which is at least an order less than typical amplitudes of monosynaptic UPSPs.

The monosynaptic nature of the recorded UPSPs and MPSPs should be checked by using common criteria. There is a special overview of such criteria for UPSP of invertebrates [120], but none of them is absolutely safe. The simplest criterion is a short (e.g. <1 ms) latency which excludes a polysynaptic transmission. However, this criterion may prove to be too strict under conditions of large distances and low conduction velocities. The distance and velocity can be estimated in principle and taken into account but the estimates can be not always precise for the CNS (see e.g. [509]). Another comparatively reliable criterion of the "monosynapticity" is the latency stability with changes in stimulus strength. This criterion is especially useful when EPSP rise-time is long enough, so that expected "utilization time" in the presumed interneurones of polysynaptic chains would be variable.

2.3 Unitary Field Potentials – "EEG Quanta"

Gutman and his colleagues [390, 400, 402, 403, 580] elaborated a comparatively simple method to analyze postsynaptic activity. The essence of the method consists in extracellular recording of FPs arising after an afferent impulse of one presynaptic unit. Such a "unitary FP" was called a "quantum of the electroencephalogram" or "EEG quantum" and defined as an extracellular reflection of numerous UPSPs (or unitary PSCs) generated synchronously in many target cells as a result of the activation of one presynaptic neurone.

The possibility of recording the EEG quantum was predicted theoretically and confirmed in experiments with spike triggered averaging in the frog visual tectum [390]. Later, EEG quantum was detected in single sweeps [402]. Amplitudes of the EEG quanta were in the range from 7 to 85 μV and they had a time course similar to that of monosynaptic EPSPs of spinal motoneurones. The EEG quanta showed a FF. Experiments with applications of Ca^{2+} and Mg^{2+} ions confirmed their postsynaptic nature [400, 402]. Recordings of the EEG quantum were used to testify the cholinergic nature of the retino-tectal connections [400, 403]. The same authors observed averaged unitary FPs with an appropriate wave form in the rabbit neocortex [400].

Postspike negative waves with amplitudes of several scores of μV were found following extracellular discharges of the mossy fibres in the turtle cerebellum [1069]. Experiments with paired stimulation and with the solution at raised Mg^{2+} concentration indicated a postsynaptic nature of such FPs which were hypothesized to be recorded from single glomeruli. Extracellular cortical potentials with an amplitude of 20–40 μV were also observed following stimulation of single Pacinian corpuscles [691]. Such waves were named "individual FPs", but activation of more than one thalamocortical fibre can be suspected if the polysynaptic character of the related pathway is taken into account. "Extracellularly recorded EPSPs" of the cat spinal motoneurones with amplitudes between 0.34 and 50 μV were also recorded after activation of a single Ia fibre [40, 708, 959, 1075].

2.4 Applications of Unitary and Minimal PSP Recordings

2.4.1 Organization of Neuronal Connections

The analysis of UPSPs is one of the basic methods to map neuronal pathways in the CNS of invertebrates (e.g. [346, 475, 481, 895]). This method has been also broadly used for a precise analysis of neuronal connections in the cat spinal cord. For example, it was shown that almost each afferent fibre of group Ia forms connections with practically all spinal motoneurones of the corresponding pool [419, 708]. It was possible to investigate the character of motoneurone innervation by "homonymous" and "heteronymous" afferent fibres [631, 708, 748, 821, 872], by different excitatory and inhibitory interneurones [169, 468, 540, 577, 578, 846, 888]. MPSP recordings were used to compare peculiarities of effects of primary afferents and cortical descending influences on spinal motoneurones of the monkey [804, 805], to investigate characteristics of different synaptic inputs (cortical and cerebellar) to the red nucleus neurones [325, 749, 982], and neocortical influences on the pontine nucleus [15, 977]. Initial information concerning the organization of excitatory and inhibitory elementary intracortical connections was obtained at both neocortical [47, 392, 883, 884, 970, 1027] and archicortical levels [90, 351, 648, 700, 716–721, 863–865, 877, 983, 1114, 1126, 1129].

2.4.2 Electrical and Chemical Synapses in the CNS

The UPSP recording is the only direct way to detect electrical transmission between neurones. It was used initially in studies on invertebrates and fishes (see reviews [113, 615, 885, 925]) and only recently it was applied to the mammalian central neurones in vitro (see references below). A dual-mode (electrical and chemical) transmission was found in the "giant" synapse of the chick ciliary ganglion [673, 674]. However, the identification of the electrical transmission in various structures of the vertebrate brain was based on indirect criteria until recently (see [113, 615, 616, 925]). Even for the classical object of the neurophysiological studies – Ia group afferent synapse on the cat spinal motoneurone – doubts existed for a long period about whether the transmission was chemical or not. In particular, some confusion resulted from demonstrations that MPSPs and UPSPs of the spinal motoneurones were not dependent upon shifts in RMP [310, 581, 758, 1081]. The dual-mode transmission at the synapse between primary afferents and spinal motoneurones was convincingly demonstrated in the frog with UPSP recordings [885, 887, 888]. Similar mixed synapses were found afterwards in the goldfish brainstem [1130]. At present, the chemical nature of the synaptic transmission at the Ia afferent-motoneurone synapse of the cat spinal cord seems to have been well established by various methods [314, 340, 811, 824, 885, 953] including measurements of the synaptic delay of the UPSPs [239, 824] and their quantal analysis (see [823] and Sect. 7.8.3). The elaboration of in vitro techniques together with staining methods opened new possibilities for more precise studies of electrical junctions in the mammalian CNS in recent years (see review [321]).

2.4.3 Spatial Organization of Synapses on the Somadendritic Membrane

It was shown [419, 708] that the number of contacts of Ia fibres with a spinal motoneurone was proportional to its surface area, so that a density of the corresponding terminals was approximately equal in different motoneurones. The use of Rall's theoretical model [454, 461, 463, 582, 817, 818, 822] provided an insight into the spatial organization of the spinal synapses on the somadendritic membrane by measuring the temporal characteristics of MPSPs and UPSPs. Numerous publications showed a general agreement between EPSP characteristics and predictions of Rall's model and indicated that, in many cases, UPSPs were generated by synapses which are concentrated within a narrow region of the neuronal surface [419], though this region itself may be localized at various electrotonic distances from the soma [708]. It has been shown [461] that the majority of synapses formed by Ia afferents are at a distance of less than the membrane length constant. MPSP analysis made it possible to suggest a more distal localization of cortical descending connections in comparison with the synapses of Ia afferents [805]. More recent recordings of UPSPs of the primary afferent synapses were complimented by the staining of both the afferent fibre and the target motoneurone [383, 825, 888].

The problem of the localization of synaptic inputs of different origin on neurones of the higher brain levels had not been so well investigated. An analysis of MPSPs by means of Rall's model showed a dendritic localization of the corticorubral synapses [981] and their collateral sprouting to proximal parts of the somadendritic membrane after the lesion of the *nucleus interpositus* [749, 982]. A proximal dendritic localization of synapses which generated high-amplitude (>3 mV) EPSPs and a more diffuse localization of synapses which mediated EPSPs with lower amplitudes were found for pontine nucleus cells [15, 16]. It was suggested that high-amplitude and low-amplitude EPSPs are mediated by synapses which were formed by collaterals of "fast" and "slow" pyramidal neurones, respectively.

On the other hand, it should be stressed that deviations of experimental data from the predictions of Rall's model were observed already in earlier publications [818]. The deviations were explained partially by generation of the PSCs at different parts of the somadendritic membrane. This explanation was supported later by some data [309, 464] including those based on staining methods [383, 825, 888].

More serious objections against the use of Rall's model were raised by Gutman's hypothesis about essential non-linear properties of the dendritic membrane [399, 401]. This hypothesis was supported by findings obtained in experiments with various neurones: see review [401] and more recent publications (e.g. [946], related to N-methyl-D-aspartate (NMDA) mediated EPSPs (see also section General Conclusions). It should be noted that the above cited studies of "unusual" effects of the RMP shifts on the EPSPs (Sect. 2.4.2) might also be explained in the frame of this hypothesis. On the other hand, it should be noted that the exact influence of dendritic non-linearity on the parameters of EPSPs generated at different parts of the neuronal membrane still remains insufficiently well investigated, which prevents precise quantitative application of the Gutman's model to the analysis of specific experimental data. No data related to the spatial organization of hippocampal synapses were obtained with recordings of MPSPs or UPSPs when we started our experiments (see Sect. 4.7 for more recent results).

2.5 Studies of Synaptic Plasticity in the CNS

2.5.1 Introductory Remarks

The important problem of the CNS developmental plasticity, memory and conditioning was under study in many laboratories and is reviewed in many publications (e.g. [6, 199, 200, 218, 328, 475, 478, 537, 636, 681, 689, 739, 815, 816, 1019, 1023, 1118, 1122]). However, the application of UPSP recording methods to study neuronal plasticity is still at its beginning and is mainly confined to the CNS of the invertebrates. The data obtained with MPSP recording after sprouting of cortico-rubral connections [749] have already been cited. An analysis of characteristics of UPSPs in chromatolyzed spinal motoneurones [573, 710] is another example of the application of this method. These studies indicated a decrease in the number of axo-somatic synapses after axotomy.

The PTP was mentioned in the Introduction as a well-known phenomenon of neuronal plasticity [102, 441, 475, 478, 534]. All studies, including recordings of UPSPs from mollusc neurones (e.g. [198, 372, 475, 548, 619, 867, 957, 1072]), as well as MPSPs and UPSPs from the cat spinal motoneurones [309, 433, 571, 633, 634], indicated a presynaptic location of PTP mechanisms (see Sect. 7.10.4).

2.5.2 Habituation and Sensitization in Invertebrate CNSs

Habituation, an interesting type of CNS plasticity which is a decrease in the amplitude of EPSP during a low-frequency (e.g. 0.1 Hz) stimulation, had been already mentioned in connection with LFD of the hippocampal FPs. This phenomenon was found in the *Aplysia* CNS [183] and described afterwards for different CNS levels in various invertebrates and vertebrates [208, 214, 326, 327, 475, 478, 789, 1066]. The LFD was shown to persist usually for only several minutes after the cessation of the low-frequency stimulation, but a long-term (several days) LFD was also found, provided repeated stimulus series were delivered [215, 476, 919]. Of especial interest is the fact that both short- and long-term LFDs possess many features characteristic of behavioural habituation [475, 968]. In most cases, LFD can be explained by homosynaptic mechanisms [437]. Two behavioural paradigms were initially subjected to detailed investigations, namely, crayfish escape behaviour [547, 1144] and the gill withdrawal reflex of *Aplysia* [475]. The results of these studies showed that the major underlying mechanism of the habituation of behavioural responses is the presynaptic LFD of EPSPs in related central neurones. In addition, a participation of changes at the peripheral nervous system [465] and an involvement of changes in postsynaptic membranes of central neurones [918, 938] were also suggested.

The neuronal organization of the gill withdrawal reflex in *Aplysia* and mechanisms of the habituation of this reflex were subjected to especially detailed studies [475]. Modifiable synaptic connections responsible for the behavioural habituation were identified by recordings from motoneurones and interneurones after intracellular stimulation of the related sensory neurones [198, 212, 475]. The time course of the LFD of the recorded UPSPs was similar to that of the behavioural habituation. Mechanisms of dishabituation or sensitization, i.e. a restoration of the habituated response as a result of a presentation of an "extrastimulus", were also studied [163, 213, 475]. The related UPSPs were found to increase as a result of the sensitization [213, 475, 895, 896]. These and other data revealed a specific mechanism in the mollusc CNS. The mechanism was termed "heterosynaptic facilitation" (HSF) and was shown to be based on special presynaptic connections [38, 95, 96, 434, 475, 478, 479, 647, 894, 895, 896, 898, 944]. Generally, the studies based on the UPSP recordings indicate mainly presynaptic mechanisms for both LFD and HSF (see Sect. 7.10).

2.5.3 Conditioning Mechanisms in the Aplysia CNS

Continuation of the studies in the same line appeared to be a clear demonstration of the possibilities of the UPSP recordings. Behavioural conditioning of the gill and siphon withdrawal reflexes was demonstrated in *Aplysia* [209, 210, 1070]. Unitary EPSPs were evoked in an identified target motoneurone by intracellular stimulation of two or three related presynaptic sensory neurones [186, 198, 204, 211, 416, 417, 1070]. One sensory neurone was induced to fire immediately prior to unconditioned ("facilitating") peripheral electric shocks; discharges of two other sensory neurones were either not evoked at all or were evoked randomly without a pairing with the unconditioned stimulus. It was demonstrated that the HSF of the unitary EPSP in the target motoneurone was selectively enhanced if the related sensory neurone was active just before the unconditioned stimulus was presented. Comparison with behavioural data strongly suggests that this "activity-dependent facilitation" is likely to contribute significantly to the conditioning of the withdrawal reflex.

Several studies indicated an existence of specialized "facilitating" or reinforcing neurones which make presynaptic connections on the modifiable presynaptic endings of the sensory neurones [5, 198, 207, 211, 375]. The facilitating neurones use serotonin and some peptides as transmitters. It might be stated that these studies confirmed experimentally the existence of a special type of modifiable synapse similar to that predicted by Konorski [519].

2.5.4 LFD in the Vertebrate CNS

Mechanisms of behavioural plasticity have not been so well investigated in vertebrates. These mechanisms are waiting for detailed studies based on UPSP and MPSP recordings. The habituation of the fish startle reflex is an example of a simple vertebrate plasticity of great potential value for future investigations [13]. Unitary IPSPs were recorded in the goldfish Mauthner cell during frequency depression [528, 529]. The statistical analysis strongly suggested a presynaptic mechanism consisting in a diminution of the probability of transmitter release (see Sect. 7.10.3). A LFD which might in principle underlie the behavioural habituation was studied with recordings of aggregate EPSPs from the frog [327] and cat [924] spinal motoneurones. The results are in agreement with the explanation of the LFD as a "self-generated homosynaptic depression" [437]. A depression of polysynaptic EPSPs at different levels of the vertebrate brain was also observed during low-frequency peripheral stimulation [801, 881, 979, 1026]. However, the underlying cellular mechanisms were not subjected to a detailed analysis.

2.6 Conclusions

The review showed that the methods of recording and analysis of MPSPs and UPSPs came into increasing use. The recordings of MPSPs and especially of UPSPs make it possible to supplement and specify data which can be obtained from the analysis of aggregate (compound) PSPs. The method provides possibilities to estimate the type of corresponding individual connections (inhibitory or excitatory, electrical or chemical) and to describe them quantitatively. The effectiveness of the methods might be exemplified by the detection of cholinergic interneurones with a double action in mollusc CNS. These neurones are able to excite some target neurones and to inhibit others [958], or even to exert a double action upon the same neurone depending on the presynaptic spike frequency [361, 1065]. This discovery, which might be considered as an original illustration of "Dale's principle" (see [299, 300]), led to a reconsideration of the idea about the obligatory existence of specialized interneurones in inhibitory pathways.

Another important problem is the existence of the electrical synaptic transmission in the vertebrate and especially mammalian CNS. The study of this problem with UPSP recordings is at its beginning (Sect. 2.4.2). Similar to many other cases, an old discussion between proponents of electrical and chemical transmission came to a compromise solution. At the same time, the role of electrical connections should not be exaggerated. Most probably, they have a rather narrow function consisting in rapid and safe transmission of signals in specialized circuits. It was found that the electrical connection in the avian ciliary ganglion did not display short-term plasticity (i.e. PTP), in contrast to the chemical synapse in the same object [674]. On the other hand, changes in the electrical transmission may participate in some forms of developmental plasticity [350]. Moreover, a kind of a LTP was recently demonstrated for both components of a mixed synapse at the goldfish brainstem [1130].

UPSP recordings can be used to estimate the distribution of synapses over the surface of the somadendritic membrane. This method enables the most precise characteristics of interneuronal connections to be obtained and it allows, in principle, to reconstruct a realistic map of the functional organization of both invertebrate and vertebrate CNSs. The UPSP recordings, in combination with the intracellular staining, seem to be especially promising for an analysis of precise morphological characteristics of individual synaptic connections [382, 383, 824, 825, 830, 888]. In addition, a method of direct observation of the neurones recorded was proposed [318], but apparently needs further improvements before it becomes more practical.

The studies of mechanisms of neuronal plasticity is a particularly promising and quickly developing field of the application of UPSP recordings. Meanwhile, the related methods are still applied mainly to neuromuscular junctions, neuroneuronal connections at peripheral nervous systems and at invertebrate CNSs. The number of the publications related to the vertebrate CNS is very limited and they are confined mainly to the spinal cord and the fish Mauthner cell (see Sect. 2.5). There were no attempts to apply methods of UPSP or MPSP recordings to analysis of the hippocampal plasticity when we started our experiments [88, 1031, 1032, 1045] and even up to now only few related papers have been published, mostly only recently [100, 343, 351, 662, 700, 1126, 1129]. Therefore, analysis of mechanisms underlying short-term and especially long-term hippocampal plasticities with the methods described in this chapter was the aim of our following studies.

2.7 Summary

Recordings of "minimal" postsynaptic potentials (PSPs), i.e. PSPs evoked by an activation of a few afferent fibres, and of "unitary" PSPs, i.e. PSPs evoked by the activation of a single afferent neurone, represent promising electrophysiological methods. Known techniques to record and detect minimal and unitary PSPs are considered. Publications are reviewed in which a less well-known method of recording of "unitary field potentials" or "electroencephalogram quantum" was used. Various applications of the recordings of minimal and unitary PSPs are analyzed. Particular attention is paid to a review of the data related to the problem of synaptic plasticity at the level of the central nervous system. The review shows that the related analysis of the analysis of plasticity mechanisms was restricted mainly to neuromuscular junctions, peripheral neuro-neuronal connections and invertebrate CNSs. There are only a few publications in which these methods were used to study synaptic plasticity in vertebrates, and these were confined until recently mainly to the cat spinal motoneurones. Of particular importance is the use of these methods for analysis of plasticity mechanisms at higher levels of the mammalian brain. Hippocampal long-term potentiation appears to be a convenient paradigm for this purpose.

3 "Minimal" Field Potentials of the Hippocampus and Their Post-Tetanic Changes

3.1 Introduction

The long-lasting changes in slow components of the FPs described in Chapter 1 were in general agreement with the hypothesis that changes in synaptic efficacy might constitute the basic mechanism of the hippocampal LTP. However, the method of FP recording is inadequate for a deeper analysis of the neural mechanisms of plasticity. It was stressed that the amplitude of aggregate FPs may vary not only because of modifications of the efficacy of the related monosynaptic excitatory connections but also due to several other reasons (Sect. 1.8).

A new method of analysis of synaptic activity has been developed by Gutman and his colleagues (Sect. 2.3). We suggested that the high amplitudes of FPs and laminar organization of the afferent inputs could facilitate a recording of the EEG quantum in the hippocampus. Therefore, we attempted to record hippocampal EEG quantum and to apply this method to analysis of LTP mechanisms [1044] before using intracellular techniques and methods of MPSP recordings, also described in Chapter 2.

3.2 Methods

In addition to the procedures described in Section 1.4, a bundle of five to seven tungsten wires (each 30 to 50 μm in diameter) was inserted into area CA1 or CA3 and fixed at the level of the pyramidal layer, as judged by the wave form and latency of the FP evoked by the septo-fimbrial stimulation. The electrodes were subsequently used both for the stimulation (to induce transhippocampal response) and for a long-term (days) recording of extracellular neuronal spikes [363] from the contralateral hippocampus at the point symmetrical to that of the FP recording. In experiments with recordings from area CA1, other stimulating electrodes were placed in CA3 or in the *stratum radiatum* between CA3 and CA1 to stimulate the Schaffer collaterals. In experiments with recordings from area CA3, additional stimulating electrodes were placed in the DG. Glass micropipettes filled with Wood's alloy and platinized afterwards were used in some experiments for recordings instead of the more common glass micropipettes.

3.3 Field Potentials Correlated with Neuronal Discharges from the Contralateral Hippocampus

Spike discharges were recorded from area CA1 or CA3 in chronic experiments. The spikes (Fig. 3.1B1) were used to trigger an analyzer for recordings the average activity from the homotopic point of the contralateral hippocampus at the level of the maximal amplitude of the negative wave of the transhippocampal FP (Fig. 3.1C, D1). From 500 to 10,000 sweeps were used for averaging. Altogether, 29 recordings of this type were made and even more averages were observed without recording. In most cases (20 out of 29 recordings), either no waves were found (Fig. 3.1A1) or their amplitudes were small (Fig. 3.1A2) and significantly not different from variations of the corresponding control potentials obtained when the triggering was not correlated with the neuronal activity (Fig. 3.1A3). In nine other cases, some fluctuations were present in the spike-triggered averages (Fig. 3.1B2, C2, D2) with amplitudes larger than that of fluctuations in respective control recordings (Fig. 3.1B3, C3). However, we hesitated to identify these as EEG quanta because the rise time of these waves (3–5 ms) was appreciably greater than we expected on the basis of the published data related to the EEG quanta [390, 400]. Even more important was their instability in time: amplitudes

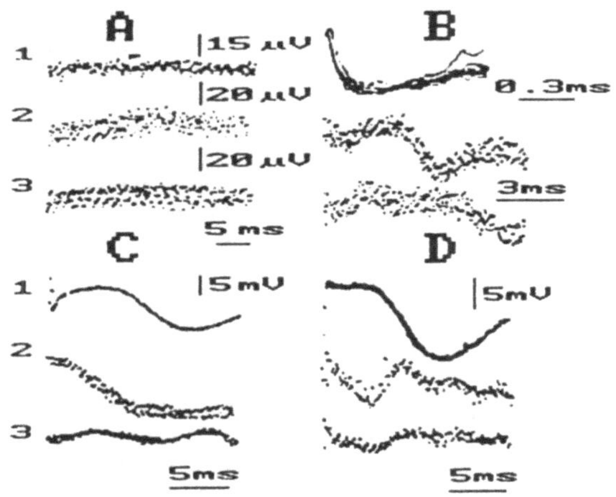

Fig. 3.1A–D. Examples of averaged post-spike field activity triggered by neuronal discharges from the symmetrical contralateral points of areas CA3 (**A**, **C**, **D**) and CA1 (**B**). **A** No field activity correlated with the triggering spikes. The field activity was triggered by spikes of two different neurones in the same experiment (*1*, *2*) or by the analyzer (*3*). **B** Examples of spikes of chronically recorded single neurone (*1*), field activity triggered by these spikes (2) and by the analyzer (*3*). **C**,**D** Averaged FP evoked by suprathreshold contralateral stimulation (*1*) and spike-triggered averages from the same recording electrode (*2*, *3*). **C3** shows a control similar to **A3** and **B3**. **D3** shows repeated averaging of the spike-triggered activity under the same conditions as for **D2**. Approximate number of realizations for **A**: 5400 (*1*), 4000 (2), 4300 (*3*); for **C1** and **D1**: 10, and for all other records about 1000 or more. [1044]

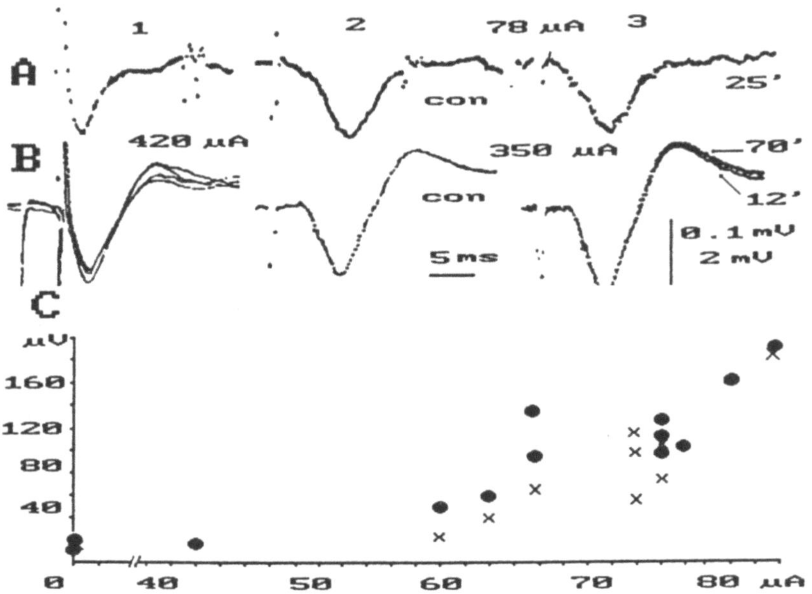

Fig. 3.2A–C. Post-spike potentials and FPs evoked by stimulation of the Schaffer collaterals. **A** Averaged potential triggered by discharges identified as spikes of a Schaffer collateral (*1*) and averaged FPs evoked by weak (near-threshold) stimuli in the same experiment before (*2*) and after (*3*) tetanization. **B** Responses of the identified Schaffer collateral to a strong stimulation of area CA3 (*1*), averaged aggregate FP before tetanization (*2*) and superposition of two averaged FPs obtained 12 and 70 min after tetanization (*3*). Stimulus current strength (in μA) and time after tetanization (in min) are shown. **C** Amplitude of averaged FPs (*ordinate*) plotted as a function of stimulus current (*abscissa*) in the same experiment before (*dots*) and after (*crosses*) tetanization. Number of averaged responses was 50 for **A** and **C**, 10 for **B2** and **B3**. Amplitude calibration 0.1 mV for **A** and 2 mV for **B**. [1044]

and time characteristics were usually substantially changed by repetition of the recordings under the same conditions or by an increase in the number of averaged sweeps. For example, the negative wave which was clearly distinguished in Fig. 3.1D2 virtually disappeared as a result of a repeated averaging of approximately the same number of realizations (Fig. 3.1D3).

3.4 Field Activity After Spikes of the Schaffer Collaterals

One of the reasons why stable recording of the EEG quantum was not obtained in the previous experiments might be an inexact matching between the target recording point and the location of the related afferent neurone. In several experiments, an attempt was made to record activity of the presynaptic axons (Schaffer collaterals). The neuronal spikes and postspike field activity were recorded with the same microelectrode in area CA1. We expected that Schaffer collaterals would respond to the CA3 stimulation by single spikes with a short (under 5 ms) and stable latency without a FF. Of several dozen units recorded, only two (both in experiments with the platinized recording electrodes) satisfied these criteria (Fig. 3.2B). All other recorded spikes exhibited a well-marked FF, i.e. latency shortening, increase in discharge probabilities and appearance of grouped discharges, typical for the postsynaptic responses in the hippocampus. Figure 3.2B shows that the spikes of the presumed Schaffer collateral appeared almost simultaneously with the beginning of the negative wave of the FP (compare also Fig. 3.2B1 with Fig. 3.2B2, which shows averaged response to a strong stimulation but minimally subthreshold for the discharge of the presumed Schaffer collateral).

In both successful experiments, the triggering of the field activity by the presumed spikes of the Schaffer collaterals produced postspike negative waves with amplitudes in the range of 40–60 μV (Fig. 3.2A1). The waves were stable in time and reproducible with repeated averaging. The amplitude of the postspike activity (Fig. 3.2A1) was similar to that of the responses evoked by weak near-threshold stimulation of area CA3 (Fig. 3.2A2).

3.5 Minimal Field Potentials

In the next series, an attempt was made to distinguish the EEG quantum by selecting a stimulus strength sufficient to evoke "minimal" averaged FP. The recording microelectrode was inserted at a depth corresponding to the maximal amplitude of the negative wave (Fig. 3.3C) evoked by the Schaffer collateral or DG stimulation with recordings from the CA1 or CA3, respectively. The dependence of the amplitude of the averaged FP (50 or 200 realizations) from the near-threshold stimulating current strength was plotted (Figs. 3.2 and 3.4).

Figure 3.3A and D shows that the current of 35 μA or less was below threshold, whereas at 35.5 μA, a FP appeared stepwise. The amplitude of the response did not increase appreciably with a further increase in the current strength (Fig. 3.3A, current 36.2 μA). Another step appeared as a result of a small additional increment in the stimulus strength (Fig. 3.3D): the response amplitude was almost doubled (Fig. 3.3A, 37 μA) and remained at approximately the same level in spite of a further increase in the stimulus current strength up to 38.5 μA (Fig. 3.3D). Similar stimulus-response curves with discrete transitions (plateaux and steps) were obtained in 10 out of 16 similar experiments with near-threshold stimuli. In the other experiments, a more smooth relationship between response amplitude and current strength was observed without clear steps and plateaux (Fig. 3.2C). It is interesting to note that in the experiments with the smooth graphs, the threshold currents were always higher (over 50–60 μA for stimulus durations of 0.2–0.3 ms) than in experiments with the stepwise stimulus-response curves (10–35 μA). One can also see that with sufficiently strong currents, the curves became smoother

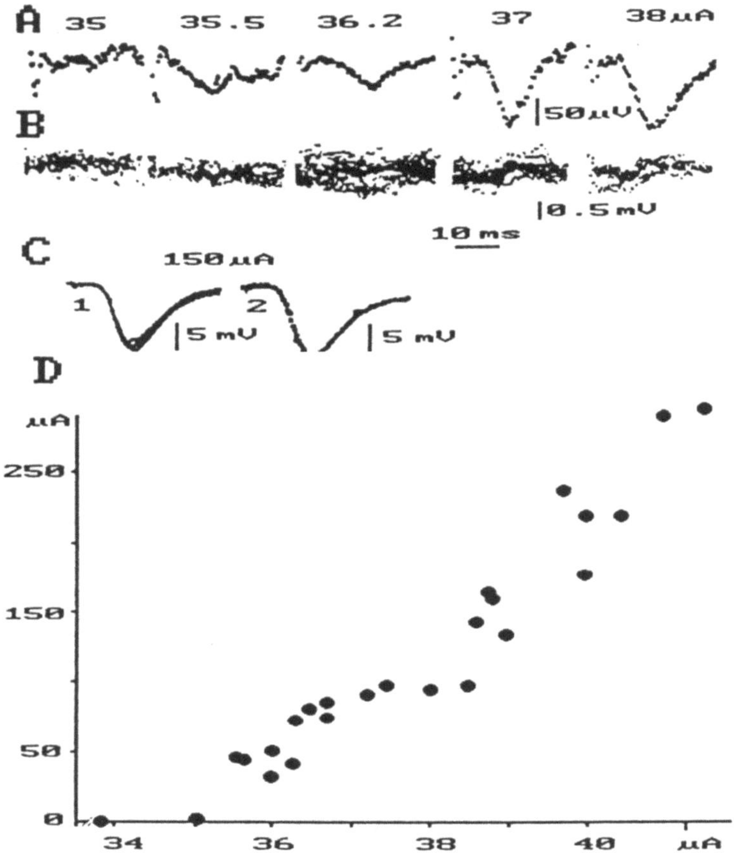

Fig. 3.3A–D. Stepwise changes in FP amplitudes with increased stimulus strength. **A** Averaged FPs (N = 50) of the CA1 region induced by stimulation of the Schaffer collaterals. **B** Corresponding superpositions of about 20 single responses. **C** Single (*1*) and averaged (N = 10) FPs (*2*) evoked by strong stimulation. Stimulus currents are shown in **A** and **C** (in µA). **D** Amplitude of averaged (N = 50) FP (*abscissa*) plotted as a function of stimulus current (*ordinate*). [1044]

(Fig. 3.3D) so that the stepwise stimulus-response relation might disappear. The FP corresponding to the first step in the stimulus-response curve we shall call "minimal" FP.

The amplitude of the averaged minimal FP, which corresponded to the first step, was from 40 to 80 µV. In most experiments, single responses were also recorded (Fig. 3.3B), but only in one experiment (with the platinized recording electrode), did the background noise appear to be low enough to measure amplitudes of single FPs evoked by weak stimulus currents. Two or three modes, with values approximately equal to integer multiples of 80 µV, could be distinguished in the amplitude distribution of the measured single responses [1044]. The experimental distribution could be fitted by the binomial distribution (with parameters $p = 0.3$, $n = 5$) convolved with the Gaussian noise distribution with the SD equal to 17 μV.

The parameters of the binomial distribution could be interpreted in this particular case as the number of excited fibres (n) and the mean probability of the response of one fibre (p). Within this interpretation, the responses corresponding to the first mode (about 80 μV) reflect the activation of one fibre, i.e. they represent EEG quanta by definition (Sect. 2.3).

3.6 Post-Tetanic Changes of Minimal FPs

Post-tetanic changes were analyzed in 13 experiments (10 and 3 experiments with recordings from areas CA1 and CA3, respectively). The tetanic stimulation (10–20 s train at 20 Hz) was delivered through the electrodes adjacent to those used for the testing stimulation. The tetanic current (0.3–0.8 mA) was chosen so that a single conditioning stimulus evoked large FP and was near the population spike threshold.

A clear LTP of the aggregate FPs was observed in only three of the ten experiments with recordings from CA1 (Fig. 3.2B2, 3). In 7 experiments, the aggregate FP showed only negligible changes or were reduced in comparison with the pretetanic control recordings. In none of the experiments was any significant prolonged increase of amplitudes of minimal FPs found (Fig. 3.2C), even when a distinct LTP of the aggregate FPs was observed (Fig. 3.2B3). In two experiments with the recordings from CA3, a significant LTP of the aggregate FPs was observed (Fig. 3.4A1–3, B1, 2). In the third experiment, the conditioning tetanization led to a prolonged depression of the recorded FP. In all cases, the minimal FPs changed accordingly (Fig. 3.4C, D). Note changes in the FPs corresponding to different steps in the stimulus-response curve (Fig. 3.4D). No significant changes in the threshold were noted after the tetanic stimulation effective for LTP induction (Fig. 3.4D).

3.7 Discussion

No reliable stable potentials that could be regarded as EEG quanta were found with the averaging of the postspike field activity when we used separate electrodes for the recording of presynaptic (spike) and postsynaptic (field) activity. One reason for this failure might be that the commissural connections between symmetrical points of area CA1 are sparse and this area receives contralateral connections mainly from area CA3 (e.g. [1000]). Another reason might be an insufficiently precise matching of the respective electrodes in the two hippocampi. An attempt

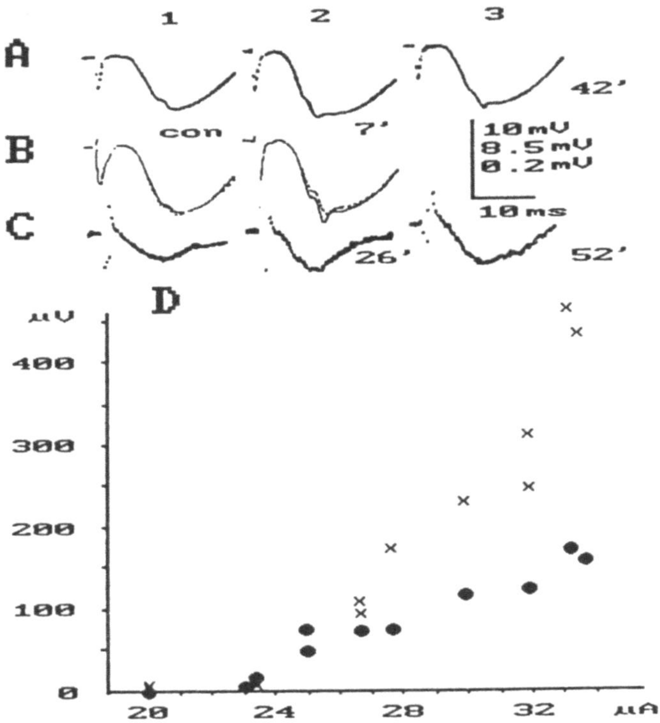

Fig. 3.4A–D. Potentiation of aggregate and minimal FPs recorded from CA3 field after the DG stimulation. **A** Averaged (N = 10) FPs evoked by a strong stimulation before (*1*), 7 min (2) and 42 min (*3*) after tetanic stimulation. **B** Examples of single FPs corresponding to averaged FPs shown in **A1** and **A2**, respectively. Responses to first (*continuous lines*) and to last (tenth) stimuli (*interrupted line*) from corresponding testing series are superimposed. Note a depression of the fast component of the last response in **B2** (*interrupted line*) in comparison with the first response (*continuous line*). C Averaged (N = 200) minimal FPs before (*1*), about 26 min (2) and 52 min (*3*) after teianization. **D** Stimulus-response curve similar to that in Figs. 3.2C and 3.3D before (dots) and after (*crosses*) tetanic stimulation. [1044]

was made to overcome this difficulty by the recordings of spikes and FPs by the same electrode following the method introduced by Gutman and his collaborators (see Sect. 2.3). However, the method by itself, without additional procedures, is not sufficient to differentiate between the EEG quantum and the proper postspike potentials. Furthermore, the cases of the successful recording of the presynaptic spike activity proved to be very infrequent. Nevertheless, the potentials observed in these rare cases could be tentatively regarded as EEG quanta. They had amplitudes similar to these of the minimal FP (see below).

A correlation between spontaneous spikes of the triggering neurone and field activity is a further possible complication which might prevent the detection of EEG quanta in our experiments with chronic neuronal recordings. The correlation of this type, which can be unstable in time, is well known for the hippocampus of unanaesthetized animals with its high-amplitude EEG [39]. Therefore, to distinguish the EEG quantum, it would be better to activate the presynaptic triggering neurone directly (see Sect. 2.2), which we were not able to achieve in these experiments.

We used here a less direct method of "minimal stimulation". Stimulus-response curves with discrete transitions were observed in most experiments with weak near-threshold stimulus currents. We suggest that the plateaux and steps are due to the activation of an increasing number of axons when the stimulus strength was increased. A similar suggestion was made independently by McNaughton [695], who studied stimulus-response curves of minimal EPSPs recorded in the DG. Similarly, we suggest that the minimal FP corresponding to the first plateau reflects a response of a single afferent fibre, i.e. that it represents the EEG quantum. This suggestion is supported by the comparison of the amplitudes of the minimal hippocampal FP (40–80 μV) and the EEG quantum (7–85 μV) of retino-tectal connections [400] and similar values of the postspike potentials of the presumed Schaffer collaterals. Our statistical analysis of single minimal responses is also in agreement with the interpretation of the minimal FPs as EEG quanta.

The rise time of the hippocampal minimal FP (4–6 ms) was substantially longer than that of the EEG quantum at the frog tectum (1.2–1.8 ms) or the duration of the monosynaptic current of the spinal motoneurones (about 2 ms). However, the results given in the following chapters show that the rise time of the hippocampal minimal EPSPs (Chap. 4) and presumed unitary PSPs (Chap. 12) is similar to these values.

A comparison of the threshold currents sufficient to evoke the minimal FPs with the radius of effective current spread [942] can be used to estimate an approximate distance of the activated fibres from the stimulating electrodes in our experiments. The estimate is about 100–150 μm for the most successful experiments (threshold currents of 10–30 μA). The distance from the recording electrode to the recorded single unit may be of about the same value [1, 613]. Accordingly, one can expect that the analysis of the EEG quanta triggered by presynaptic spikes recorded by a separate electrode may find a practical application in future. This suggestion was confirmed by more recent findings (see Chap. 12 with references related to recordings of hippocampal UPSPs).

No LTP of minimal FPs in area CA1 was found. Partially, it might be explained by the above discussed relative weak reproducibility of LTP in CA1 of intact animals (Sect. 1.10). In cases when a significant LTP of the aggregate FP was observed (Fig. 3.2), the minimal FP might presumably reflect the activity of those fibres in which LTP was not manifested. The other reason for the lack of LTP of the minimal FPs might be a LFD due to a great number of stimuli necessary for the averaging. In contrast, a pronounced LTP of the minimal FPs was found in the CA3 region. This supports the notion that LTP of the hippocampal aggregate FP is not explained by changes in processes of the transformation of the currents produced by single synapses into aggregate FPs. The contribution of changes in spike responses or in polysynaptic and inhibitory activity is also not supported by these observations. In general, the data of the present chapter are in agreement with the synaptic location of basic LTP mechanisms.

3.8 Summary

An attempt was made to record hippocampal electrical field activity induced by the activation of a single presynaptic fibre (so-called EEG quantum). In experiments on unanaesthetized rabbits, the averaging of the field activity triggered by spontaneous neuronal spikes from the homotopic point of the contralateral hippocampus revealed no reproducible waves. In two cases, spikes identified as responses of the Schaffer collaterals were recorded in area CA1. Averaging of postspike field activity recorded by the same microelectrode showed negative waves with amplitudes of 40–60 μV. The waves were interpreted as EEG quanta. Stimulus-response curves constructed for the averaged field potentials (FP) showed discrete transitions (steps and plateaux) in most experiments. We suggest that the minimal FP, corresponding to the first plateau (with amplitude of 40–80 μV) is evoked by the activity of one fibre and, therefore, represents the EEG quantum. No long-term potentiation (LTP) of the minimal FP was found in area CA1 after above-threshold tetanization of the Schaffer collaterals. The minimal FPs recorded in area CA3 following the DG stimulation demonstrated a clear LTP. The results are in agreement with a presumed synaptic location of basic LTP mechanisms.

4 Excitatory Postsynaptic Potentials of Hippocampal Neurones and Their Low-Frequency Depression

4.1 Introduction

The results obtained with recordings of aggregate (Chap. 1) and minimal FPs (Chap. 3) were considered as a support of the hypothesis suggesting that the basic LTP mechanism is a persistent increase in the efficacy of excitatory synapses. Nevertheless, it was necessary to test this hypothesis more directly with intracellular recordings, including recordings of minimal PSPs (see Chap. 2). The hippocampus seemed to be a convenient structure for this kind of studies because of the possibility of stimulating morphologically identifiable monosynaptic inputs. However, as it was noted (Sect. 1.2), distinct hippocampal EPSPs were very infrequently encountered in the in vivo experiments. Single intracerebral stimuli evoked predominantly primary IPSPs (see Sect. 1.2 for references). Another factor interfering with the intracellular recordings in the hippocampus is the presence of high-amplitude FPs, which can mask cellular synaptic responses. We met such difficulties from the very beginning of our experiments with intracellular recordings from the in vivo hippocampus [891, 1030, 1031, 1042, 1045]. Therefore, it was important to find a method which would allow to overcome these difficulties and to record distinct hippocampal EPSPs following stimulation of related monosynaptic pathways.

The aim of the present chapter is to describe this method together with some properties of hippocampal EPSPs revealed by this method [1031]. Intracellular recordings of minimal EPSPs (Chap. 2) allow a more precise analysis of plasticity mechanisms than the methods discussed in previous chapters. As a first step, we studied neuronal mechanisms underlying LFD described with FP recordings in Chapter 1. Numerous studies of LFD and behavioural habituation were reviewed in Section 2.5. It was mentioned that no special analysis of LFD mechanisms at the higher levels of the mammalian CNS had been previously undertaken.

4.2 Methods

Intracellular and so-called quasi-intracellular [690] or partially intracellular [911] recordings were made in experiments described here and in the following chapter. Glass recording microelectrodes were inserted in areas CA3 or CA1 as described in Section 1.5. Testing stimulations were applied to the SFA, DG or Schaffer collaterals. The spike amplitude varied from 7 to 30 mV and the steady potential shift was in the range of 10 to 40 mV in the quasi-intracellular recordings. The spikes were from 30 to 70 mV at resting membrane potential (RMP) from 40 to 60 mV in healthy neurones with intracellular recordings. However, essential parts of the stable intracellular recordings were obtained from neurones with lower RMP (20–40 mV). The spike generation in these cells was inactivated due to depolarization. All three types of record, i.e. the intracellular recordings with a relatively high RMP and preserved spikes, intracellular recordings with inactivated mechanism of the spike generation and quasi-intracellular recordings, provided similar results and were mutually complementary. Exceptional cases with different results will be indicated below. In general, the results obtained with all three kinds of recordings will be considered together. The main criterion for the inclusion of a record into analysis was the stability of the steady potential level, wave form and amplitude of spikes and PSPs throughout the recording period. The duration of the stable recording period varied from 1 to 40 min for various neurones.

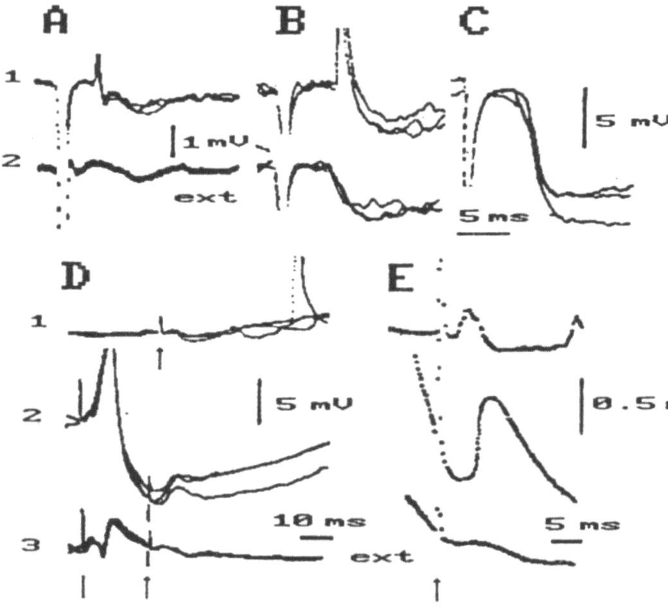

Fig. 4.1A–E. Responses of two neurones (**A–C** and **D–E**) to the septo-fimbrial stimulation. **A** Extracellular discharges before penetration (*1*) and control FPs after the cell was lost (*2*). **B** Responses of the same neurone recorded intracellularly with (*1*) and without (*2*) evoked spike discharges. **C** Responses of the same neurone after a depolarization for 10–15 mV as a result of injury. **D** Single superimposed responses of another neurone to weak near-threshold (55 μA) stimulus current (*1*), to paired pulses of 200 and 55 μA (*2*) and control extracellular FPs (*3*). **E** Averaged responses to weak pulses (55 μA) corresponding to **D1–D3**, respectively. Note that only a small hyperpolarization was evoked by the weak testing stimulus (**D1**); the initial depolarization in **E1** is accounted for by accidental spontaneous discharges. Here and in other figures, the strong (conditioning) stimuli and weak (testing) stimuli are marked by *bars* and *arrows*, respectively. [1031]

4.3 Responses to Single Stimuli

Well-marked IPSPs were observed in most neurones in the present experiments, similarly to the experiments of other authors performed on anaesthetized animals and on animals with removed neocortex (see Sect. 1.2 for references). Primary EPSPs were not always distinct even when spike discharges were evoked: either the prespike depolarization was absent (Fig. 4.1B1, 2) or its amplitude (Fig. 4.1D2) did not exceed the FP recorded immediately after the neurone was lost (Fig. 4.1D3). In some cases, paradoxical observations were made: if the FP recorded by the same microelectrode later (Fig. 4.1D3, first stimulus) was subtracted from the intracellular record (Fig. 4.1D2, first stimulus), the spikes would arise from a hyperpolarized level.

Stimulus evoked spikes were observed infrequently (Fig. 4.1B1, D2) and intracellular responses typically started with primary IPSPs (Fig. 4.2A2). The occurrence of primary IPSPs might be partially explained by the depolarization and traumatic effects of the recording microelectrode. In fact, we sometimes observed a decrease in the probability of spike responses and an increase in their latency after the cell was penetrated (compare Figs. 4.1A1 and 4.1B1, 2) and with time passing after the cell penetration (Fig. 4.1C). Attempts were made to reduce the influence of the IPSPs and to avoid the high-amplitude FPs by delivering near-threshold stimuli of low intensity (under 0.1 mA). However, no initial depolarizing responses were usually observed which would significantly exceed control extracellular FPs (Figs. 4.1D1; 4.2A1, B1).

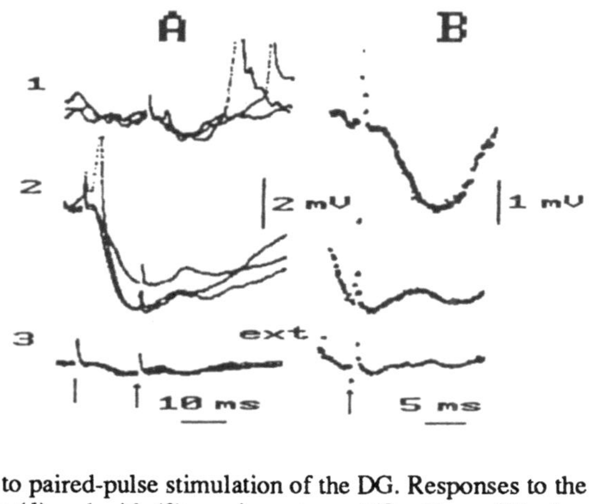

Fig. 4.2A,B. Intracellular responses to paired-pulse stimulation of the DG. Responses to the weak testing stimuli (39 μA) without (*1*) and with (2) previous strong (68 μA) conditioning stimuli and also respective extracellular controls (*3*) are shown in **A** and **B** similarly to Fig. 4.1D and E, respectively. Note the reversion of the hyperpolarizing potential (**A1**, **B1**) into the depolarizing one (**A2**, **B2**) due to the conditioning procedure

4.4 Responses to Paired Stimuli of Different Amplitudes

With recordings of 62 neurones, a weak near-threshold (usually below 0.1 mA) testing stimulus was delivered with an interval of 15–30 ms following a strong suprathreshold conditioning stimulus (up to 1 mA). Under these conditions, the response to the testing stimulus appeared against the background of the IPSP evoked by the conditioning stimulus. The primary hyperpolarizing response to the near threshold testing stimulation converted into a pure depolarization in 45 out of 62 neurones tested with this conditioning procedure (Figs. 4.1D2 and 4.2A2, second stimulus; Figs. 4.1E2; 4.2B2). The depolarization substantially exceeded the corresponding extracellular FP in 34 cells (compare Fig. 4.1E2 and 3). Primary hyperpolarizing responses persisted in 17 neurones in spite of the conditioning procedure. However, the testing stimuli were usually applied during the descending phase of the conditioning IPSP in these cases. The latency of the averaged depolarizing response evoked by the testing stimulus during the conditioning proce-

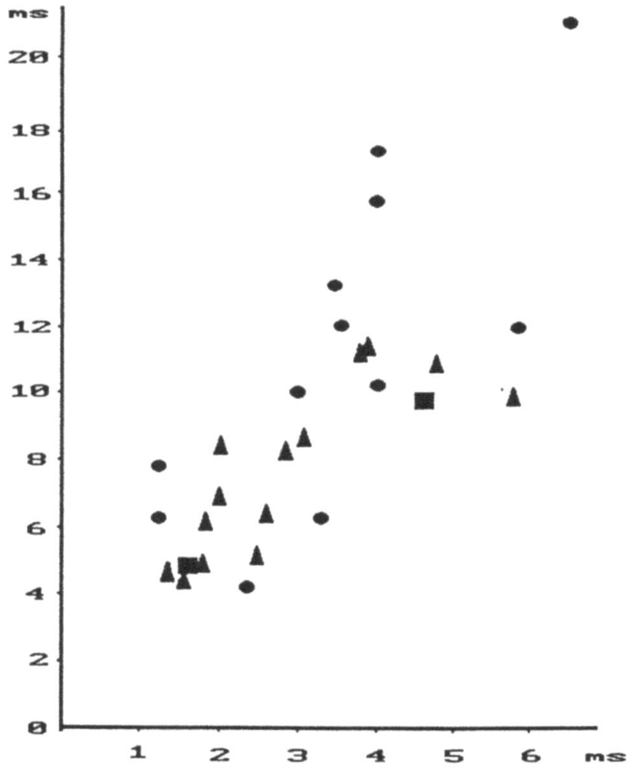

Fig. 4.3. Relationship between the half-width (*ordinate*) and rise time (*abscissa*) of minimal EPSPs. *Dots* and *triangles* represent responses of CA3 neurones evoked by stimulation of the DG and septo-fimbrial stimulation, respectively. Responses to stimulation of the Schaffer collaterals in two CA1 neurones are shown by *squares*. [87]

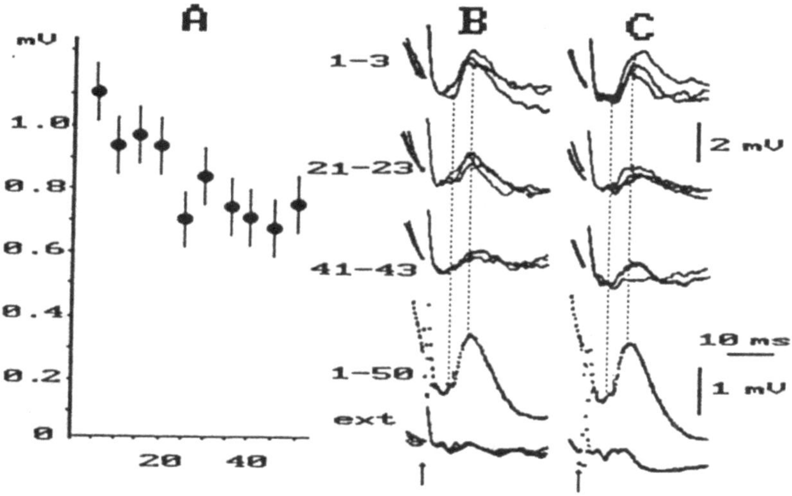

Fig. 4.4A–C. Low-frequency depression of minimal EPSPs evoked in CA1 by the septo-fimbrial stimuli delivered with intervals of 2–3 s. **A** Mean amplitudes of five consecutive EPSPs measured from three stimulus series. *Abscissa* stimulus number; *ordinate* mean amplitude of 15 EPSP. *Error bars* represent ±SEM. **B** and **C** Examples of responses from two series. Sequential numbers of superimposed responses are indicated to the *left* of the records. 1–50 marks averaged responses (N = 50). *Bottom records* represent single (**B**) and averaged (**C**) control extracellular responses. [1031]

dure was usually close to the latency of the hyperpolarizing response evoked by the same stimulus without the conditioning (compare Fig. 4.2B1 and 2). However, deviations in the range of 1–3 ms to either direction were observed in some cases.

To determine the nature of the slow depolarizing waves induced by the testing stimuli (see Sect. 4.7), a correlation was calculated between their amplitudes and that of the IPSP evoked by the conditioning stimuli in 4 neurones with at least 50 measurements. The correlation was significant in no case: the coefficient of correlation ranged from –0.18 to 0.15 ($P>0.05$).

On the basis of our results, the depolarizing responses will be regarded as EPSPs (Sect. 4.7). The amplitude of the averaged EPSPs varied from 0.2 to 1.7 mV, and it increased with an increase in stimulus strength. The amplitudes of single non-averaged EPSPs evoked by testing stimuli of constant strength were varied, and failures might appear. These effects indicate that only a small number of presynaptic fibres were stimulated, and, therefore, the majority of the responses could be regarded as "minimal" EPSPs (Sect. 2.2.4). Figure 4.3 shows shape indices of the minimal EPSPs: the half-width, i.e. the duration at the half-amplitude, and the rise-time, i.e. the time taken to rise from 10 to 90% of the peak amplitude. The latencies of the minimal EPSPs varied from 1.5 to 6.2 ms.

4.5 Changes of EPSPs During Low-Frequency Stimulation

Effects of low-frequency stimulation with intervals from 1 to 5 s were studied in 15 neurones with distinct minimal EPSPs recorded from area CA3. The SFA (13 neurones) or DG (2 neurones) were stimulated with paired pulses (conditioning and testing) of two different amplitudes as described in the previous section. Responses of 3 out of 15 neurones were recorded after preliminary tetanizations of the same pathway (see the next chapter). The amplitudes of single responses were measured between two time points corresponding to the beginning and maximum of the averaged EPSP, respectively (Fig. 4.4B, C, broken lines). The measurements were appropriately corrected whenever the amplitude of the control extracellular FP was not negligible in comparison with the EPSP amplitude. The amplitudes of single EPSPs varied (Figs. 4.4, 4.5) but the general tendency was the amplitude diminution in the course of the stimulation. The diminution was evident with comparisons of both single EPSPs (Figs. 4.4B, C) and their mean amplitudes calculated from several consecutive responses (Figs. 4.4A and 4.5). The LFD was statistically significant in the majority of neurones (11 out of 15) tested with trains consisted of at least 40 testing stimuli. The rate and the degree of the LFD varied in various neurones (compare Figs. 4.4 and 4.5) and even in the same neurone with stimulations of different intensities. Similar to the LFD of the FPs

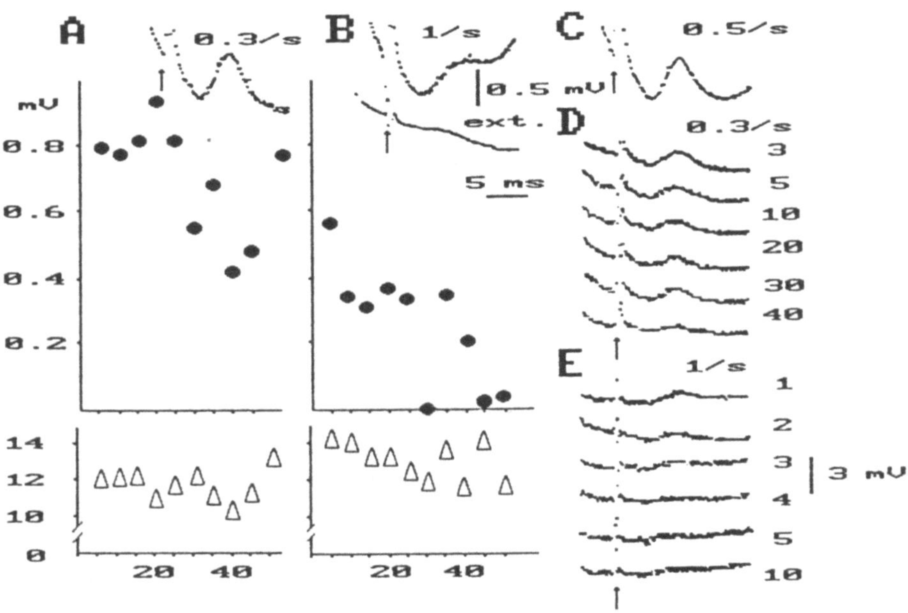

Fig. 4.5A–C. Dependence of LFD on the stimulus frequency. Examples of averaged (**A–C**) and single (**D, E**) responses to stimulation with various mean frequencies are shown. Amplitudes of the testing EPSPs (*dots*) and conditioning IPSPs (*triangles*) are plotted in **A** and **B** against stimulus numbers. See Figs. 4.1 and 4.4 for other explanations. [1031]

(Chap. 1), the LFD of the EPSPs was especially prominent in neurones recorded after the tetanic stimulation (see Fig. 5.4C, D).

After an interval of several minutes without testing stimulation, a partial or complete recovery of the initial EPSP amplitude took place (Figs. 4.4C). When the low-frequency stimulation was repeated, the EPSP amplitude decreased gradually again (Figs. 4.4C and 4.5B). The repetition usually led to a faster and a more profound LFD (compare Fig. 4.5A and B). There was a dependency between the rate of the depression and stimulus frequency in the frequency range tested (0.2–1 Hz). Examples are given in Fig. 4.5, where a rather mild depression was seen at 0.3 Hz stimulation (Fig. 4.5A, D), while it was more prominent at 1 Hz stimulation (Fig. 4.5B, E). However, with a further increase in the frequency, a FF was observed [25] (see Sects. 7.10.2, 9.7.4 for more references and discussion).

4.6 Changes of IPSPs During Low-Frequency Stimulation

Changes in IPSP amplitudes during low-frequency stimulation were analyzed in seven neurones. The IPSPs were evoked by strong conditioning stimuli in four neurones and by minimal near-threshold testing stimuli in three neurones. A decrease in IPSP amplitude was found during the low-frequency stimulation but it was not as prominent as that of EPSP (Fig. 4.5A, B, triangles). Only in one case was a tendency to an increase in the IPSP amplitude observed during the low-frequency train, but this increase was not statistically significant.

4.7 Discussion

The absence of well-marked EPSPs in the majority of hippocampal pyramidal cells recorded in the in vivo experiments was explained by Andersen and Lømo [25] on the basis of a remote dendritic location of the corresponding synapses. However, more recent in vitro experiments indicated that it is hardly likely to be an important reason. These experiments demonstrated clear-cut initial EPSPs, sometimes without significant IPSPs (see, e.g. initial publications [252, 253, 910, 912]). This does not mean, however, that hippocampal inhibitory connections are completely suppressed in the slice preparations because distinct IPSPs can also be recorded (see reviews [193, 784] and below). Our experiments with paired stimulation suggest that a mixed synaptic input which consists of excitatory (EPSP) and inhibitory (IPSP) components is typical for in vivo preparation and that the inhibitory component is usually predominant in this mixed response. The predominant activation of the inhibitory connections may be explained by the location of the stimulating and recording electrodes in different hippocampal laminae. Traumatic effects of the recording microelectrode should additionally diminish the EPSP amplitude due to a membrane depolarization and a decrease in the input resistance (R_{in}). The depolarization can also augment the IPSP amplitude. These effects might be more essential in in vivo as compared to more stable in vitro preparations.

We found a method to reveal EPSPs from the presumed mixed synaptic input. The predominant hyperpolarizing responses were converted into depolarizing ones following a strong conditioning stimulation. The alternative explanation of this conversion is a reversion of the sign of the testing IPSP due to the conditioning hyperpolarization. This seems to be unlikely because of the lack of a significant correlation between amplitudes of the testing depolarization and conditioning IPSP. Occasional differences between the latencies of the responses recorded without and with conditioning are also more compatible with the interpretation of the depolarization recorded during paired stimulation as an EPSP. Moreover, no inversion was found in recent in vitro studies of paired-pulse depression of pure IPSPs [250]. The detection of the EPSP in the mixed synaptic responses during the conditioning procedure can be explained, firstly, by a suppression of the earlier inhibitory components and, secondly, by a PPF of the excitatory component. The suppression of the inhibitory component could be due to several mechanisms which have been under intensive study during recent years in connection with two related problems, firstly, an IPSP suppression during repetitive stimulation [25, 256, 687, 967] and, secondly, a suppression of IPSP evoked by a stimulation of one pathway following a "priming" stimulation of another pathway [512, 594, 934]. In addition to the postsynaptic membrane hyperpolarization during conditioning stimulation, several other mechanisms might be responsible for the IPSP suppression, such as changes in K^+ and Cl^- gradients [687, 967], a desensitization of subsynaptic gamma-aminobutyric acid (GABA) receptors [770, 1113] (see, however, [967]), a release of "modulating" substances (e.g. [649, 878]) and, especially, GABA feedback on its own release which includes $GABA_B$ receptor mediated presynaptic inhibition (e.g. [250, 256, 753]). In a more general sense, the term "presynaptic inhibition" can include also an inhibition of inhibitory interneurones. In fact, clear IPSPs were evoked by the strong stimuli in practically all neurones recorded in our studies both within and outside the pyramidal layer. The inhibition of inhibitory interneurones seems to be a major mechanism of the IPSP suppression under our conditions, especially in view of recent observations [250] that pharmacological blockade of excitatory synaptic transmission completely suppressed IPSPs evoked by a remote stimulation in hippocampal slices.

The inhibition in presynaptic pathways should favour recordings of monosynaptic responses during paired stimulation. An important advantage of the method of paired stimulation is that not only averaged responses but also single responses to the near-threshold stimuli can be detected and measured. The detection of single responses is favoured by a significant reduction in the background synaptic noise during the conditioning IPSP. Such a synaptic noise of large amplitude is typical for CNS neurones of unanaesthetized animals [1036]. The method of the paired stimulation is technically more simple and more effective in EPSP detection than the more common method of IPSP suppression by intracellular injection of hyperpolarizing current.

Both averaged and single EPSPs were measured in the present study. The rise time and half-width of the averaged minimal hippocampal EPSPs (Fig. 4.3) were found to be appreciably larger than the corresponding values for minimal and unitary EPSPs of spinal motoneurones (see Chaps. 2, 7). The longer rise time of the hippocampal EPSP could hardly be explained by asynchronous presynaptic inputs, especially when the occurrence of failures indicated the activation of only a few fibres (see Sect. 2.2.4). Stability of the temporal characteristics in experiments with changes in the stimulus strength is further evidence against this explanation. More recent recordings of minimal [864, 865, 983] and unitary hippocampal EPSPs in both slices (e.g. [864, 865]) and in vivo preparations (Chap. 12) showed similar time characteristics. The long rise times might be explained by a remote dendritic location of the related excitatory synapses. The explanation is supported by the lack of a significant correlation between the amplitude of the testing EPSP and that of the conditioning IPSP and by related morphological studies [840, 1000, 1079]. However, we did not find any significant difference between temporal characteristics of EPSPs evoked by the stimulation of the DG and SFA. The differences are expected on the basis of morphological data showing that the corresponding synapses are located on the proximal and more distal parts of the apical dendrites, respectively [410, 1000, 1079]. Similarly, very small differences were found in later recordings of EPSPs following activation of distal and proximal synapses in CA1 neurones in vitro [587]. However, a more accurate analysis with recordings of minimal EPSPs [983] did reveal some differences predicted on the basis of Rall's model (see Sect. 2.4.3).

Some discrepancies with the model might be attributed, in principle, to the various durations of the synaptic currents or to multiple generation of these currents in several parts of the dendritic membrane [818]. For our case, a possible explanation may be the presence of specific giant spines with thin necks [410, 1079] which might increase the electrotonic distance from the proximal DG–CA3 (mossy fibres) synapses to the cell body. A quantitative estimate of the influence of the spine structure on temporal characteristics of the EPSP is necessary to check this explanation. A possible activation of recurrent collaterals of pyramidal cells with both types of the stimulation used in our studies might complicate our results and contribute to the similarity of EPSP time-courses (see the end of this section).

The lack of the correlation between the temporal characteristics of EPSPs and presumed location of corresponding synapses might also be explained by non-linear properties of the dendritic membrane suggested by Gutman (Sect. 2.4.3). However, the above cited recent study [983] indicated that at least for minimal EPSPs in CA1, non-linearity was hardly an essential factor. It should be noted that analysis of temporal characteristics of hippocampal EPSPs was not the special aim of the present study.

The method of paired stimulation was used here to study changes in minimal EPSPs during low-frequency stimulation. A decrease in the EPSP amplitude during stimulation exhibited some characteristic features of behavioural habituation [968], such as the asymptotic decay of the response, its recovery after an interruption, a more rapid decay during repeated stimulus series and a dependence of the decay rate on the stimulus frequency within certain limits. An increase in EPSP amplitude following a higher frequency stimuli (FF and especially LTP, described in the next chapter) can be considered as an analogue of the "dishabituation" or "sensitization" of the habituated behavioural response. Not all parametric characteristics of behavioural habituation [968] were studied here for the LFD of the hippocampal EPSPs. Nevertheless, the hippocampal LFD can be regarded as a neuronal analogue of behavioural habituation. The habituation of hippocampal spike responses to repeated peripheral stimulations [1000] can be explained by the LFD of the related hippocampal EPSPs.

The data described here are in agreement with the results obtained at other CNS levels of various species (Sect. 2.5), and they support the hypothesis which explains behavioural habituation on the basis of homosynaptic depression at excitatory synapses. An alternative explanation, based on a potentiation of the postsynaptic inhibition, seems unlikely for the present case. In fact, presumably pure EPSPs without IPSP components were recorded in most cases. Moreover, our data showed that IPSP amplitudes also decreased rather than increased during low-frequency stimulation.

It should be noted that a precise analysis of the mechanisms of synaptic plasticity demands recordings of monosynaptic responses. The stimuli used here could not only activate the related afferent inputs, but also induce antidromic excitation of the pyramidal cells and activate recurrent excitatory connections. The following arguments suggest that mainly monosynaptic EPSPs were recorded in our experiments: (1) known morphological results demonstrate that the related pathways are monosynaptic; (2) the method used which favours the activation of the monosynaptic EPSPs and a suppression of polysynaptic responses; (3) the latencies (1.5–6.2 ms) which are compatible with the monosynaptic transmission, taking into account the distances between the stimulating and recording electrodes (more than 2–4 mm) and low, under 0.5 m/s [1125] conduction velocities; (4) the stability of the latencies of the averaged EPSPs when the stimulus strength was increased including the cases when EPSP amplitude was modified as much as several times (Fig. 8.1). There was practically no doubt about the monosynaptic nature of the responses with the shortest latency (about 2 ms and below) which also exhibited LFD (see Fig. 5.4).

4.8 Summary

A method to detect "minimal" excitatory postsynaptic potentials (EPSPs) in hippocampal neurones of the unanaesthetized rabbit was proposed. The method consists in presenting a strong (up to 1 mA) conditioning stimulus which induced a distinct inhibitory postsynaptic potential (IPSP) before a near-threshold (0.03–0.1 mA) testing stimulation of the same intrahippocampal pathway. The response to the testing stimulus, evoked against the background of conditioning

IPSP, appeared to be purely depolarizing in most cases. It represented a pure EPSP, as judged by both occasional latency differences in comparison with the unconditioned hyperpolarizing response and the absence of the correlation between its amplitude and that of the conditioning IPSP. A low-frequency depression (LFD) of the EPSPs was observed when the paired stimuli were repeated with intervals of 1 to 5 s. The LFD exhibited some of the characteristic properties of the behavioural habituation, including a recovery of the depressed response after an interval without stimulation and a more rapid amplitude diminution with repeated stimulus series. The amplitude of the IPSP also diminished or showed no significant change during the low-frequency stimulation. It is concluded that the LFD of the hippocampal field potentials described in Chapter 1 is explained by the LFD of excitatory hippocampal synapses (homosynaptic depression). The results are also considered as support for the hypothesis that the major mechanism of behavioural habituation is a homosynaptic depression in related excitatory synaptic connections.

5 Responses of Hippocampal Neurones During Long-Term Potentiation

5.1 Introduction

Experiments described in the previous chapters created a basis for the analysis of LTP mechanisms at the level of single neurones. Before we started our experiments, long-term facilitation of hippocampal neuronal discharges had been described only after repeated ("chronic") tetanic stimulations [161]. After we started our experiments, a few intracellular recordings from the DG neurones in vitro were published [252, 287]. Long-lasting changes (>90 s duration) were not found, and the conclusion was that "LTP is not an intracellular event". The aim of our following experiments [1030, 1032] was an attempt to analyze LTP mechanisms with extra- and intracellular recordings of neuronal activity from the intact hippocampus of unanaesthetized animals.

5.2 Methods

Extracellular, intracellular and quasi-intracellular recordings (Sect. 4.2) were performed mainly in CA3 with stimulation of the SFA or the DG. The testing stimulation was usually applied in series (ten stimuli in each series) with 3–10 s intervals between the stimuli and 2–10 min between the series. The experimental procedure consisted in two to five control series, a conditioning tetanization (through the same or adjacent electrodes) and a subsequent post-tetanic testing. The duration of the tetanic train (10 or 20 Hz) was 10–20 s. The method of the paired-pulse stimulation with two different current strengths was used to detect EPSPs in some cases as described in the previous chapter.

5.3 General Description of Neuronal Responses

The results presented in this chapter are based on recordings from 47 neurones, which include 39 neurones recorded intracellularly or quasi-intracellularly. However, the activity of almost half of them was lost (or the neurones were damaged) during the tetanization or immediately after it. The recordings were especially unstable with the septo-fimbrial tetanization. The probable explanation might be

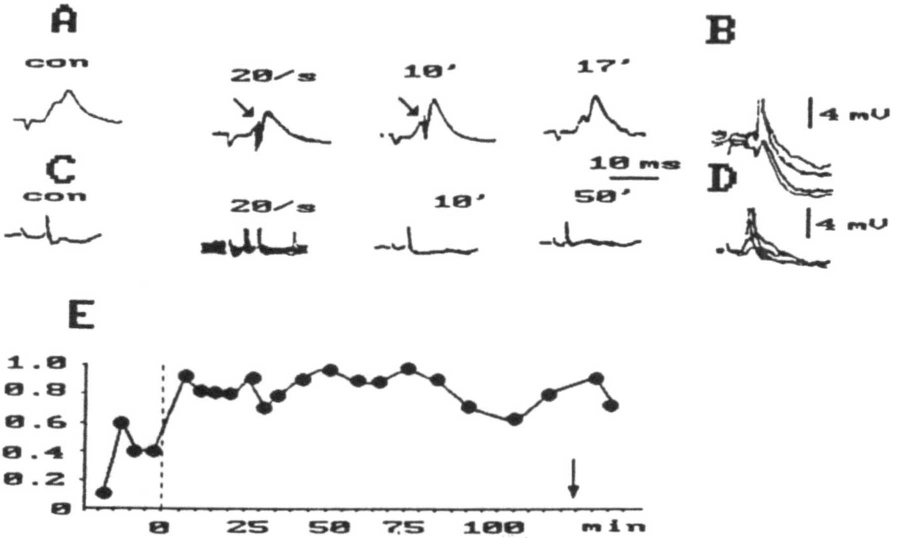

Fig. 5.1A–E. Long-term potentiation of field and neuronal responses evoked by stimulation of the SFA. **A** Examples of FPs recorded from CA3 pyramidal layer before (*con*), during (20/s) and after (10 and 17 min) tetanization. *Oblique arrows* mark neuronal discharges superimposed on the FPs. **B** Superposition of responses recorded after the penetration of the same neurone. **C** Examples of extracellular neuronal responses before (*con*), during (20/s) and after (10 and 50 min) tetanization. **D** The same neurone recorded "quasi-intracellularly". Top parts of spikes in **B** and **D** are not shown. **E** Probability of the spike discharge (*ordinate*) calculated from 10 consecutive responses plotted against time (*abscissa*). Measurements from the neurone illustrated in C and D were used. *Vertical broken line* marks tetanization. *Vertical arrow* marks a shift to the "quasi-intracellular" recording shown in **D**. [1042]

changes in blood pressure during the septal stimulation [240]. Therefore, the stimulation of the DG was used in more recent experiments. With extracellular recordings, single testing stimuli often produced spikes with a latency in the range of 2 to 10 ms (Fig. 5.1C). The neuronal spikes were often superimposed on the population spike (Fig. 5.1A, arrows) and this was a criterion for the recording from the pyramidal cell layer. With intracellular recordings, short-latency spikes were less common (Fig. 5.3A2). Clear-cut EPSPs were evoked by single pulse stimuli only in rare cases (Fig. 5.2C1). As a rule, these neurones were recorded outside the pyramidal layer as judged by FP wave form (Fig. 5.2C7). When the paired stimulation was used (see previous chapter), the second (testing) stimulus usually evoked a clear-cut EPSP on the background of the conditioning IPSP produced by the first stimulus (Fig. 5.4A).

5.4 Short-Term Post-Tetanic Changes

Changes in cellular responses during and immediately after tetanization were described by many authors [25, 161, 912]. Intracellularly, a suppression of hyperpolarizing components and an augmentation of the depolarizing components were observed during the tetanic stimulation (Figs. 5.2A2 and 5.3F). IPSPs remained suppressed and EPSPs augmented during initial 10 to 120 s following the tetanus (Fig. 5.2A3). This potentiation was analogous to the previously described short-term neocortical PTP following tetanic stimulation of the neocortical surface under similar recording conditions [1007, 1008]. The hippocampal PTP was often accompanied by seizure activity which continued for about 5 to 15 s after tetanic stimulation similarly to the observations in the in vivo neocortex. A depression of neuronal and field responses could be observed after the short-term PTP and seizure activity. According to our and literary data [25, 145], the general period of the short-term PTP and depression does not usually exceed 1–2 min. Therefore, changes in neuronal responses which persisted for more than 3 min after the tetanus will be conventionally called "long-term changes".

5.5 Long-Term Changes in Spike Responses after Tetanization

Statistically significant long-term post-tetanic changes in spike responses were found in 13 out of 28 neurones which were recorded during the period from 3 min to 2.5 h after tetanic stimulation. A post-tetanic depression was observed in three neurones and a facilitation of spike responses was found in ten cells. The facilitation was manifested as an increase in the probability (Fig. 5.1A, C) and a reduction in the spike latency (Figs. 5.1C, D). It lasted usually throughout the whole recording period (Fig. 5.1E). Figure 5.1 illustrates an example of such changes in a neurone recorded in the pyramidal layer (Fig. 5.1A) as judged by the shape of the FPs (Fig. 5.1A), and in a neurone recorded outside the pyramidal layer (Fig. 5.1C). Quasi-intracellular recordings were made in both cases (Fig. 5.1B, D) after the initial period of the extracellular recordings which lasted 30 and 130 min, respectively.

5.6 Post-Tetanic Changes of Postsynaptic Potentials

Responses of 24 neurones were recorded intracellularly and quasi-intracellularly before and after tetanic stimulation. No substantial long-term (above 3 min) post-tetanic changes of evoked PSPs were observed in 13 of them. PSP amplitudes were reduced in three neurones. Post-tetanic enhancement was observed in eight neurones. Six out of the eight neurones were recorded at the level of the pyramidal layer as judged by the FP wave form (Fig. 5.2A7) and two neurones were

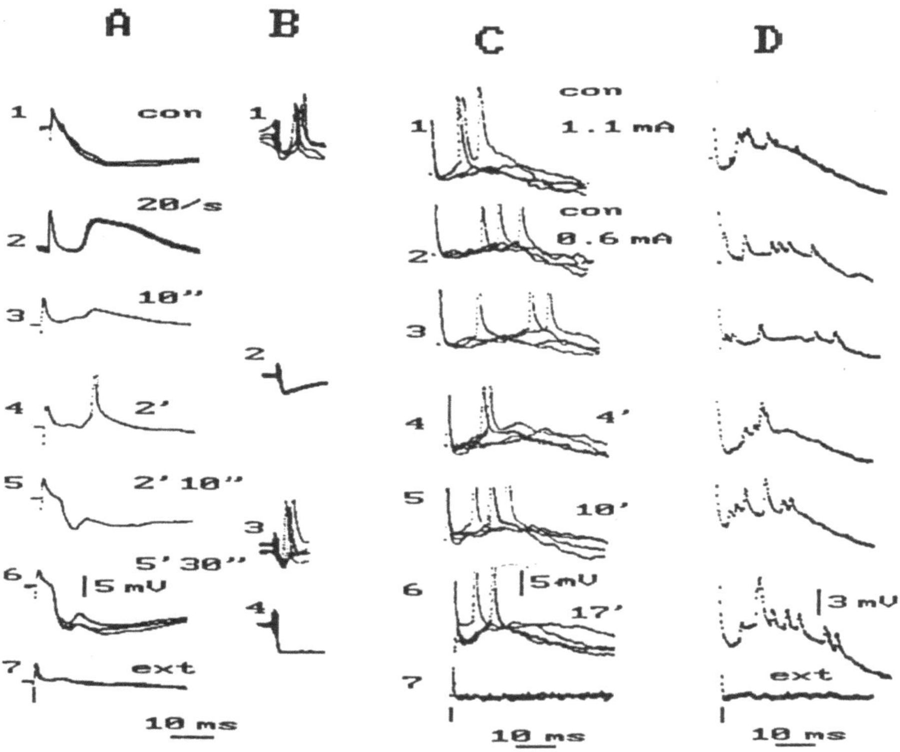

Fig. 5.2A–D. Changes in responses of two hippocampal neurones after tetanization of the DG (**A, B**) and SFA (**C, D**). Neurones shown in **A, B** and **C, D** were recorded at the level of the pyramidal layer and above it, respectively. **A** Responses before (*1*), during (*2*) and after (*3–6*) tetanization; time after tetanization is indicated. **B** Testing the postsynaptic excitability in the same neurone by current pulses delivered through the recording electrode. Note the absence of spikes to the intracellular current pulses immediately after tetanization (*2*) and the recovery of the excitability in 2 min (*3*). Control extracellular record (*4*) indicates bridge disbalance which should be taken into account for estimates of the RMP shifts during intracellular pulses in **B1–B3**. **C, D** Single (**C**) and corresponding averaged (**D**) responses of the second neurone recorded before (*1–3*) and after (*4–6*) tetanization. Stimulus strength was 1.1 mA for **C1, D1** and 0.6 mA for **C2–C7, D2–D7**. Recordings **A7, B4, C7** and **D7** were obtained extracellularly after the neurones were lost. [1032]

recorded outside the pyramidal layer (Fig. 5.2C, D). In six neurones, comparatively long-latency (4–10 ms) EPSPs either increased after tetanic stimulation or appeared in neurones without essential evoked EPSPs before the tetanus. In two neurones, EPSPs of a short (about 2 ms) and stable latency were potentiated.

An example of a neurone with a comparatively long-latency EPSP is shown in Fig. 5.2A. A short-latency (about 2 ms) IPSP (Fig. 5.2A1) was suppressed in this neurone during the tetanic stimulation and an EPSP with a longer latency (8–9 ms) appeared (Fig. 5.2A2). The EPSP persisted during the period corresponding to the short-term PTP (Fig. 5.2A3). IPSP recovered after about 2 min

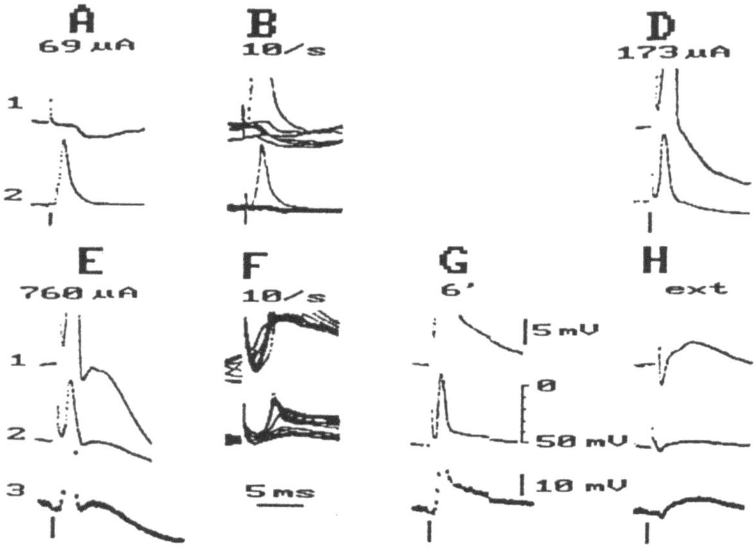

Fig. 5.3A–H. Responses of a neurone from CA3 pyramidal layer to the septo-fimbrial stimulation. Responses to stimuli of three different strengths (**A–C**, **D** and **E–H**) are shown before (**A**), during (**B**) and after the first tetanization (**C–E**); before (**E**), during (**F**) and after the second tetanization (**G**). **H** Control extracellular responses after the neurone was lost. *1* AC recordings with a higher amplification; *2* DC recordings with a lower amplification; *3* averaged responses (N = 10). Responses to two successive near threshold stimuli are shown in **A1** and **A2**; note that spike was induced only in one case (**A2**). The same responses with different amplifications are illustrated in other cases (**A–H**). Stimulus currents are indicated in **A** (valid for **A–C**), **D** and **E** (valid for **E–H**). Stimulus duration was 0.15 ms for **A–D** and 0.05 ms for **E–H**. 10/s marks the first (**B**) and second (**F**) tetanic stimulations with different current pulses (69 μA, 0.15 ms and 760 μA, 0.05 ms, respectively). Time (in min) after respective tetanizations is shown in **C** and **G**. [1032]

(Fig. 5.2A5, 6) with an increased slope as compared to the pretetanic IPSP (Fig. 5.2A1). The EPSP which appeared after the tetanus (Fig. 5.2A3) persisted during the whole post-tetanic recording period (Fig. 5.2A5, 6).

Figure 5.2C and D illustrates post-tetanic changes which might be interpreted as an increase in the amplitude of an EPSP with an intermediate latency (3–4 ms): compare pretetanic (Fig. 5.2C2, 3, D2, 3) and post-tetanic (Fig. 5.2C4–6, D4–6) responses. Note that a sufficiently strong septo-fimbrial stimulation before the tetanus (Fig. 5.2C1, D1) produced EPSP similar to the post-tetanic one: compare averaged responses in Fig. 5.2D1 and D4–6.

Appearance of short-latency (about 2 ms) EPSPs after tetanization or a significant increase in the amplitude of such EPSPs was found in two neurones recorded during the septo-fimbrial and DG stimulation, respectively (Figs. 5.3 and 5.4). Initial IPSPs (Fig. 5.3A1) occasionally preceded by a spike (Fig. 5.3A2) were evoked in one of these neurones by the septo-fimbrial stimulation before the conditioning tetanus (see also responses to a stronger stimulation, Fig. 5.3D, E).

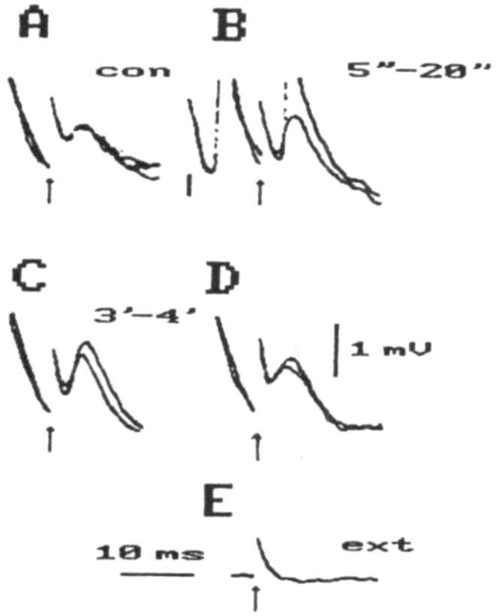

Fig. 5.4A–E. Post-tetanic changes of minimal EPSPs evoked by a weak DG stimulation in a CA3 neurone. Responses to both conditioning (0.5 mA, marked by *bar*) and testing stimuli (85 μA, marked by *arrows*) are shown in **B** as recorded soon after tetanic stimulation. Only the last part of the response to the conditioning stimulus and responses to the weak testing stimuli are shown before (**A**) and 3–4 min after the tetanization (**C**, **D**). Control extracellular record obtained after the neurone was lost is shown in **E**. [1016]

The spike was most probably antidromic, since it was characterized by a stable and short latency and corresponded in time to the short-latency (presumably antidromic) population spike (Fig. 5.3H). A weak tetanic stimulation produced neither FF (Fig. 5.3B) nor LTP (Fig. 5.3C). During the tetanic stimulation with the higher strength (Fig. 5.3F), the antidromic spike was fragmented and blocked. EPSPs with a latency of less than 2 ms appeared (Fig. 5.3F1) and occasionally evoked orthodromic spikes (Fig. 5.3F2). The IPSP was absent during the tetanic stimulation (Fig. 5.3F) and afterwards (Fig. 5.3G). The antidromic spike prevented a precise latency measurement of the post-tetanic depolarizing response. Nevertheless, a comparison of the responses recorded before (Fig. 5.3E), during (Fig. 5.3F) and after tetanization (Fig. 5.3G, see also control FPs, Fig. 5.3H) strongly suggested an appearance of EPSPs with a latency less than 2 ms after the tetanic stimulation.

An increase in the amplitude of a short-latency EPSP was clearly seen in another neurone after the DG tetanic stimulation (Fig. 5.4). The method of the paired-pulse stimulation was used so that the second (testing) stimulus (Fig. 5.4A, arrow) was delivered during the IPSP evoked by the first (conditioning) stimulus (Fig. 5.4B, bar). Note a short (<2 ms) and stable latency of the testing EPSP (Fig. 5.4A). Both conditioning and testing stimuli evoked EPSPs with occasional spikes shortly after tetanic stimulation (Fig. 5.4B). A post-tetanic increase in EPSPs was clearly seen during the whole period of the stable recording (compare Fig. 5.4A and C). A clear-cut gradual decrease in the EPSP amplitude was observed during the post-tetanic testing series (compare Fig. 5.4C and D). This was analogous to the LFD of the potentiated FPs described in Chapter 1.

IPSPs also usually increased after tetanic stimulation. The increase in the slope of the early IPSP component in Fig. 5.2 has already been mentioned. Note also post-tetanic increases in the hyperpolarization following EPSPs in Figs. 5.2C, D and 5.4C. Post-tetanic IPSP suppression during and after tetanus was observed in only one neurone (Fig. 5.3F, G).

5.7 Post-Tetanic Changes in Excitability and Resting Membrane Potential

Attempts were made to measure neuronal excitability by injecting current pulses through the intracellular microelectrode in several experiments. The current effects were measured before and after tetanization in two neurones. In the neurone illustrated (Fig. 5.2B), the excitability was decreased during the period of the short-term PTP. The decrease was similar to that found in neocortical neurones after tetanic stimulation of the cortical surface [1007]. Such a short-term post-tetanic depression of neuronal excitability might be explained at least partially by the post-tetanic hyperpolarization. Both the excitability and RMP returned to the control pretetanic level within 1–2 min after tetanic stimulation. During the period corresponding to the LTP of intracellular responses (Fig. 5.2B3), the threshold current strength was not significantly different from the control pretetanic level (Fig. 5.2B1).

It was possible to evaluate the R_{in} of this neurone by measuring the bridge disbalance (Fig. 5.2B1–4). The estimates gave a value of about 10 MOhm, which is within the range of the input resistances found in the cat hippocampal pyramidal cells in vivo [922]. The mean value reported in the former publication was 13 MOhm and the range reported in the latter was from 2 to 10 MOhm. Note that significantly higher values of 30 MOhm or more were reported in studies of hippocampal CA3 neurones in vitro (see e.g. [514]). Even higher values were found in more recent experiments using whole cell recordings [34].

No significant post-tetanic changes in the input resistance (R_{in}) were found in the experiment shown in Fig. 5.2B. It should be noted, however, that the accuracy of the estimation was not better than about 3 MOhm. Long-term post-tetanic shifts in RMP were observed in several neurones: usually it was a hyperpolarization. However, attempts to quantify such shifts also encountered difficulties associated, firstly, with changes in recording conditions and, secondly, with variable and uncontrollable changes in the extracellular steady potential level. For the most stable record in this series (Fig. 5.3), the mean value of the RMP before the second tetanic stimulation was 49±1.4 mV (mean ±SEM; the number of measurements, N = 10), and it amounted to 56±1.7 mV (N = 12) between 6 and 7 min after tetanic stimulation which induced the post-tetanic changes in evoked PSPs described in the previous section. The difference between the pre- and post-tetanic values of the RMP was statistically significant ($P<0.01$). Having in mind possible shifts in the extracellular steady potential, the changes in the intracellular spike amplitude were regarded as being a more reliable index of the true RMP shifts. These changes indicated even stronger hyperpolarization: 68±2.8 mV (after tetanic stimulation) versus 53±2.1 mV (before the tetanus; P<0.01). In no other neurone recorded, did similar estimates give statistically significant differences between pre- and post-tetanic values.

5.8 Discussion

The analysis of single neurone responses demonstrates that LTP consists not only in FP changes, but also in an enhancement of spike and synaptic responses of a considerable number of hippocampal neurones. This is consistent with the data obtained with recordings of single neurones in area CA3 during repeated "chronic" DG stimulations [161]. The potentiation of neuronal responses could also be expected on the basis of post-tetanic changes in the population spike (Chap. 1). Nevertheless, it was of importance to evaluate a possibility to produce long-term changes in responses of single hippocampal neurones after one tetanic stimulation of the afferent input. It is interesting to note that the LTP of spike responses appeared to be more prominent in neurones recorded presumably outside the pyramidal layer (3 out of 3 neurones) than in the pyramidal cells (7 out of 25 neurones). Neuronal discharges recorded both at the dendritic level and at the pyramidal layer approximately corresponded in time to the population spike (see Sect. 1.2).

As was noted previously, LTP had been initially described in the DG [145]. The finding of a similar LTP in the pyramidal layer (Chap. 1) facilitated the analysis of membrane and synaptic mechanisms of LTP. Initial data of such an analysis are given in the present chapter. Our demonstration of the lasting post-tetanic changes of intracellular EPSPs in the experiments with unanaesthetized animals was at variance with the negative results of initial related experiments with intracellular recordings in in vitro preparations (see Sect. 5.1). However, they are in general agreement with more recent results of several other in vitro studies published almost simultaneously [31, 723, 910, 1127] with our initial publications [912, 1030, 1032] or more recently (see reviews cited in the Introduction).

In spite of a comparatively small number of neurones which were recorded under stable conditions in our in vivo experiments, an increase in the amplitude of EPSPs was clearly shown. Both long-latency, probably polysynaptic, EPSPs and short-latency (less than 5 ms, sometimes less than 2 ms), most probably monosynaptic, EPSPs were enhanced. These data indicate that the major mechanism of LTP consists in an increase of the efficacy of excitatory synapses. It might be suggested that the mechanism is similar to that underlying the short-term PTP described for neuromuscular junctions and many other preparations (see Introduction for references). A presynaptic change in transmitter release is the most probable candidate to explain the short-term PTP in simplified preparations (see Introduction and Sect. 7.10). However, the participation of some other mechanisms could not be excluded in more complicated structures (Sects. 1.10, 6.1). For polysynaptic chains, changes in presynaptic neuronal elements and in inhibitory connections should be taken into account. Some evidences of modifications of this type were found in our study. In fact, long-latency, presumably polysynaptic EPSPs were often increased after tetanic stimulation. A suppression of IPSPs was found in all neurones both during and shortly after the tetanus. The short-lasting IPSP suppression had been known for many years [25] and its mechanisms were listed in Section 4.7. The late IPSP, in its turn, would be highly sensitive to the high-frequency stimulation which transiently elevates the extracellular K^+ concentration. A more persistent IPSP suppression might participate in mechanisms of some epileptic states such as the state induced by multiple tetanic stimulations [914] in a paradigm similar to "kindling" (see Section General Conclusions for definition and references). However, the lasting IPSP suppression is not typical after a single tetanization: it was observed clearly in only one neurone. It remains to be resolved whether this suppression resulted from a true blockade of the postsynaptic inhibition or was due to the membrane hyperpolarization [12, 912] or "masking" effects of increased EPSPs. It should be stressed that, according to our data, more typical post-tetanic changes in IPSPs were an enhancement rather than a suppression (see also Chap. 12 and General Conclusions). Therefore, the IPSP changes seem not to be an essential mechanism of LTP maintenance.

As has been already noted, postsynaptic (membrane) mechanisms might, in principle, explain long-lasting plasticities [1008, 1019]. However, we did not observe any significant changes which could indicate the participation of such membrane mechanisms in either short- or long-lasting post-tetanic potentiation of cellular responses. This result is in general agreement with later studies of our group and several other research groups using in vitro preparations (see General Conclusions). Note that post-tetanic changes in the RMP tended to be hyperpolarizing rather than depolarizing. The membrane hyperpolarization and a decrease in neuronal excitability, naturally, cannot explain the observed post-tetanic increase in the probability of spike discharges. Nevertheless, it is interesting to note that the hyperpolarizing shifts may actually lead to an increase in the population spike amplitude. This can result from an increase in the amplitude of the intracellular spike, which was occasionally observed in our experiments and from a larger synchronization of spikes in a population of neurones, for example, due to a suppression of the discharges to near-threshold EPSPs. It can thus be suggested that a post-tetanic increase in the amplitude of the short-latency (probably antidromic) population spike mentioned previously (Sect. 1.6) might reflect such a hyperpolarization. The above-mentioned divergence in the degree of the potentiation of slow and fast FP components (Sect. 1.10) can be partially explained by the same mechanism.

5.9 Summary

Long-lasting (up to 2.5 h) post-tetanic enhancement of cellular responses to the stimulation of the SFA or DG was found in 10 out of 28 CA3 hippocampal neurones of the unanaesthetized rabbit. The facilitatory effect included a reduction in the latency and an increase in the probability of single neurone discharges. An increase in the amplitude of excitatory postsynaptic potentials (EPSPs) which persisted during the whole post-tetanic period, was observed in 8 out of 24 neurones recorded intracellularly or quasi-intracellularly from 3 to 20 min after tetanic stimulation. EPSP latency was short (2 ms) and stable in two neurones, strongly indicating the monosynaptic nature of these potentiated EPSPs. IPSP suppression lasting more than 2 min was observed in only one neurone. Otherwise, the IPSPs tended to be increased after tetanic stimulation. An evaluation of the membrane excitability in two neurones by current pulse injections through the recording microelectrode showed a short-term (up to 2 min) increase in spike threshold after tetanic stimulation. No significant prolonged (more than 3 min) shifts in the RMP or a tendency to a post-tetanic membrane hyperpolarization were found after tetanic stimulation, but these measurements encountered difficulties resulting from changes in the extracellular steady potential level and the low stability of the intracellular recordings from the unanaesthetized animals.

Therefore, the data demonstrated long-term potentiation (LTP) at the level of hippocampal neurones after a single afferent tetanization. There were no significant changes in either IPSPs or the membrane characteristics which could underlie LTP. We conclude that an increase in the efficacy of excitatory synaptic connections is the most probable candidate to explain the major mechanism of the hippocampal LTP.

6 Changes in Acetylcholine Sensitivity During Long-Term Potentiation

6.1 Introduction

The data obtained in the previous chapters indicate that the major mechanism of hippocampal LTP is an increase in the efficacy of excitatory synaptic connections. The aim of the following series of experiments is an attempt to perform a more detailed analysis of mechanisms which underlie these changes in synaptic efficacy. To analyze these mechanisms in more detail, it is convenient to use the terminology of the quantum hypothesis of synaptic transmission [260, 261, 262, 485]. This hypothesis and its application to central synapses will be considered in Chapter 7. Here, we shall use only the simplest equations of this hypothesis.

The mean amplitude of an aggregate PSP, which is the measure of the synaptic efficacy of the stimulated pathway, could be represented as follows:

$$V = \sum_{i=1}^{N} E_i \qquad (6.1\text{–}1)$$

where E_i are amplitudes of UPSPs and N is the number of excited fibres. Assuming all $E_i = E$, which could be approximated in the hippocampus when a small number of identical adjacent fibres are activated, (6.1–1) gives

$$V = NE. \qquad (6.1\text{–}2)$$

From the quantum hypothesis (Chap. 7), $E = mv$, where m is the mean quantal content, i.e. the mean number of transmitter packages ("quanta") released per presynaptic volley, and v is quantal size, i.e. the mean membrane potential shift produced by one transmitter quantum. From (6.1–2) and the relation $E = mv$

$$V = Nmv. \qquad (6.1\text{–}3)$$

The estimate of changes in these three parameters (N, m and v) makes it possible, in principle, to resolve the problem of pre- versus postsynaptic location of LTP mechanisms. On the basis of the data discussed in Section 1.10, one might exclude changes in N after conditioning tetanic stimulation. The mean number of quanta released (m) depends on purely presynaptic mechanisms, while the quantal size (v) is dependent on several factors. The latter dependence could be expressed by the following formula (from [495] and [572], with modifications):

$$v = K * R_{in} * G * V, \tag{6.1–4}$$

where K is a coefficient dependent on the electrotonic distance from the locus of the generation of the synaptic current to the recording site; R_{in} is the input resistance; G is the conductivity change evoked by one quantum; V is the difference between the RMP and the equilibrium potential for the respective PSP. All variables in Eq. (6.1–4) are measurable, in principle, with the intracellular technique and all of them, apart from G, are of postsynaptic nature. The variable G depends on both presynaptic factors (the amount of the transmitter in one package) and postsynaptic, or more exactly "subsynaptic" (sensitivity of subsynaptic receptors) factors. In addition, "intermediate" factors can influence G [493, 494]. The "intermediate" factors determine which part of the released transmitter reaches the subsynaptic receptors. They include characteristics of the synaptic cleft and activity of re-uptake mechanisms or related enzymes.

It should be stressed that the above analysis is not purely theoretical. For every factor considered, related experimental results can be found which suggest a real involvement of the particular factor into modifications of postsynaptic responses under various conditions. Changes in the subsynaptic sensitivity can be cited as examples. The changes were observed in various junctions following electrical stimulation [965, 1065], transmitter application [496, 605] or lesions of afferent pathways [129, 874]. A decrease in the concentration of the enzyme (acetylcholinesterase, AChE) that hydrolyses a synaptic transmitter in the hippocampus (acetylcholine, ACh) has been suggested as an explanation for the increased sensitivity of hippocampal neurones to ACh after septal lesions [129]. The other related examples are post-tetanic changes (swelling) of dendritic spines in the DG [335] and CA3 [737, 738] which can lead to post-tetanic response modifications [177, 1089]. Some data [1091] might be interpreted as a participation of a decrease in the efficiency of the uptake of transmitter in LTP, but other studies denied uptake changes [381, 645]. Similar factors can be evaluated by drug applications, provided the transmitter in the tested pathway is identified. The septo-hippocampal pathway is one of the first pathways in the vertebrate CNS with an identified transmitter [765]. ACh has been suggested to be the transmitter since pioneer studies of Shute and Lewis [899]. Therefore, the LTP of the responses evoked by the septo-fimbrial stimulation appeared to be a convenient paradigm to analyze a possible involvement of subsynaptic and "intermediate" factors underlying LTP. Of particular interest seemed to be the role of changes in sensitivity of subsynaptic ACh receptors which was proposed to explain memory formation by several authors [267, 561, 869, 880].

In this connection, our next step was an attempt to study the participation of subsynaptic factors in LTP maintenance by evaluation of changes in ACh sensitivity with microiontophoretic techniques [545, 546, 1014, 1048].

6.2 Effects of Acetylcholine Iontophoresis on Cellular and Field Responses

Although there was no doubt about the existence of the excitatory septo-hippocampal cholinergic pathway [193, 765, 784] when we started our experiments, the studies of effects of ACh microiontophoresis in the hippocampus were not in complete agreement with each other. A high sensitivity of hippocampal neurones to ACh was reported by some authors [129], in contrast to a moderate [131] or rather low sensitivity [423] described in earlier publications. Moreover, inhibitory effects of ACh on FPs in area CA1 of hippocampal slices had been reported [439]. Most of the data cited were obtained on anaesthetized animals, often after severe surgical interventions. The influence of ACh on evoked cellular activity had not been well investigated [936]. Therefore, our first aim was to study ACh effects on spontaneous and evoked responses of hippocampal neurones of unanaesthetized rabbits. In an attempt to diminish the possible involvement of non-cholinergic pathways, stimulating electrodes were inserted not into the SFA but into the *nucleus septalis medialis* according to coordinates AP 1.5–3; L 0.5; H 5–8.5 [336]. Their more precise position was adjusted to evoke population spikes in response

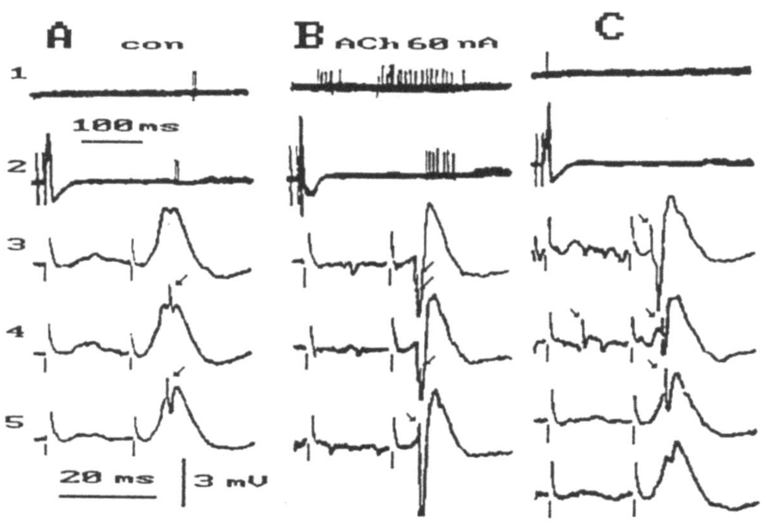

Fig. 6.1A–C. Effects of ACh iontophoresis on hippocampal cellular discharges and field potentials. Spontaneous single neurone activity (*1*), responses to paired septal stimuli at a low sweep speed (*2*) and at a higher sweep speed (*3–5*) are shown before (**A**), during (**B**) and after (**C**) ACh iontophoresis. Responses before the first, second and third ACh applications are shown in **A3**, **A4** and **A5**, respectively. Responses during the third application of the iontophoretic current are shown in **B3–B5** and responses recorded immediately after cessation of the iontophoresis are shown in **C3–C5**. *Bottom record* in **C** was obtained 1 min after switching off the iontophoretic current. Cellular discharges are marked by *oblique arrows* in **A–C** (records *3–5*), bottom parts of the population spikes in **B2–B5** are not shown. [545]

Table 6.1. Distribution of recorded neurones according to changes in spontaneous and evoked spike activity during ACh iontophoresis

Type of activity	Excitation	Inhibition	No effect	Total number of neurones
Spontaneous	13	8	15	36
"Late" response	9	6	19	34
"Early" response	7	1	18	26

to paired (10–50 ms interval) pulses with duration of 0.1–0.5 ms and with equal or different current strength in the range from 0.05 to 2 mA. The stimuli were delivered with 2 to 10 s intervals between the pairs. The recording barrel of the two- or five-barrel glass microelectrode was filled with 2M NaCl. One of the barrels was filled with 1 M ACh solution (pH 4). A holding current of 16 nA was used to prevent an ACh leakage between injections. From 10 to 50 responses were collected before, during and after ACh iontophoresis.

Altogether, 54 neurones were recorded extracellularly. Effects of ACh on the spontaneous activity were examined in 36 neurones with iontophoretic currents from 20 to 60 nA (Table 6.1). Statistically significant changes (predominantly activation) were observed in 60% of the neurones recorded (Fig. 6.1A1–C1). Generally, these moderate changes did not differ significantly from those described in most publications [131, 137, 935, 936].

Delayed discharges with latencies from 90 to 340 ms were evoked by septal stimulation in 34 neurones (Fig. 6.1A2). In neurones with spontaneous spiking, the delayed discharges were preceded by an inhibition. Table 6.1 shows that ACh effects on the delayed discharges were similar to its effects on spontaneous activity (Fig. 6.1A2, B2). Short-latency (2 to 10 ms) responses were recorded in 26 neurones (Fig. 6.1A4, 5, arrows). In 8 of them, significant ACh effects on the short-latency responses were found (Table 6.1). The effect was facilitatory with the exception of one neurone. The facilitation consisted in an increase in discharge probability and in latency reduction (Fig. 6.1B3–5, arrows). Bursts of spikes appeared in some cases during the ACh iontophoresis (Fig. 6.1B3).

It unexpectedly turned out that ACh iontophoresis strongly effected the fast component of the evoked FP (population spike). The increase in the population spike amplitude was clearly seen in recordings of neuronal activity (Fig. 6.1) and also in a special series of experiments with FP recordings (Fig. 6.2). In general, statistically significant increases in the population spike amplitude during ACh electrophoresis with 60 nA current was observed in 24 out of 30 experiments. The amplitude could be increased as much as several times (compare Fig. 6.1A3–5 and B3–5). In some cases, the population spike was absent in control responses and appeared only during the ACh application (Fig. 6.2). The increase in the amplitude during the ACh iontophoresis was often accompanied by a latency diminution. In some experiments, the testing stimulus evoked two population spikes; the

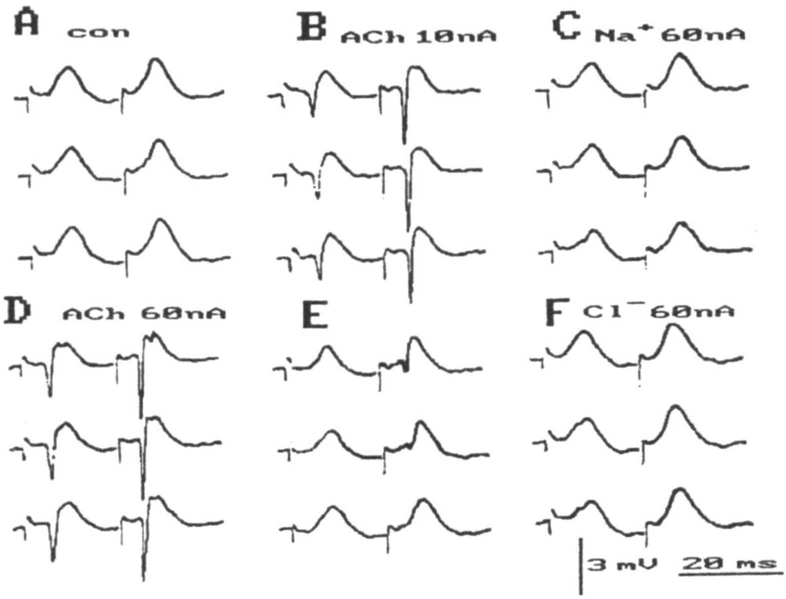

Fig. 6.2A–F. Changes in field responses to paired septal stimuli during ACh iontophoresis. **A** Control before iontophoresis. **B** Responses during ACh application with 10 nA current. **C** Control application of Na^+ ions. **D** ACh iontophoresis with 60 nA current. **E** Immediately after switching off the iontophoretic current. **F** Control obtained during electrophoresis of Cl^- ions. [545]

second spike was usually more sensitive to the ACh iontophoresis than the first one.

The characteristics of the FPs as well as of spontaneous and evoked neuronal discharges returned to the control level within scores of seconds following the cessation of the electrophoresis (Figs. 6.1C and 6.2E). It is interesting to note that the aftereffects (Figs. 6.1C and 6.2E) were superficially similar to the short-term PTP after moderate tetanization (see Sect. 5.4).

We did not observe substantial changes in the slow FP components similar to those reported by other authors with glutamate (Glu) iontophoresis [371, 455]. The positive wave (component 2 of the FP) decreased in some experiments (compare Fig. 6.2A and B). However, the decrease might be interpreted as resulting from the increase in the amplitude of the population spikes. No significant FP changes were found during control applications of Na^+ ions with currents of the same strength and direction as during ACh iontophoresis through the barrel filled with either 2 M NaCl or sodium Glu in eight control experiments (Fig. 6.2C).

Comparisons of cellular discharges with FPs confirmed that the generation of a population spike is the result of a summation of discharges of a large number of hippocampal neurones [28] (Sects. 1.2, 1.10, see also Fig. 6.1). Therefore, the increase in the amplitude and area of the population spike can be naturally explained by an increase in the probability and synchronization of neuronal discharges. This explanation agrees with the data of neuronal recordings (Fig. 6.1, Table 6.1).

In general, the results obtained with the recordings of both single neurones and population spikes showed a well-marked sensitivity of the hippocampal neurones of unanaesthetized animals to ACh. The prominent ACh effects on the population spike amplitude was used in the next series of our experiments to answer the question whether the subsynaptic sensitivity was changed as a result of LTP induction in the hippocampus [1048, 546].

6.3 Changes in ACh Sensitivity During LTP

To quantify the comparison of the sensitivity to ACh before and after tetanic stimulation, a "relative sensitivity" was calculated according to the following equation

$$RS = (A_{ACh} - A_{con})/A_{con}, \qquad (6.3\text{–}1)$$

where A_{con} and A_{ACh} are the mean amplitudes of the population spike before and during ACh microiontophoresis, respectively. The values were taken as half-sums of the following two measurements: first, from the point preceding the beginning of the population spike to its peak, and the second, from the peak to the top of the positive wave [10]. From 10 to 15 responses were measured in every testing series and averages were used for the estimates of ACh sensitivity. Similarly, to measure LTP magnitude, a relative change in the FP amplitude was calculated as

$$RA = (A - A_0)/A_0, \qquad (6.3\text{–}2)$$

where A_0 and A are population spike amplitudes before and after tetanic stimulation, respectively.

Results of 12 experiments with 14 septal tetanizations in 6 rabbits were analyzed. Figure 6.3 shows the time-courses of LTP (Fig. 6.3B) and ACh sensitivity (Fig. 6.3C). The first tetanization caused no significant changes in the population spike amplitude in this experiment (Fig. 6.3AI, BI), but the "relative sensitivity" to ACh [Eq. (6.3–1)] decreased in comparison with the pretetanic control (Fig. 6.3C). The second tetanization (Fig. 6.3AII) produced LTP which persisted during the whole experiment (Fig. 6.3B). The "relative sensitivity" was further reduced after the second tetanization and remained below the pretetanic level during the rest of the experiment (Fig. 6.3C). There was a statistically significant negative correlation between the relative sensitivity to ACh and the degree of potentiation estimated from Eq. (6.3–2) in this experiment ($r = -0.44$, $P<0.05$).

Because the amplitude of the population spike increased after tetanization, the reduction in relative ACh sensitivity might, in principle, be explained by the dif-

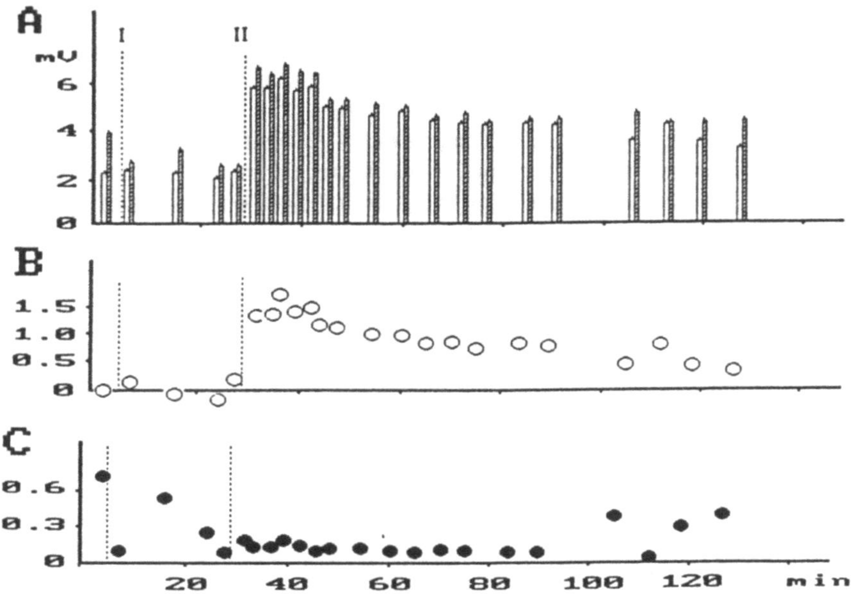

Fig. 6.3A–C. Comparison of time-courses of LTP and relative ACh sensitivity of septo-hippocampal responses. *Ordinates* amplitude of the population spike before (*unshaded columns*) and during (*shaded columns*) ACh application (**A**), the relative value of potentiation (**B**) and the relative sensitivity to ACh (**C**). *Abscissa* time in min. *Interrupted vertical lines* mark the first (*I*) and second (*II*) tetanic stimulations. Note greater ACh sensitivity (**C**) after the first tetanization (*first dotted line marked I*) which produced no LTP (**B**) in comparison with the sensitivity after the second tetanization (*marked II*). [1048]

ferences in pre- and post-tetanic amplitudes, for example because of a "saturation" effect. To evaluate this possibility, several stimulus strengths were used in the majority of the reported experiments. Comparison of responses evoked by different currents but with similar population spike amplitudes before and after tetanic stimulation showed a significant post-tetanic reduction in ACh sensitivity as well. Statistically significant potentiation of the population spike which persisted for more than 10 min was observed in 7 out of 12 experiments. The relative sensitivity was reduced after seven effective tetanizations in six out of the seven experiments. The mean relative sensitivity decreased from the pretetanic level of 0.46±0.05 (mean ±SEM) down to 0.17±0.06 as calculated 10 min following the conditioning tetanus (N = 7). In 3 out of the 12 experiments (including one experiment with LTP), the relative sensitivity increased for 1 to 2 min following the tetanization. A more prolonged increase in ACh sensitivity was observed in one experiment. This increase persisted <30 min after conditioning tetanic stimulation. Following this period, the relative sensitivity to ACh diminished and became not significantly different from the control despite the persistence of the LTP. In five experiments, conditioning tetanization led to a prolonged (>5 min) post-tetanic depression (PTD). The relative ACh sensitivity was increased in two out of the five cases, and it was decreased in the other three cases. For the pooled data of all 12 experiments which included both LTP and PTD, average relative ACh sensitivity decreased from 0.57 (control) to 0.40 (after tetanization). The decrease was statistically significant ($P<0.05$, Wilcoxon test for matched pairs). A weak but statistically significant negative correlation ($r = -0.41$, $P<0.01$) was found between the relative post-tetanic amplitude [Eq. (6.3–2)] and ACh sensitivity for measurements made after at least 10 minutes following tetanization. It should be stressed that both a definite decrease or increase in the relative sensitivity to ACh could be observed in different experiments irrespective of the direction of the post-tetanic changes of the population spike amplitude, i.e. LTP or PTD.

In a special series [546], stimulation of non-cholinergic input from the DG was used in addition to septal stimulation. The effect of ACh iontophoresis on the dento-hippocampal FPs was similar to that on the septo-hippocampal responses (compare Fig. 6.4A1 and B1, columns). Changes in ACh sensitivity studied with 24 tetanic stimulations of the medial septum and 7 tetanizations of the DG were also similar to those described above. The average relative ACh sensitivity of the responses from both inputs decreased during LTP following the tetanization of the corresponding pathway (compare unshaded and shaded columns in Fig. 6.4 before and after tetanization). The lack of a significant increase in ACh sensitivity is clearly seen when control responses (Fig. 6.4A1, B1) are compared with the respective post-tetanic responses evoked either at the same (Fig. 6.4A2, 3, B2, 3, continuous columns) or at a lower current strength (Fig. 6.4A2, B3, dotted columns).

A heterosynaptic PTD at a control input (either from the septum or DG) was observed in 10 out of 26 cases with tetanizations of the other input (Fig. 6.4A3). In cases with the PTD, the post-tetanic changes in ACh sensitivity were also variable similarly to the cases with LTP, and a decrease in the relative ACh sensitivity was the most common observation. In several additional experiments with Dr. R.G.Kozhedub, we tested changes in sensitivity to iontophoretic Glu application after either DG or septal tetanization. The results were qualitatively similar to the results of our experiments with ACh iontophoresis and with results of other authors obtained with Glu application to CA1 neurones in vitro (see Sect. 6.4). We found no changes or a tendency to a decreased Glu sensitivity during LTP irrespective of the pathway tetanized or tested.

6.4 Discussion

The results of the microiontophoretic study revealed well-marked changes in the population spike amplitude during ACh application. These data, obtained in experiments on unanaesthetized animals, were in agreement with the suggestion about the cholinergic nature of the septo-hippocampal pathway and allowed us to make an attempt to study changes in ACh sensitivity after LTP induction. The result of these experiments was negative by showing no increase in ACh sensitivity

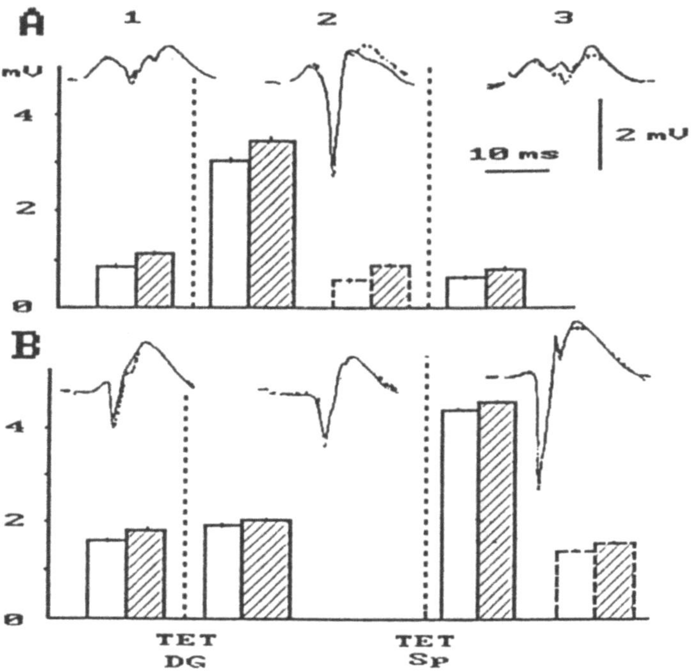

Fig. 6.4A,B. Specificity of long-term potentiation and effects of ACh iontophoresis on FPs evoked by stimulation of the DG (**A**) and medial septum (**B**). Oscillograms show examples of FPs evoked by the DG (**A**) and septal (**B**) stimulation before (*continuous records*) and during (*interrupted records*) ACh iontophoresis. The population spike amplitudes before and during ACh iontophoresis are presented by *empty* and *shaded columns*, respectively. Columns *bounded by broken lines* correspond to responses to testing stimulation of reduced intensity. Control responses (*1*) as well as responses after tetanization of the DG (*2*) and septum (*3*) are shown. Differences between amplitudes under normal conditions and during ACh application are significant for all cases: P<0.01 (for **A**) and P<0.05 (for **B**, Student's t-test)

during LTP. Actually, even a slight decrease in sensitivity was found in most cases. Therefore, no confirmation was found for the participation of changes in subsynaptic receptor sensitivity in LTP expression. Our findings are in agreement with the results of Lynch et al. [638], who described a decrease in sensitivity of CA1 neurones to Glu during LTP in hippocampal slices. No increase in the sensitivity to Glu and its agonists was found also in our more recent observations in vivo (Kozhedub and Voronin, unpubl.) as well as in more recent in vitro experiments [955] at least during the initial 30 min after LTP induction [251]. We found that the changes in transmitter sensitivity during both LTP and LTD were not specific either for the particular transmitter (ACh or Glu) or pathway stimulated. The diminution in sensitivity during LTP might be related to the heterosynaptic PTD revealed by stimulation of a non-tetanized pathway (see Sect. 1.10) and to the prolonged post-tetanic hyperpolarization found in some cases (Chap. 5). The decrease in both postsynaptic excitability and transmitter sensitivity should result in a depression of responses mediated by non-tetanized pathways. This depression might be regarded as an analogue of behavioural "conditioned inhibition" ("differentiation") which is a depression of responses to unpaired stimuli after elaboration of a CR [797].

Our observations of mainly excitatory responses to the ACh application are in agreement with the most of the literary data cited (Sect. 6.2). More recently, a similar increase in the amplitude of population spikes evoked by stimulation of various inputs was reproduced by other authors in experiments with ACh iontophoresis to the pyramidal layer [552, 554, 838]. It is interesting to note that similar facilitation of population spikes was reported during a weak septal stimulation which did not evoke any significant FP [125, 552]. The decrease in the amplitude of the slow component of the hippocampal FP cited in Section 6.2 [439] was confirmed in more recent publications (see review [784]): it was obtained with ACh application not to the pyramidal layer but to the dendritic region. The decrease was explained by presynaptic inhibitory effects [274, 439, 876, 989]. Various complicated ACh effects were found recently with intracellular recordings from hippocampal slices (see reviews [765, 784]). These observations can be explained by several ACh actions recently established, including the direct muscarinic depolarization, increase in R_{in}, as well as direct or indirect hyperpolarization and "disinhibition" (i.e. IPSP suppression) probably with involvement of presynaptic mechanisms. Direct ACh effects were shown to be mediated by the blockade of several types of K^+ currents [650, 765, 766, 752, 784]. Participation of muscarinic receptor-mediated modulations of Ca^{2+} conductances was proposed recently as a mechanism of a long-lasting facilitation of hippocampal EPSP in slices with suppressed inhibition [667]. In addition, ACh can modulate effects of other neurotransmitters (e.g. [712]) probably via some other mechanisms like modification of intracellular messenger metabolism.

This variety of possible mechanisms make both the interpretation of the observed changes in transmitter sensitivity in our experiments and an increased sensitivity of neocortical neurones to ACh after the acquisition of a behavioural CR [561] ambiguous. Moreover, in view of more recent data [552], it is probable that in our experiments, the testing responses were evoked not due to the activation of the proper cholinergic pathways but rather to an involvement of non-cholinergic fibres passing through the septum or adjacent structures. In fact, the existence of LTP in the proper cholinergic septo-hippocampal pathway is still to be demonstrated, and effects of the septal stimulation are complicated (e.g. [760]). On the other hand, it should be stressed that these reservations do not reject the basic conclusion of the present chapter because our additional experiments with the DG stimulation and measurements of the sensitivity to Glu, which is a probable transmitter in the DG–CA3 pathway, as well as the above-cited observations of other authors for area CA1 [251, 638, 955] showed either no significant changes or a decrease in transmitter sensitivity, at least during initial stage of LTP maintenance.

In parallel histochemical experiments [713], we made an attempt to study changes in AChE activity as one of the putative "intermediate" factors of the synaptic modification (see Sect. 6.1). In addition, in the same experiments, the activity of monoamine oxidase (MAO), which is an enzyme of monoamine metabolism, was measured in view of a possible participation of a "reinforcing" brain system in LTP (Sects. 1.9, 1.10). This system might involve monoaminergic pathways [185, 367, 875, 990, 1107]. Monoaminergic fibres, passing through the septum, can be activated during the tetanus, which produces LTP of the septo-hippocampal responses. The histochemical measurements revealed rather complicated and unexpected changes in AChE during LTP. A decrease in AChE level was found after 20 min and an increase was found after 60 min post-tetanically, which is hard to interpret in the light of a participation of "intermediate" mechanisms of synaptic plasticity (Sect. 6.1). In general, these results did not confirm any essential role of the diminution of the enzyme activity in LTP maintenance. An increase in MAO activity was found in the same experiments [713]. It may be regarded as an indirect support to the hypothesis suggesting a participation of monoaminergic "reinforcing" systems in LTP formation (see Section General Conclusions).

On the whole, the results of our experiments did not provide justification for either subsynaptic (i.e. an increase in receptor sensitivity) or "intermediate" mechanism (i.e. a decrease in enzyme activity). In the light of the results obtained, some other mechanisms seem to be more attractive to explain hippocampal LTP, with presynaptic mechanisms especially challenging.

6.5. Summary

Principal mechanisms which might underlie changes in synaptic efficacy were discussed. One possible mechanism, an increase in the sensitivity of subsynaptic receptors, was chosen for initial experimental analysis of the increased synaptic efficacy during long-term potentiation (LTP). Sensitivity of single hippocampal neurones to acetylcholine (ACh) and glutamate microiontophoresis was tested. The iontophoresis produced significant changes in spontaneous and evoked activity of hippocampal neurones in intact unanaesthetized rabbits and a prominent increase in the amplitude of the population spike evoked by stimulation of the medial septum or dentate gyrus. Such an increase was used as an index of transmitter sensitivity. A tendency to a decrease in sensitivity was found during LTP, with a negative correlation between sensitivity to ACh and LTP magnitude. An additional series of experiments is discussed which revealed no significant post-tetanic decreases in monoamine oxidase level determined histochemically. In general, our iontophoretic and histochemical studies did not confirm the participation of either subsynaptic or "intermediate" mechanisms in LTP maintenance. In the light of the results considered in this chapter, a more attractive explanation of LTP mechanisms seems to be changes in presynaptic release.

7 Quantal Analysis of Postsynaptic Potentials (Literature Review)

7.1 Introduction

The results of the previous chapter did not allow elucidion of the mechanisms responsible for changes in synaptic efficacy during LTP. Therefore, more precise methods were necessary. The quantum hypothesis of synaptic transmission was mentioned in Section 6.1. The quantum hypothesis was formulated by Katz and his colleagues [260, 261, 331] on the basis of end-plate potential (EPP) recordings from the frog neuromuscular junction. Initial attempts to apply the related methods to neuronal junctions were made in early 1960s with recordings from sympathetic ganglia [133, 134, 673, 674] and the spinal cord [488, 570, 571]. The application of the quantal analysis to neurones of the CNS met many difficulties, in particular, because of their multiple innervation. However, the development of methods related to the recording of MPSPs and UPSPs (Chap. 2) made it possible to overcome some difficulties. The object of the present chapter is to systematize methods used in the context of the quantum hypothesis and some results obtained by these methods in studies of neuronal synapses and their plasticity. Many of the following considerations were strongly influenced by our own results which will be discussed in the following chapters. Additional explanations as well as historical surveys can be found in several reviews [53, 55, 446, 474, 475, 485, 486, 525, 572, 670, 672, 684, 823, 928, 1084]. These publications with rare exceptions [525, 572, 823] consider data obtained mainly in studies at neuromuscular junctions.

7.2 Principles of the Quantum Hypothesis

The theory is based on the assumption that the generation of a PSP at a synapse with chemical transmission results from the action of one or several fixed transmitter portions ("quanta") so that the postsynaptic effect of each quantum is constant under constant conditions but the number of quanta released varies from trial to trial in a stochastic manner. On the basis of this assumption, the amplitude (E_i) of the PSP recorded after one trial with sequential number i

$$E_i = xv, \tag{7.2–1}$$

where v is "quantal size" or "unit PSP", i.e. the membrane potential shift produced by one quantum and x (x = 0, 1, 2, and so on) is the number of quanta in a given PSP (or "quantal content"). An approximate fulfilment of this relation has been confirmed many times with recordings of responses from neuromuscular junctions and neuro-neuronal synapses.

The simplest model [260, 261] assumes that a synaptic transmitter is kept in the presynaptic ending in the form of n discrete portions (quanta). If the mean probability of a release of one quantum is p, the mean number of quanta released per one presynaptic impulse, i.e. the mean quantal content, is

$$m = np, \tag{7.2–2}$$

and the mean PSP amplitude is

$$E = mv. \tag{7.2–3}$$

The value of m is given by

$$m = \Sigma\, xN_x/N, \tag{7.2–4}$$

where summation is done from 1 to n, N_x is the number of PSPs with quantal content equal to x and N is the number of responses recorded. If p is constant and independent for each member of the population n, the binomial statistics can be applied so that

$$N_x = N(nx)p^x(1 - p)^{n-x}, \tag{7.2–5}$$

where (nx) = n!/[(n − x)!x!] is the number of combinations of n elements x at a time. For example, the number of "zero responses" or "failures" (E = 0) is

$$N_0 = N(1 - p)^n, \tag{7.2–6}$$

and the number of "unit PSPs" (i.e. responses evoked by one quantum) is $N_1 = Nnp(1 - p)^{n-1}$. With a small p ($p \ll 1$), the binomial distribution approximates to the Poisson distribution and Eq. (7.2–5) might be substituted by

$$N_x = Nm^x * \exp(-m/x!) \tag{7.2–7}$$

In this case, the number of "unit PSPs", $N_1 = Nm * \exp(-m) = mN_0$. The Poissonian approach was used in most of the initial publications to calculate the "basic" quantal parameters m and v. With the development of these studies (see below) it became clear that p is typically not small and the binomial approach might be more adequate. Moreover, the postulate of a constant p is not generally true. In this case, more complicated models should be used. Nevertheless, the simplest models based on the binomial or Poisson laws were used almost exclusively when we started our experiments and are still widely used at the present.

7.3 Methods of Determination of the Mean Quantal Content

7.3.1 Direct Counting of Quantal Contents of Single PSPs

If one can determine quantal content of every separate PSP (N_x) directly, it is possible to calculate m from Eq. (7.2–4). Theoretically, it is the most accurate method because it does not depend on the character of PSP amplitude distribution. However, in practice, the adequate situations are very rare and may occur at a low temperature in some neuromuscular junctions when synaptic delays are large and variable [485, 469, 489, 490, 1145]. The counting of N_x may be facilitated by electronic differentiation of the PSP [136, 246, 941]. In central synapses, the asynchronous release was observed only exceptionally [55].

7.3.2 The Method of "Miniatures"

If one assumes [261] that the amplitude (μ) of a "miniature" PSP (mPSP), i.e. of a spontaneous unit PSP [329, 331] is equal to v, it follows from Eq. (7.2–3)

$$m = E/\mu. \qquad (7.3.2\text{–}1)$$

This method is often called a "direct method" [446, 475]. Similar to the previous method (Sect. 7.3.1), it does not depend on the type of PSP amplitude distribution, but relies on the equality of μ and v. If the innervation is spread over a wide area of the postsynaptic cell, a number of restrictions must be taken into account [683, 684]. If a presynaptic fibre forms multiple contacts or contacts of a large area, the preference should be given to extracellular recording of PSPs generated at local sites [122, 263, 489, 490]. This technique has been used in practice until now only for neuromuscular junctions. The method of the miniatures is hard to apply to most of central neurones under normal conditions because of their multiple innervation. Only in rare cases is it possible to distinguish mPSPs related to a particular synapse under study, using certain criteria, for example, changes in spontaneous PSP frequency after presynaptic activation [570, 692], during depolarization of the presynaptic element [56, 408] or local application of hypertonic solution [99]. An approach based on the identification of particular wave forms was applied recently to a central inhibitory synapse [530, 839]. Computer algorithms are promising for the separation of mPSPs of different origin by their wave forms.

A variant of this method is the noise analysis introduced by Katz and Miledi [492] for analysis of ionic channels and based on Campbells's theorem [873]. If a shift in the membrane potential, dV, is accompanied by a change in variance, dS^2, then

$$\mu = 2S^2/dV. \tag{7.3.2–2}$$

7.3.3 Histogram Method

The method was applied when regular modes (peaks and troughs) were clearly visible in the experimental amplitude histogram [212, 475]. The average distance between the modes was taken as an estimate of v; m was determined from Eq. (7.2–3). In the simplest variant, the distance from zero to the first peak is taken as v ("initial peak method"). Generally, this method is simple and universal. However, the appearance of clear regular peaks demands a sufficient number of observations (N) and, especially, a sufficiently low ratio of the standard deviation (SD) of noise (S_n) to v. The SD of v (S_v) also expands the peaks in the histogram. The method becomes especially unreliable when S_n/v is comparable to 0.5 or larger. There are several approaches to make the histogram method more objective by using the autocorrelation [655] or Fourier transformation [244, 700] but these variants were not commonly used. The so-called deconvolution technique [308–310] will be considered in Section 7.3.7, after more traditional methods.

The three methods described above are suitable for distributions of any kind. For the following two methods, formulae based on the Poisson law were used.

7.3.4 Method of Failures

From Eq. (7.2–7), it is easy to obtain [261]

$$m = \ln(N/N_0). \tag{7.3.4–1}$$

The ratio v/S_n should be high enough to distinguish the failures. Averaging of sweeps taken as failures can be used to facilitate their separation [212]. An objective method is to determine N_0 as the double number of negative amplitudes [763, 798]. The method should overestimate N_0 at high noise levels.

For binomial distribution, Eq. (7.3.4–1) overestimates m. The overestimate is equal to 5% for p = 0.1 and reaches 39% for p = 0.5. From Eq. (7.2–6):

$$m = -p * \ln(N/N_0)/\ln(1 - p) \tag{7.3.4–2}$$

and it is necessary to know p (Sect. 7.6) to determine m.

7.3.5 Method of the Coefficient of Variation

The method [55, 261, 670] is based on the equality of the mean value and its variance for the Poisson distribution. Therefore, the coefficient of variation of m is equal to the square root of m. Since the coefficient of variation does not depend on the unit of measurement, one can obtain:

$$m = [1 + (cv)^2]/(CV)^2, \qquad (7.3.5\text{–}1)$$

where CV and cv are the coefficients of variation of E and v (or μ), respectively. Since cv<0.2–0.3 [55, 312], one can take $(cv)^2 = 0.1$ [134], or ignore it. Therefore, the Eq.

$$m = 1/(CV)^2 \qquad (7.3.5\text{–}2)$$

can be used. If noise is taken into account

$$m = E^2[1 + (S_v^2/v^2)]/(S^2 - S_n^2), \qquad (7.3.5\text{–}3)$$

where S^2 is the variances of E. Respective expressions for the binomial distribution are:

$$m = (1 - p)/(CV)^2, \qquad (7.3.5\text{–}4)$$

$$m = E^2[1 + (S_v^2/v^2)](1 - p)/(S^2 - S_n^2). \qquad (7.3.5\text{–}5)$$

In cases of binomial distributions, Eqs. (7.3.5–1) and (7.3.5–2) overestimate m. The overestimate is larger than for the method of failures: for Eq. (7.3.5–4) it is equal to 11% at p = 0.1 and reaches 100% at p = 0.5.

7.3.6 Analysis of Dispersions

The method is a variant of the variance method and it is based on the assumption that v and n are constant during modifications of transmitter release. Initially the method was used by Christensen and Martin [225]. They calculated p_1, p_2, n and v by solving a system of four equations composed from equations E = npv and Eq. (7.3.5–4) with two sets of E and CV measured at two different Ca^{2+} concentrations. It should be noted that the assumption of n constancy is at variance with studies which demonstrated that changes in Ca^{2+} concentration could essentially influence n [105, 107, 111, 166, 635, 725]. In this connection, a more recent variant applied to spinal cord synapses in tissue culture [755, 806] seems to be more reliable. Assuming that p is negligibly low under conditions of suppressed transmitter release at low ratio of Ca^{2+} to Mg^{2+} concentrations, they calculated p after combining Eqs. (7.3.5–2) and (7.3.5–4) for low and normal ratios of Ca^{2+} to Mg^{2+} concentrations, respectively.

Another variant of this method [229] is based on a relationship between S^2 and E which are modified, e.g. during PTP. The relationship can be approximated by a parabola with parameters c_1 and c_2:

$$S^2 = c_1E - c_2E^2. \tag{7.3.6–1}$$

After finding the best estimates of c_1 and c_2 with the least-squares criterion, Clamann et al. [229] took $v = c_1$ and $n = 1/c_2$. It is worth stressing again that all the variants assume that v and n are constant during E modifications. Therefore, they are not suitable for distinguishing between changes in m and v.

7.3.7 Deconvolution Technique

This method [81, 308, 464, 568, 823, 1112] represents a development of the histogram method and aims at using all information contained in the amplitude histograms. The method is based on a computer optimization algorithm. The algorithm searches for a discrete "noise-free" amplitude distribution which after being convoluted with a known noise histogram gives the best fit to the observed amplitudes. The deconvolution procedure, in principle, allows the "noise-free" discrete amplitude distributions to be obtained from the experimental noisy histograms without assumption about the distribution of underlying release probabilities (p_i) and allows both non-uniform p and different v at different release sites. Theoretically, no constraints for parameters of the PSP amplitude distribution can be assumed, except common assumptions that the distribution is discrete, that PSPs add linearly and that noise and PSPs are statistically independent. Initial publications [308, 309] demonstrated "non-quantal" fluctuations at Ia afferent synapses on spinal motoneurones. However, with a refinement of the technique [464, 1067, 1068, 1112], fluctuations of EPSP amplitudes with regular increments between successive components were obtained for spinal neurones. The quantal nature of synaptic release was confirmed, but it was found that the amplitudes of spinal EPSPs were not distributed according to the Poisson or (in general) binomial law (Sect. 7.7). A "compound" binomial model was proposed with a uniform v but with different p_i associated with different release sites [823, 1068].

Nevertheless, more restrained deconvolution procedures were also used assuming the traditional binomial model for synaptic release [81, 528, 1110]. These variants have the advantage of a less severe demand of v/S_n and N to obtain statistically reliable estimates. Therefore, the following variants of the deconvolution can be listed with a different number of free parameters: "unconstrained discrete model", "unconstrained quantal model", "compound binomial model", convolution of two or several binomial distributions, simple binomial model, Poisson model [823]. Various statistical criteria for the optimization can be used, like the Chi-square [528], least-squares [1112], Kolmogorov-Smirnov test and maximum likelihood estimator (MLE). The MLE is preferred in the majority of recent studies [823]. In particular, the fitting with the MLE does not depend on binning.

When "unconstrained" models are adopted, the most critical problem is the uniqueness of the solution obtained, especially in the case of noisy experimental histograms without clear multiple peaks. Monte Carlo simulations are usually used to evaluate the reliability of the estimates. Clamann et al. [230] demonstrated recently that with large S_n, the deconvolution overestimated v, giving values which are proportional to S_n (Fig. 15.2). Therefore, estimates of v from histograms without clear peaks might be incorrect.

7.3.8 Other Methods

Variants and combinations of various methods will be considered here. Values of m calculated by method 7.3.2 (m_1) and 7.3.5 (m_2) are usually underestimated and overestimated, respectively (Sect. 7.5). Therefore, a weighted mean might be preferred [103, 670]:

$$m = \sqrt[3]{m_1^2 m_2}. \qquad (7.3.8\text{–}1)$$

Mendell and Weiner [709] used a combination of methods 7.3.4 and 7.3.7 to analyze unitary EPSPs of spinal motoneurones. Using three amplitude levels (0, 50 and 100 μV) as borders to select failures, they plotted theoretical distributions for various m values, determined by Eq. (7.3.4–1), and various S_v, chosen by the method of trial and error. For most cases (22 out of 33), the computer algorithm found Poisson distributions. In half of the remaining cases, binomial distributions were fitted (Sect. 7.6.4).

Blankenship and Kuno [138] measured changes in the amplitudes of spontaneous PSPs after pharmacological challenges to estimate approximate mean m for excitatory synapses on spinal motoneurones. A system of two equations was written and solved relative to N and m with one Eq. (7.2–7) for normal conditions and the other one for application of tetrodotoxin or excessive K^+. One shortcoming of this method is the suggestion that neither of the two variables (N and m) was modified.

7.4 Corrections for the Methods of Estimating m

7.4.1 Non-Linear Summation

With sufficiently high m (or v), the summation of unit PSPs may be non-linear [55, 669, 670] because of the approximation to the equilibrium potential for the PSC. To allow for this "non-linear summation", correlated values E' = DE can be used instead of the measured amplitudes E [669, 670]. "Martin's correction" is given by

$$D = V_0/(V_0 - E), \qquad (7.4.1\text{–}1)$$

where V_0 is the difference between the RMP and the equilibrium potential. This correction ignores the membrane capacity [446], allowance for which requires more complicated equations [671, 694, 808, 939, 994]. According to some estimates [111, 1103], 0.5D can be used as a first approximation instead of D. Sometimes, it is convenient to use Eq. (7.3.8–1) which is not affected by the non-linear summation. To avoid the use of the corrections, the RMP can be clamped. These corrections, in theory, might be essential for neuro-neuronal synapses. The reason is that UPSPs can reach several mV at the site of their generation because of a high R_{in} of the dendrites [75, 76, 485, 488, 572, 670, 822]. However, more recent studies on central neurones indicated their relative electrical compactness [983]. Therefore, v at the site of the generation might not be significantly different from the value estimated from somatic recordings. However dendritic spines can influence both the electrotonic spread and non-linear summation but this problem is not resolved (e.g. [1090]). Another complication is related to possible non-linear properties of dendritic membrane (Sect. 2.4.3). In the case of essential non-linearity, Eq. (7.4.1–1) should be modified. The corrections would depend on the voltage-current characteristic which may vary under various conditions [401].

7.4.2 Control of Steady State

Generally speaking, the application of statistical methods demands a "stationarity" of the experimental data. Without discussing this complicated matter in detail [64, 175, 1124], some methods to control a "quasi-stationarity" may be mentioned. The simplest one is to compare mean amplitudes [212] or quantal parameters [1145] for small consecutive samples (e.g. $N = 20$). In this way, a systematic trend can be detected. Regions without any significant trend can be selected or corrections can be introduced [171, 725].

7.4.3 "Non-Specific" Activity

FPs resulting from the activation of adjacent postsynaptic elements [330, 490, 607] or small PSPs of distant synapses [1145] are understood by "non-specific" activity. It can modify E, CV [446] and distort the amplitude histograms. Comparison of amplitudes for different stimulus strengths, analysis of failures and evaluation of FPs recorded extracellularly are used to control the "non-specific" activity.

7.4.4 Multicomponent Miniature PSPs

When some methods described previously (7.3.2; 7.3.3; 7.3.5) are used, it is necessary or advisable to know the values of μ and S_v. In practice, these values are calculated from amplitudes of mPSPs which are usually distributed according to the Gaussian law [157, 261]. However, the appearance of both "giant" [515, 606, 607] and "subminiature" EPPs [123, 549, 993] were noted. Multimodal amplitude distributions of mPSP were observed also in autonomic ganglia [156, 265, 673]. The problem was studied in many publications (e.g. [156, 315, 549, 550, 655, 740, 996, 1086, 1134, 1135], see reviews [551, 975]). Recordings from muscle fibres of small diameters (and with large μ, up to several mV) were used in most of these studies to facilitate the analysis. Polymodal amplitude distributions (2–15 modes) were found. These results compelled a fresh look at the basic postulates of the quantum hypothesis and at the interpretation of the quantal parameters. The idea that a single quantum is composed of several subunits was formulated (see Sect. 7.11). This idea was questioned [655, 722, 1134] and the multimodal distributions of μ were explained either by random statistical variations due to small N [722] or by the existence of spatially separated transmitter release sites [153, 1134–1136]. However, the strong spatial attenuation of mPSP claimed in the latter publications was not supported by other studies [124, 975, 994]. A harmonic analysis indicated the appearance of subminiature PSPs in the neuromuscular junction, at least in some cases [244]. A simple physical model (a drop formation process) was proposed which simulates mPSP substructures [551].

The problem of corrections that, in view of the hypothesis of "subminiature" PSPs, should be introduced into methods of determination of quantal parameters is out of our scope. However, it can be tentatively suggested that such corrections might not necessarily be substantial. In fact, with a sufficiently large number of subunits (n') and the probability of their release (p'), the conditions $n'p'>4$ and $n'(1-p')>4$ are satisfied. Then, the amplitude distribution can be replaced with a Gaussian one. With a smaller value of n', the histogram becomes skewed and, as has been shown for autonomic ganglion neurones [692], it can be described by the gamma-distribution. In this case, as well as in cases of a high noise [607], it is better to use the mode of distribution of mPSP amplitudes (instead of the mean) as an estimate of μ for the determination of m from Eq. (7.3.2–1).

7.4.5 Additional Remarks

These corrections do not exhaust all possible ways to increase the accuracy of the methods described above [444, 999, 1145]. The introduction of the corrections depends on the conditions and aims of the investigation. If the only purpose is an approximate evaluation of quantal parameters, the simplest approximations can be used. However, if synaptic mechanisms of various challenges are to be analyzed, corrections can be important [57]. The type of the amplitude distribution may play a special role.

7.5 Poisson and Binomial Distributions of PSP Amplitudes

As was stated above, the binomial distribution of PSP amplitudes followed from the original model [261]. The binomial distribution can be approximated by a Poisson distribution when $p \ll 1$. In fact, it was the similarity between experimental data and those predicted by the Poisson law that provided initial support for the quantum hypothesis [55, 158, 261, 288–290, 485, 486, 490, 669, 670]. In most of these studies, synaptic release was suppressed by changes in extracellular Ca^{2+} and Mg^{2+} concentrations. First substantial deviations from the Poisson law were found with recordings from crustacean [51, 132] and insect [987] neuromuscular junctions. Agreements with the binomial law were found for crayfish and frog neuromuscular junctions when recordings were taken from local zones [469, 1082, 1083, 1145] and from the whole end-plates in the presence of a normal or only slightly reduced Ca^{2+} concentration [106, 165, 166, 725, 1084, 1103]. The binomial distribution is described by parameters n and p. Methods of p determination will be presented below. Parameter n is either estimated simultaneously with p or calculated from Eq. (7.2.2).

7.6 Methods of Estimating Binomial Parameter p

7.6.1 Variance of Quantal Distribution

The method [833, 1084, 1145] is based on the fact that the variance of the binomial distribution is equal to $np(1 - p)$. From this and Eq. (7.2–2),

$$p = 1 - (S_1^2/m). \qquad (7.6.1\text{–}1)$$

The value of m is calculated by Eq. (7.2–4), and the variance of m is given by $S_1^2 = \sum N_x(x - m)^2/(N - 1)$ with summation from $x = 1$ to $x = n$. The method is suitable in those rare cases (Sect. 7.3.1) when N_x is known.

7.6.2 Variances of Amplitudes of PSPs and mPSPs

This method [106, 833] is a variant of the method of moments 7.6.1. According to this method

$$p = 1 - [S^2/(E\mu)] + (S_v^2/\mu^2). \qquad (7.6.2\text{–}1)$$

When S_n can not be neglected, the equation assumes the form

$$p = 1 - [(S^2 - S_n^2)/(E\mu)] + (S_v^2/\mu^2). \qquad (7.6.2\text{–}2)$$

7.6.3 Division of Amplitude Histogram According to μ

This method is another variant of method 7.6.1 [109]. According to a known μ, amplitude histograms can be divided into classes corresponding to different quantal contents (N_x); m and p can be calculated from Eqs. (7.3.2–1) and (7.6.1–1) or (7.6.2–1).

7.6.4 Automatic Computation when v Is Known

This approach was used by Belluzzi [103], who applied several algorithms which gave the best fit of a theoretical binomial distribution to the observed amplitude histogram [Eq. (8.7–1)]. As a control of the accuracy of the calculation, he compared the p by n product with m calculated by Eq. (7.3.7–1). In Monte Carlo simulations, the low sensitivity of this method was demonstrated to an increase in S_n but not to variations in v: a 30% error in v evaluation led to almost 300% error in the calculation of p. The disadvantage of this method as well as other methods described in the previous sections is the necessity of preliminary estimation of μ, which is usually problematic for neuro-neuronal synapses.

7.6.5 Automatic Analysis of Amplitude Distributions

Variants of the deconvolution techniques based on the binomial assumption (Sect. 7.3.7) can be used to avoid the necessity to estimate v independently [528, 529, 709, 823]. It has been noted (Sect. 7.3) that the problem of the uniqueness of the solution might be a limitation of the deconvolution technique, especially when "unconstrained" variants are used. The problem is aggravated if one wishes to estimate p and n. In many cases, unrealistically large N had to be used to distinguish between several possible distributions of p_i.

7.6.6 Combination of Variance and Failures Methods

If E and μ are known, m can be determined from Eq. (7.3.2–1). Then one can calculate p from Eq. (7.3.5–4) or (7.3.5–5), which is equivalent to method 7.6.2. The other variant [1084] is estimating p from Eq. (7.3.4–2). If μ is unknown, which is a common case for CNS synapses, but failures are present, Eq. (7.3.4–2) can be equated with m obtained by the histogram method 7.3.3 [475]. It is also possible to solve the system of the two Eqs. (7.3.4–2) and (7.3.5–4) relative to p and m. The first of these two methods is a variant of method 7.6.2. The second was used in practice to estimate the mean value of p for several neurones [570, 578]. These authors combined the equations as $1/(CV)^2 = -p * \ln(N/N_0)/[(1-p)\ln(1-p)]$ and plotted experimental data as $1/(CV)^2$ (abscissa) versus $\ln(N/N_0)$ (ordinate). They calculated p from the coefficient K of the obtained linear approximation $1/(CV)^2$

= K $*$ ln(N/N_0). Recently, a similar method was used by Miles [716] to obtain initial estimates of p, n and v for a deconvolution histogram analysis assuming the binomial model.

7.6.7 Other Methods

Two other methods are based on the assumption that one of the binomial parameters (either n or p) is constant during procedures used to change transmitter release. One of these methods assumes the constancy of p during frequency depression. It is based on a so-called model of transmitter depletion [608] and will be considered in Chapter 9. The second method (analysis of dispersions) was considered in Section 7.3.6.

7.7 Additional Explanations to Methods of Computing Parameters

7.7.1 Non-Stationarity and Non-Uniformity

The best variant is to use several methods to control the accuracy of the estimates. An effective control procedure is a comparison of predicted (theoretical) distributions with the observed (experimental) histograms. The corrections mentioned in Section 7.4 had to be taken into account for estimating p and n. Non-stationarity in time may have a particularly strong influence [175].

In cases when MPSPs (Chap. 2) rather than UPSPs are recorded, a "non-quantal" interpretation of the parameters estimated by the above described methods can be valid in principle (Sect. 8.8). The calculated binomial p gives proper evaluation of release probability only under the condition that probabilities of activation are close to 1 and similar for all fibres. Since this can hardly always be the case, the calculated p must be regarded as one of the possible approximations of the binomial parameter p (see below the discussion of non-uniform p). A more detailed analysis of influences of recordings of MPSP rather than UPSP and of p non-uniformity (see below) on method 7.3.5 was made recently [323].

A presynaptic block is another situation in which deviations from the binomial distribution can be expected. The presynaptic block is known for peripheral and central axons in invertebrates [227, 414, 916, 1076]. It was suggested for the primary afferents in the cat spinal cord [634], but denied by other authors on the basis of analysis of latency distributions and analysis of amplitude fluctuations [823–825]. The absence of presynaptic blocks was directly evidenced by measurements of electrical components of EPSPs in the frog spinal motoneurones [888] and goldfish Mauthner cell [609]. Nevertheless, failure of synaptic transmission might occur at high stimulus frequencies (e.g. 40 Hz) in both peripheral and central junctions [415, 420, 916].

Most described methods assumed that p is uniform for all release sites so that PSP amplitudes are binomially distributed. A spatial non-uniformity of p, i.e. different probabilities of transmitter release for different release sites, might be a more general situation (Sect. 7.3.7). In fact, in their initial publication, Del Castillo and Katz [261] found essential deviations from the binomial model at m>10. At present, the spatial non-uniformity of both spontaneous and evoked releases with proximal-distal gradient of p along terminal branches was shown in numerous publications ([52, 107, 108, 835, 1085, 1087, 1134, 1135], see review [834]). Respective experimental data for mammalian neuro-neuronal synapses are less direct. Only a small part of the presynaptic varicosities of sympathetic nerves was shown to participate in transmitter release [246, 941]. In distinction from the data on the fish Mauthner cell [528] and on descending fibres on the frog spinal motoneurones [81], the number of presynaptic boutons appeared to be larger than binomial n for amphibian [81, 383, 888] and mammalian [541, 754, 755, 806] spinal neurones. A probable explanation was the non-uniformity of p and existence of functionally inactive (silent) synapses [81, 82, 806]. The data obtained with the deconvolution method also indicated the non-uniformity of p for both frog [81] and cat [464, 823, 825, 1067, 1068] spinal cord synapses. A comparison of n estimated by deconvolution procedures with the number of stained boutons also indicated the existence of ineffective synapses [81, 823, 825].

When p' is the mean and S_p is the SD of p_i, the binomial p [175, 261] is

$$p = p' + S_p^2/p', \tag{7.1–1}$$

and Eq. (7.6.2–1) assumes the following form $p + S_p^2/p = 1-[S^2/(\mu E)] + (S_v^2/\mu^2)$ [111]. Therefore, p calculated from the binomial model overestimates average value of non-uniform p and underestimates n. It has been noted [111] that a good agreement between experimental distributions and theoretical binomial ones based on parameter p from Eq. (7.6.2–1) should indicate a small value of S_p^2/p. However, computer simulations demonstrated that even with high values of S_p^2/p, simulated histograms might remain statistically indistinguishable from those predicted by the binomial law [84, 175]. In these cases, parameter p calculated by Eq. (7.6.2–1) deviates from the mean simulated p' [Eq. (7.1–1)]. The sense of the deviations is that the estimated parameter preferentially reflects release sites with higher p, exaggerating average p and attributing it to correspondingly fewer n, i.e. effective number of release sites [175]. This effect might be especially essential for cases with a small N, since the sites with small p would contribute only insignificantly to the population of the recorded responses.

7.7.2 Models with Non-Uniform p

Several approaches were attempted to evaluate non-uniform p_i [82, 273, 415, 726, 823, 1067, 1137, 1138]. An initial approach was to use a convolution of two or several binomial distributions [64, 415]. With an analysis of spinal cord unitary

EPSPs [82], p was found to be close to 1 for one of the two best fit binomial distributions and in the range of 0.12–0.3 for the second distribution. A development of this approach for cases with large n is a description of p_i distributions by beta-distributions [107, 273, 1137, 1138]. Miyamoto [726] proposed a model which described the spatial non-uniformity of p by a symmetric distribution with three major parameters: n, average p and its variance (S_p). However, an analysis of this model using both physiological and simulated data showed its inferiority as compared to the beta model [273]. Distributions of p_i for Ia synapses on the neurones of the spinocerebellar tract assuming compound binomial model could also be described by a beta-distribution [823, 1068].

Summarizing, the assumption of the compound binomial model seems at present to be the most general approach to estimate the non-uniform p_i, especially when n is not too large. When n is large (several tens), the beta model may give a better description.

7.7.3 Other Modifications

Korn and Mallet [527] modified the simple binomial model to explain a greater central tendency of the observed distributions than that predicted by the best-fitted binomial distribution [823]. These authors suggested an interaction of postsynaptic loci in a non-linear manner, i.e. proportionally to the number of sites at which transmitter was released [324, 527]. A model of a generalized binomial distribution based on the "reflection principle" was formulated recently [83] and tested using physiological (EPSP amplitude distributions) and morphological (number of boutons) data.

7.8 Basic Quantal Parameters for Various Junctions

7.8.1 Neuromuscular Junctions and Autonomic Ganglia

The major aim of Sections 7.8 and 7.9 is a consideration of the parameters of central synapses known when we started quantal analysis. More recent data will also be mentioned in this and the following chapters. A brief overview of initial results obtained on peripheral structures will be given for comparison. Under normal conditions, the amplitude of the EPP reaches dozens of mV [330, 485]. Initial estimates of m assuming the Poisson distribution gave values of 100–300 [157, 261, 607, 670]. Somewhat lower m estimates (about 100, on average) were reported later assuming binomial models [165, 166, 725, 1084, 1103]. The classical publication of Del Castillo and Katz [261] gave μ (and v) values of 0.3–0.5 mV for the frog sartorius muscle. A more detailed study [495] showed that μ is a linear function of R_{in} with a slope of about 1 mV/MOhm. The amplitude of recorded mPSP varied from 0.2 to 2.7 mV, but theoretical estimates for thin muscle fibres gave a value of about 6 mV. Values of this order were indeed recorded more recently [549, 550, 1085]. It should be stressed that most earlier data had been obtained from preparations treated with an anticholinesterase. This treatment increases μ by as much as two to three times [331, 606]. On the other hand, increased Mg^{2+} concentration in the same experiments might diminish μ by about two times [260, 606].

The values of m for synapses of the peripheral nervous system were smaller, especially for the mammalian autonomic ganglia: e.g. 8–20 for the thoracic ganglion [134] and 0.6–12 for the superior cervical ganglion [111, 692, 851]. The data given correspond to a low frequency stimulation (1–2 Hz). The influence of stimulus frequency will be discussed in Section 7.10. According to McLachlan [692], μ and v usually varied between 0.5 and 1.0 mV. Other authors mentioned higher v for autonomic ganglia, namely, 0.9–3.0 mV [133–135, 767, 851] or even up to 5 mV [673]. The differences may be explained by different R_{in}.

7.8.2 Invertebrate CNS

There were surprisingly few publications related to the quantal analysis at the invertebrate CNS when we started our experiments. This may be explained partially by the fact that the low R_{in} of the invertebrate neurones and low μ made the quantal analysis difficult. In the squid stellate ganglion treated with tetrodotoxin, μ estimates were usually <0.1 mV, sometimes up to 0.2 mV [714] and they depended linearly on R_{in} with a slope of 0.3 mV/MOhm. Estimates on the basis of this functional dependence gave μ for the giant synapse of the squid stellate ganglion equal to about 10–20 μV. With E = 50 mV, this corresponds to m of about 10,000, allowing for the non-linear summation. Although this value seems to be very large, it had been noted [714] that the calculated density of the transmitter release is of the same order as for the neuromuscular junction (about one quantum per square micrometre).

For the synapse between an identified sensory neurone and a motor neurone of the abdominal ganglion of *Aplysia* [475], m was reported to be similar to that for the frog neuromuscular junctions (100–300). To obtain this estimate, v values in the range of 14–70 μV [212] and 20–90 μV [213] were used. The authors applied the Poisson equation [Eq. (7.3.4–1)] and the histogram method under conditions of elevated external Mg^{2+} concentration. A similar mean value of m (180) and a slightly larger value of v (190 μV) were given more recently for cholinergic IPSPs recorded in the buccal ganglion of *Aplysia* [901]. Similar v values were found in another identified excitatory molluscian synapse [798] with comparatively small m from 1 to 22. A manifold increase in m during development was shown. In contrast, v declined with age in parallel with a decrease in R_{in}.

Table 7.1. Mean amplitudes (E) and quantal parameters (v and m) of unitary and minimal Ia EPSPs of spinal motoneurones of frogs (## 1 and 2) and cats (## 3–8)

#	E(mV)	v(mV)	m	Method	Reference
1	0.14–0.24	0.1–0.2	1–3	7.3.3	[488]
2	0.01–0.27	0.06–0.09	0.8–17.7	7.3.5	[382]
3	0.10–0.45	0.12–0.24	0.63–5.4	7.3.4	[570]
4	0.68–1.57	–	5.6	7.3.4	[190, 191]
5	0.06–2	0.11	0.7–15	7.3.4	[575]
	0.27		2.5	7.3.5	
6	0.02–0.6	–	–	–	[708]
	0.1				
7	0.05–0.6	0.05–0.02	0.2–25	7.3.6	[709]
	0.1	0.1	1.7	7.3.5	
8	0.004–0.28	–	–	–	[1075]

Note. Reported ranges and mean values are given.

7.8.3 Spinal Cord

Initial estimates of the basic quantal parameters for the synapses of the primary afferents on spinal motoneurones are summarized in Table 7.1. The amplitudes of the UPSPs were significantly smaller in comparison with peripheral synapses. The differences can be explained mainly by differences in m, which is often close to 1 for the central synapses. It should be noted that in initial studies, m and v were estimated mainly on the basis of the Poisson equations. However, later estimates assuming either binomial or "unconstrained" models led to similar values. Shapovalov and his collaborators made extensive studies of synaptic connections to motoneurones in isolated spinal cord preparations of amphibians and fishes [59, 58, 383, 885–888, 897] assuming both Poisson and binomial statistics. One of these statistics conformed better to the experimental data in some neurones, while approximately equal fit was found in other cases. In addition to Table 7.1, responses to stimulation of ventrolateral tract fibres showed m of 1.1–16.4 and v of 18–113 µV [59] and 50–116 µV according to a more recent analysis with the deconvolution technique [81].

According to Table 7.1, v for the cat spinal motoneurones is about 100 µV. Similar values were reported in more recent studies using the "unconstrained" deconvolution technique [464]. An unexpected result was that v did not depend on the calculated electrotonic distances from the cell body [464]. However, these estimates were criticized recently, and smaller v values for some synapses were suggested [230]. It is tempting to assume that the quantal unit of central synapses might correspond to the "subunit" of the neuromuscular junctions (Sect. 7.4.4). Recent demonstrations of multiquantal amplitudes of miniature hippocampal IPSPs and especially relatively small conductance changes mediating one quantal event [307, 839] are consistent with this suggestion.

The small m (Table 7.1) can be naturally explained by the small size of central junctions. In fact, a strong correlation of m with junctional area was found for neuromuscular preparations [579]. Similarly, the large m (20–50) in dorsal spino-cerebellar tract neurones [311] corresponds to large (up to 20 μm) diameters of the related synapses [311, 574, 973]. Large m (1–28) for these synapses can be estimated from the data obtained with the deconvolution technique [1068]. Values of m might be influenced by the state of the preparation, in particular by the anaesthesia level [1077]. This may partially explain the different results in Table 7.1. The method of PSP detection may also be relevant: the authors who used averaging with large N revealed smaller UPSPs [1075] in comparison with the authors who did not use averaging [190, 191].

Larger v values were reported for cultured mammalian spinal cord [754, 755]: 738±655 μV (mean±SD, here and below) for EPSPs evoked by stimulation of dorsal root ganglion neurones and 165±45 μV by stimulation of other spinal neurones; corresponding values for m were 11±7 and 79±52 (methods 7.3.5 and 7.6.6, binomial statistics). For comparison, amplitudes obtained in spinal motoneurones with activation of other pathways may be mentioned: 0.09–0.23 mV (minimal IPSPs with intracerebral near threshold stimulation of the chronically deafferented spinal cord [576]); 0.03–1.1 mV (minimal corticospinal EPSPs [805]); 68–220 μV (unitary IPSPs evoked by spikes of Ia inhibitory interneurones [468]); 26–116 μV (UPSPs evoked by activation of interneurones of the reticulo-spinal system [540, 888]); 0.07–1.1 mV (unitary and minimal reticulo-spinal EPSPs [482]); m was 1.3–4.6 for the most reliable cases in the latter publication.

7.8.4 Cerebrum

Detailed data for the synapses of the higher levels of the vertebrate CNS were obtained only recently. The most extensively studied connection is an inhibitory synapse on the goldfish Mauthner cell [319, 525, 528, 529]. The unitary IPSPs appeared as depolarizing responses due to the injection of Cl^-- ions into the cell. Binomial statistics were shown to give a more adequate description of the experimental histograms than the Poisson statistics. Deconvolution estimated m (binomial product np) to be 1.1–26.6 and v = 70–550 μV or about 1% of the full-sized IPSP [528]. Similar estimates (180–300 μV) gave a recent analysis of synaptic noise [530].

Somewhat larger v values (300–700 μV) were estimated by the method of miniatures (with m in the range of 6–20 at 2 Hz stimulation) for excitatory synapses between the Mauthner cell and two identified neurones [408]. On the contrary, v values for EPSPs evoked at a single club ending in the goldfish brainstem were only 30–60 μV [609]. These estimates were criticized recently [823] because S_n was too large (50–70 μV) in comparison with the estimated v.

Purkinje cell EPSPs evoked by stimulation of single climbing fibres in vitro were found to have a low CV in the normal ionic medium [407]. This indicates a high m which corresponds to the great number of synaptic boutons. Failures appeared at low external Ca^{2+} concentration. Equation (7.3.4–1) gave v as 0.3–0.4 mV. Amplitudes of spontaneous EPSPs were of the same order [407, 773]. Therefore, m should be between 15 and 30 for EPSP of 5–10 mV under normal ionic conditions.

When we started our experiments and published our first results on hippocampus and neocortex [88, 89, 1045–1047], only very rough estimates of the quantal parameters of mammalian brain synapses were available on the basis of several published distributions of amplitudes of MPSPs [15, 16, 460, 749]. For example, amplitudes of thalamo-cortical MPSPs were 0.2–0.7 mV [1074] and unitary rubral EPSPs about 0.75 mV [972]. The latter did not fluctuate much. Parameter p should be close to 1 and m = 1 (Sect. 7.9.3). In that case, v = E = 0.75 mV. More recent estimates of amplitudes of unitary EPSPs of rubrospinal neurones gave values of 0.22–0.65 mV [325].

7.8.5 Hippocampus

An initial attempt at quantal analysis at the hippocampus was made with recordings of spontaneous PSPs from CA3 neurones in vitro [176]. The authors reported amplitudes up to 1–2 mV. However, a high noise level (>0.3 mV) and a multiple innervation could lead to ignoring synapses with smaller v. In more recent current- and voltage-clamp studies of CA3 neurones, the same authors found even larger spontaneous PSPs (up to 12 mV) when the cells were hyperpolarized [174]. Characteristics of hippocampal synapses known at the present time are listed in Table 7.2. Practically all authors of the initial publications used method 7.3.5 with the Poisson approximation [Eq. (7.3.5–2)]. Note also that mainly IPSPs were recorded in initial experiments with simultaneous intracellular recordings from two hippocampal neurones. Excitatory connections were studied initially with MPSP recordings, and related data should be considered with reservations. For example, the large m (Table 7.2, #2) might be explained by excitation of several presynaptic fibres. A more careful selection of minimal EPSPs of the DG granular cells was done by McNaughton et al. [700]. They constructed stimulus-response curves similar to those described in Chapter 3 for minimal FPs [1044]. The MPSPs corresponding to the first plateau, and presumably resulting from single fibre activation, were of 0.1 mV, on average. They also applied methods 7.3.3 with Fourier analysis and 7.3.5 and found similar mean estimates (0.08 and 0.14 mV, respectively).

An original method to record presumably unitary EPSPs was used by Yamamoto [1126], who found that homocysteate iontophoresis into local sites of the DG was accompanied by the appearance of regular bursts of EPSPs in CA3 neurones (Table 7.2, #4). However, a synchronous activation of several afferent neurones was not excluded. This method may favour selection of the most effective synaptic connections. Later on, the method was improved by recording of extracellular spikes from the DG [430] and the number of recorded neurones was increased (Table 7.1, #5). Note that the new study gave similar m and even larger v. In initial attempts to apply the "unconstrained" deconvolution techniques to mini-

Table 7.2. Mean amplitudes (E) and quantal parameters estimated by method 7.3.5 in experiments on slices (## 1–5, 7–10) and tissue culture (#6) with recordings of minimal (## 1, 2, 9) and unitary (## 3–8, 10) hippocampal PSPs

#	E (mV)	v (mV)	m	N	Structure and PSP	Reference
1	0.1	0.1	0.9	15	DG, EPSP	[700]
2	–	0.29	46	4	--"--	[243]
3	2–10	–	–	12	CA1, EPSP	[513]
	2–4	–	–	13	CA1, IPSP	
4	0.9–4	0.19–0.50	4–12	5	CA3, EPSP	[1126]
	0.32		8.3			
5	–	0.61	4.4	11	--"--	[430]
6	2.2	0.46	4.7	1	CA3, IPSP	[717]
7	361*	11.6±2.7*	31±5	21	Culture, IPSP	[877]
8	1.9±0.8	0.9±0.2	2.1	5	CA3, EPSP	[716]
9	0.1–0.8	0.1–0.4	0.4–3	33	CA1, minimal EPSP	[343]
	0.26	0.25	1.3			
10	0.1–0.7	0.1–0.4	0.5–2	10	CA1, unitary EPSP	[343]
	0.29	0.22	1.2			

Note. N is the number of neurones. Reported ranges or (and) mean values (±SEM) are given. Values marked by the asterisk are given in volts * s integrated over a period of 0.5–2.0 ms.

mal EPSPs of CA1 neurones evoked by stimulation of the Schaffer collaterals [864, 865], only three results were considered to be reliable with v about 0.2 [864] and 0.3 mV [865]. More recent estimates in the context of the binomial model [343] gave almost identical parameters for minimal (Table 7.2, #9) and unitary EPSPs (Table 7.2, #10) of CA1 neurones calculated by the variance method. Similar values gave the method of failures [Poisson Eq. (7.3.4–1)]. The deconvolution technique with the binomial assumption was used also for analysis of unitary EPSPs (Table 7.2, #8) evoked in inhibitory interneurones by intracellular stimulation of CA3 pyramidal cells [716].

Spontaneous inhibitory PSCs [839] and spontaneous as well as evoked, presumably unitary, inhibitory PSCs [307] were analyzed recently in area CA1 and DG, respectively. The quantal nature of the PSCs was demonstrated, and it was shown that one quantal current represented simultaneous opening of <20–30 GABA-receptor chloride channels.

7.9 Binomial Parameters n and p for Various Junctions

7.9.1 Neuromuscular Junctions and Autonomic Ganglia

Parameter p for the frog neuromuscular junctions with partially suppressed release was 0.25–0.62 [165, 1082] with n within 9–124 [1082] or within 54–116 [165]. In neuromuscular preparations treated with glycerol to prevent muscle contractions, mean p and n (±SD, here and below) were 0.65±0.86 and 170±42, respectively [725]. In the dissected diaphragm muscle of the rat, p was 0.7–0.9 and n was 44–111 with respective averages equal to 0.82 and 70 [1103]. Estimates of the binomial parameters for a limited number of "active zones" of the end-plate [263] were obtained with intracellular recordings from frog muscles after focal depolarization in the presence of tetrodotoxin [1084] and also with extracellular recordings from crayfish neuromuscular preparations [469, 1084, 1145]. In these cases, n was 3.7–22.0 [1084] or 2–10 [469, 1084], and p was 0.04–0.49 [1084] or 0.1–0.49 [469, 1084] for the frog and crayfish, respectively.

Parameter p (method 7.6.2) was 0.35–0.9 in the guinea pig superior cervical ganglion but it was about 0 for some neurones [111, 692]. Parameter n ranged from 1 to 28, usually from 3 to 10.

7.9.2 Invertebrate CNS

The mean p for the identified central synapse of *Aplysia* (Sects. 2.5, 7.8.2) was 0.16 [212, 475]; method 7.6.5 was used. However, our estimates from the published amplitude histograms show that at the beginning of the low-frequency stimulation, p might usually be higher (about 0.5–0.6), but fell to 0.3 or below as a result of the LFD. Typical initial values of n might be in the range of 4–6. All related experiments were done in the presence of increased Mg^{2+} concentration. Our observations (Chap. 11) indicate that for the CNS of *Helix pomatia*, p might be as high as 0.7–0.9 in normal ionic medium. This value may fall to 0.5–0.6 or even below during low-frequency stimulation. Parameter n varied between 10 and 20 (see Sect. 11.4 for more details).

7.9.3 Spinal Cord

The authors of the earlier publications cited in Table 7.1 used mostly Poisson approximations. Mendell and Weiner [709] estimated p in the range of 0.2–0.69 in 6 of 33 neurones. The estimate of the mean p for minimal IPSPs of the deafferented spinal cord [578] gave the value of 0.5, with n below 4 (method 7.6.6). Our application of the same method to individual cases showed that p might vary between 0.3 and 0.8 according to the data of these authors. Similar estimates can be made on the basis of some other studies on the spinal motoneurones. For instance, in addition to the mean estimate ($p = 0.25$ by method 7.6.6) made by Kuno [570] for cases with $m<3$, the same method gives parameter p close to 1 for EPSPs with low CV and small N_0. Similar high p values might also be attributed to large EPSPs (Table 7.1) with small CV in other publications [191, 709]. Binomial p varied from 0.15–0.69 and n ranged from 5–24 for six stained primary afferents in the frog spinal cord [383]. Similarly, n ranged from 3–54 in reticulomotoneuronal EPSPs of the rat [482]. Binomial p was from 0.13 to 0.71. Studies of Redman and his coworkers with the deconvolution technique (Sects. 7.3.7, 7.6.5 and 7.8.3) indicated non-uniform p (Sect. 7.7) with values from 0 to 1 and with effective n ranged from 2 to 5. Larger values of n (6–19) were determined for six stained descending fibres on frog motoneurones [81]. In distinction from the primary sensory afferents, in only one case was n for the descending fibres less than the number of the stained boutons. A detailed analysis of the Ia afferent synapses on the neurones of the dorsal spinocerebellar tract was performed by Walmsley et al. [1068]. Estimates in the frame of the compound binomial model showed average p = 0.52–0.98 for different neurones with the mean 0.76 and with effective n = 3–30.

7.9.4 Cerebrum

As shown in Section 7.8.4, detailed quantal analysis at higher levels of the vertebrate brain has been performed only for synaptic connections at the fish brainstem. For the inhibitory synapse on the Mauthner cell, p was found to vary from 0.17–0.68 averaging 0.37 [528] (Sect. 7.8.4). Estimates for 18 cases with $m<7$ gave binomial n between 3 and 28 which were in practicaly every case equal to the number of stained presynaptic boutons (see also Sect. 7.11). More recently, two identified neurones of the goldfish brainstem were recorded after activation of the Mauthner axon [408]. Similar n (8–35 at 2 Hz stimulation) were found with higher p of 0.6–0.8 for EPSPs and about 1 for IPSPs.

Note that the authors of the most initial studies on the mammalian in vitro hippocampus (Table 7.2) used the Poisson approximation and, therefore, they did not calculate binomial parameters. One of the published histograms for the IPSP recorded in CA3 [717] can be fitted by the binomial distribution with n = 8 and p = 0.58. More recently (Table 7.2, #8), the deconvolution technique was used and binomial parameters of EPSPs evoked in inhibitory interneurones by activation of CA3 pyramidal cells were determined. Parameter n ranged from 2–6 in five different neurones with average p equal to 0.53 ± 0.14. In a detailed study on CA1 neurones (Table 7.1, ## 10, 9), mean p was equal to 0.24 (range 0.17–0.46) and 0.29 (0.09–0.41) and mean n was 3.8 (2–5) and 5.3 (2–9) for unitary and minimal EPSPs, respectively.

7.10 Quantal Analysis of Synaptic Plasticity

7.10.1 Various Applications of Quantal Analysis

The quantal analysis allows not only to determine quantitative characteristics of various synaptic junctions but also to study a number of problems related to synaptic function, e.g. presynaptic inhibition and modulation [52, 53, 231, 284, 289, 341, 571, 894]; "functional feedback connections" [682, 684]. Numerous publications are related to the effects of various ions and drugs. The quantum hypothesis stimulated multidisciplinary approaches, in particular a combination of electrophysiological, histological and electron microscopical techniques [58, 59, 81, 82, 223, 383, 428, 449, 483, 541, 528, 529, 730, 754, 756, 806, 823–825, 887, 888, 1139]. The application of the quantum hypothesis seems to be especially promising for studies of synaptic plasticity, for example development, degeneration and regeneration [104, 106, 108, 109, 170, 269, 421, 747, 798, 971, 982, 988, 1131, 1140], including synapse development in cell cultures [755, 756]. Some results related to analysis of postactivation plasticities will be described below.

7.10.2 Frequency Facilitation (FF)

Two basic forms of FF can be extracted. The first form is the so-called "paired-pulse facilitation" (PPF) which is an amplitude increase of the response to the second (testing) stimulus delivered with a short interval (typically from several ms to several scores of ms) after the first (conditioning) stimulus. The second form is a proper FF which is a similar increase during repetitive stimulation (e.g. at frequency of 10–100 Hz) in comparison with the response observed during low-frequency stimulation (e.g. at 0.3–1 Hz). In an initial analysis of the FF at the frog neuromuscular junction, equations of the Poisson distribution (methods 7.3.2 and 7.3.4) were used [262]. The FF was explained mainly or exclusively by changes in m. Therefore, the basic mechanisms of the FF should be localized presynaptically (Sect. 6.1). More recently, the presynaptic location of FF mechanisms was confirmed [653] on the basis of calculations of m from Eq. (7.3.5–2). These early studies established the idea of the constancy of v during short-term modifications of synaptic efficacy [485] so that some investigators, in fact, postulated a constant v during FF (e.g. [289]). Further supports of the presynaptic location of FF mechanisms were obtained in detailed investigations based on the binomial model and direct estimates of m (Sect. 7.3.1) at the crayfish neuromuscular junction [414,, 1082, 1145] (see reviews [50, 53]). The results of the analysis of changes in binomial parameters appeared to be more controversial: no significant changes in n were found during FF at the crayfish neuromuscular junction by some authors [415, 1145], while increases in n during stimulation at relatively low frequencies with predominantly increased p at higher frequencies (5–20 Hz) were found by others [915, 1082]. Mainly an increase in n was also found during frequency stimulation at raised Mg^{2+} concentration [105, 112] and in newly formed amphibian neuromuscular junctions [106]. They used method 7.6.2. Parameter p remained constant or was increased only slightly [105]. More complicated changes in both n and p at the frog neuromuscular junction have been described more recently [1133].

The quantal analysis of FF at neuro-neuronal synapses was initiated at the frog sympathetic ganglion [134]. An increase in m at a constant v (method 7.3.5) was found in one experiment. Postulating constancy of v and calculating n and p by method 7.6.2, McLachlan [692] found that an increase in m in the guinea pig superior cervical ganglion was mainly due to an increase in p when initial p was low but mainly due to an increase in n when p was >0.5 initially. McLachlan [693] summarized this and other similar observations by suggesting that p was always affected by repetitive stimulation, but it saturated. When this happened, an increase in n might occur. Quantal analysis of FF at the mammalian CNS was performed with recordings of minimal and unitary EPSPs evoked in cat spinal motoneurones by stimulation of the primary muscle afferents [571, 709] or by intracerebral microstimulation in deafferented spinal cord [577]. The authors used equations based on the Poisson distribution, and found an increase in m without changes in v. These data, as well as the lack of correlation between the amplitudes of the second (facilitated) and first (control) responses during PPF [577], indicated a presynaptic localization of underlying mechanisms. More recently, a similar conclusion was reached for several in vitro preparations of invertebrate and vertebrate CNSs (see review [823]), e.g. for DG–CA3 synapses in mammalian slices [1126], excitatory synapse on the goldfish Mauthner cell [609] and serotoninergic connections formed in tissue culture of identified central neurones of the leech [761]. However, the authors were not able to resolve n and p for the control and facilitated responses.

7.10.3 Frequency Depression (FD)

A diminution of the response amplitude instead of FF and PPF might occur under certain conditions, depending on the type of a synaptic junction, its functional state and stimulus parameters. In addition to paired-pulse depression, two types of FD can be separated: FD which occurs during short stimulus trains or at the beginning of a long train, and FD which results from "long-term" (minutes, at least) repetitive stimulation [693]. The second type of FD is very common, even if FF takes place initially, and it is attributed to inability of transmitter synthesis to maintain the stores of the transmitter depleted during the repetitive stimulation [312, 313, 693]. The so-called transmitter depletion model (Chap. 9) was used in numerous studies to analyze FD at neuromuscular junctions (Sect. 9.7). In general, the results are in agreement with a presynaptic location of basic FD mechanisms. Quantal analysis showed a decrease in m and also suggested a presynaptic location of FD mechanisms [262, 313]. Method 7.6.1 revealed a decrease in n at a constant p during short-term FD at the regenerated frog neuromuscular junction [106].

In several studies on various peripheral junctions, FD during lasting repetitive stimulation was shown to be also due to a decrease in m [415, 470, 692]. However, in contrast to the above data [106], a considerable diminution in p with only a slight tendency to a reduction in n (method 7.6.1) was found during long-lasting frequency stimulation at the crayfish neuromuscular junction [415]. In addition to an essential reduction in m (for about 50%, on average), a decrease in μ was observed in this study. The amplitude of mPSPs could fall to 80% of an initial value, which actually had been also noted in an earlier study [262] and in several other publications [313, 376, 692]. Similar changes were described more recently also after a brief (100 stimuli at 2–20 Hz) stimulation [275]. On the other hand, more prolonged stimulation was reported to have no effect on v [803] or even to increase it [995] in some neuromuscular preparations. A substantial decrease in v (methods 7.3.2 and 7.3.5) was found in addition to a reduction in m during FD at the rat neuromuscular junction treated with hemicholinium-3, which presumably prevented the filling of the synaptic vesicles [470]. No substantial change in v was found at the rat superior cervical ganglion under similar conditions [850–852]. A decrease in v during long-lasting stimulation without hemicholinium pretreatment was also indicated by recordings from neurones at the goldfish brainstem [728] but without a special quantal analysis. A more recent quantal analysis showed that at least at lower frequencies (< 10 Hz), the FD of EPSPs and IPSPs in the same neurones was accompanied by a decrease in m and n [408]. Parameter p decreased only rarely or even increased, while v was constant [408]. In the Mauthner cell, similar FD of IPSPs could be attributed solely to a decline in p [529]. Parameter n remained constant and equal to the number of the stained boutons. It should be stressed that the diversity of the changes in n and p might be explained not only by different preparations and experimental conditions but also by the inadequacy of the simple binomial model (Sect. 7.7). Actual increase in p at initially low-efficacy sites could give an increase in calculated binomial n with variable changes in binomial p (increase or decrease depending on the initial distribution of p_i and their changes). Similarly, an essential drop in p might appear as an apparent decrease in n during FD (e.g. [358]).

Analysis of the LFD at invertebrate CNSs is of special interest (Sect. 2.5). The first attempt of quantal analysis, according to data of one experiment, showed that a decrease in the amplitudes of aggregate EPSPs from the crayfish CNS was accompanied by an increase in their CV [1144]. However, the increase in the CV was unproportionally large, which was explained by a non-uniform reduction in m at different synapses [1144]. In an attempt to check this suggestion, Krasne [547] analyzed three unitary EPSPs, and unexpectedly found decrease in the CV. Krasne [547] inclined to explain these data by p heterogeneity (Sect. 7.7) and its non-uniform changes. A more detailed investigation of LFD mechanisms was conducted for the CNS of *Aplysia* [212, 475]. Poisson statistics (method 7.3.4) and method 7.3.3 were used. No changes in v were found and presynaptic mechanisms of both LFD and behavioural habituation were supported. A similar conclusion in relation to LFD of EPSPs evoked by descending fibres in motoneurones of the frog spinal cord was reached on the basis of mPSP recordings [374].

7.10.4 Post-Tetanic Potentiation (PTP) and Heterosynaptic Facilitation (HSF)

An increase in m during PTP was found at the rat neuromuscular junction under conditions of suppressed transmitter release [110]. The increase was due to changes in p. Parameter n was reduced in the majority of cases. In the avian ciliary ganglion, a post-tetanic increase in mPSPs frequency was observed without considerable changes in μ [674]. However, the estimates of the quantal parameters were not given. Kuno [571] found a decrease in N_0 for minimal EPSPs of spinal motoneurones and suggested presynaptic mechanisms. A constancy of the distances between discrete components of the deconvolved distributions of amplitudes of motoneuronal EPSPs (Sect. 7.3.7) indicated a presynaptic mechanism for the PTP [309, 433].

The same methods as those used in studies of the LFD (Sect. 7.10.3) were employed to investigate the HSF in *Aplysia* CNS (Sect. 2.5). An increase in m was found without considerable changes in v [213, 475].

7.10.5 Long-Term Facilitation (LTF)

LTF after a comparatively long-lasting (e.g. 20 min) repetitive activation of the crayfish neuromuscular preparation [50, 53, 892, 1109] was already mentioned (see Introduction). It was found more recently that a shorter train (50 Hz, 15 s) can also induce LTF [98]. This phenomenon looked superficially like hippocampal LTP. Three methods (7.3.2, 7.3.4, 7.3.5) of quantal analysis with Poisson equations were applied [98] and no effect on v was found after LTF induction, while m increased. This finding supports a presynaptic mechanism for the LTF. Similar conclusions were made in other studies on the basis of PSP [1110] and PSC [1111] recordings. Binomial deconvolution showed increases in both n and p. A long-lasting increase in m was also found in bullfrog sympathetic ganglion following tetanic stimulation of the preganglionic fibres [539]. In addition, an increase in v was seen in some experiments, particularly in cases where the tetanization evoked spikes. Another form of LTF was revealed in the bullfrog sympathetic ganglion [569]: a facilitation of the fast EPSP recorded in a low Ca^{2+} and high Mg^{2+} solution after antidromic stimulation of the preganglionic nerve trunk. A significant increase in v was found with only a small rise in m, which indicates predominantly postsynaptic involvement. No significant changes were found in R_{in}, while the increased sensitivity of the subsynaptic membrane to ACh was indicated by an increase in nicotinic ACh potential. It is interesting that this "postsynaptic" LTF was not observed in the rat sympathetic ganglion [167]. A more detailed review of "pre-" and "postsynaptic" types of LTF in sympathetic ganglia was made recently [562].

Summarizing, most data obtained indicate presynaptic localization of mechanisms of short- and long-term plasticities at both neuromuscular and neuro-neuronal synapses studied so far. Nevertheless, adequate methods of quantal analysis were not used in all studies cited. Inaccuracies might result from using Poisson instead of binomial statistics. Only a small number of studies at the vertebrate CNS were performed, and a few very recent studies dealt with phenomena of long-term plasticities. Related studies at the mammalian forebrain appeared only recently and they will be considered in Chapters 12–15.

7.11 Conclusions

One can see from the literature reviewed in this chapter that the quantum hypothesis possesses an effective apparatus for both quantitative description of the synaptic transmission and analysis of changes in its efficacy. These techniques are under continuous refinement and their field of application is steadily expanding. However, it is clear at present that some important aspects of the quantum hypothesis require further clarification. Some observations (Sect. 7.4.4) questioned one of the basic postulates of the quantum hypothesis and suggested that one quantum might consist of several subunits. This hypothesis rejects a more customary "one quantum-one vesicle" principle [293, 494, 975] but might provide an easier explanation for changes in v, which were sometimes reported as a result of action of various ionic [261, 549, 606], pharmacological [312, 448] and temperature challenges [445, 549] as well as frequency stimulation (Sect. 7.10.3). The interpretation "one quantum-several vesicles" corresponds to the data, indicating that there are several sites at which vesicles can fuse with the membrane in each presynaptic "active zone" [8, 286, 428, 976]. Within this interpretation, the number of the "active zones" or "synaptic complexes" might be related to the maximal quantal content, i.e. to n [111, 528, 976, 1085, 1145]. One problem of the more customary principle "one quantum-one vesicle" is the explanation of the fact that only one vesicle is released out of many available [976]. On the other hand, the vesicular hypothesis itself, i.e. the view that the transmitter release is preceded by its packaging and storage in the form of vesicles [359, 444, 485, 486, 487, 1141] encountered difficulties in interpretation of some experimental results [17, 293, 458, 551, 666, 956, 977, 991, 992, 1003]. One example of the difficulties is the fact that 99% of spontaneously released ACh at the frog neuromuscular junctions is due to a continuous "non-quantal release". Recently, an alternative to the vesicular hypothesis was proposed on the basis of consideration of the above mentioned process of drop formation [551]. The packet size of the transmitter was suggested to be determined at the moment of release.

Note that the present day interpretation of parameter n as the number of "active zones" involved in transmitter release is distinct from the original interpretation of this parameter as the number of available portions of transmitter (Sect. 7.2). Zucker [1145] first paid attention to the small values of n (2–4) for the local release sites at the neuromuscular junction and suggested that n corresponded rather to a number of active release sites. The mean distance between the release sites in the neuromuscular junction was found to be of about 1–2 μm [1085]. In view of the estimated values of n for neuromuscular junctions, one should expect values of n of about 1–2 to be typical for central synapses which consisted of one bouton or only a few small boutons. Every such bouton was shown to contain typically only one active zone [790, 976]. On the basis of studies at neuromuscular junctions, it has been suggested that normally at most only one quantum is released from a release site by one presynaptic impulse [1084, 1085, 693]. The modern scheme of PSP generation at central synapse [306, 307, 525, 823, 1068] presumes all-or-nothing release of one transmitter quantum at each release site with generally non-uniform p_i ranging from 0 (silent release sites) to 1. The physical correlate of n estimated in the frame of the compound binomial model is the number of effective release sites with $p_i>0$. It may be equal to the number of synaptic boutons if every bouton contains one effective release site.

Definite complications in the simple apparatus of binomial (or Poisson) statistics were introduced by the understanding the fact that spatial heterogeneity of p cannot be ignored in many situations (Sect. 7.7.2). This problem awaits more direct studies with recordings from separate release sites. For many central synapses, for which n is small, this problem either does not exist (when n = 1) or may be resolved much more easily than for junctions with multiple release sites. Recent studies, which used a combination of physiological and anatomical techniques (Sect. 7.10.1 and below), made important contributions to the understanding of the physical correlate for parameters n and p in the vertebrate brain. A striking equivalence between the binomial n and the number of stained presynaptic boutons was found at an inhibitory synapse on the Mauthner cell [528]. More complicated relations were found for spinal neurones [383, 483, 541, 806, 824, 825, 1068], where the spatial non-uniformity of p and the existence of non-effective (silent) boutons were suggested.

Summarizing, it can be stated, that despite some unsettled problems related to both methods of calculation of the quantal parameters and interpretation of their physical nature, the quantum hypothesis provides effective methods of quantitative analysis of synaptic transmission. When we started our analysis of the mechanisms of hippocampal plasticity, only very little information was available about statistical parameters of CNS synapses. Moreover, serious difficulties for the quantum hypothesis had been created by demonstrations of irregular amplitude fluctuations [308, 309]. Edwards et al. [308, 309] inclined to reject the applicability of the quantum hypothesis to the central synapses. At the same time, there were practically no data on quantal analysis on synapses of mammalian forebrain. A modification and testing of the methods of the quantal analysis for their application to studies of vertebrate central synapses seemed to be an urgent task necessary to start the quantitative analysis of mechanisms of firstly short- (PPF, LFD) and afterwards long-term (LTP) synaptic plasticities at the vertebrate brain.

7.12 Summary

Basic assumptions of the quantum hypothesis of synaptic transmission are discussed. Known statistical methods are systematized, which allow estimation of the mean quantal content (m), quantal size (v) and binomial parameters (n and p) of the distributions of amplitudes of unitary and minimal postsynaptic potentials. Estimates of the quantal parameters of various junctions are presented. It is shown that before we started our experiments, only estimates for peripheral synapses and synapses at the spinal cord were available for mammalians. Literary data are reviewed, which studied phenomena of synaptic plasticities (frequency facilitation, frequency depression, short-term post-tetanic potentiation and heterosynaptic facilitation) at neuromuscular junctions and invertebrate neuro-neuronal synapses. Most of these studies revealed only (or predominantly) changes in m and, therefore, indicated a presynaptic localization of basic mechanisms of plasticities studied. In some cases, changes in v were also noted. The following seems to be urgent: (1) modifications of methods of quantal analysis for application to synapses of the vertebrate CNS; (2) testing these methods and their application to study mechanisms of short-term CNS plasticity (frequency facilitation and depression); (3) application of these methods to studies of mechanisms of hippocampal long-term potentiation.

8 Application of Quantal Analysis to Central Synapses

8.1 Introduction

The data reviewed in Chapter 7 showed that, despite the informative power of the quantum hypothesis, its application to the analysis of synaptic mechanisms at the vertebrate CNS was carried out only at the beginning when we started our studies of hippocampal plasticity. Therefore, our initial aim was to test the applicability of the related methods to responses of hippocampal neurones in vivo. We planned to choose the most adequate methods of measurements and computations, to modify them in order to increase their accuracy, and to compare the results obtained by different methods. In addition, unitary EPSPs of the snail central neurones were analyzed for comparison with results obtained with minimal hippocampal EPSPs.

8.2 Methods

Recordings from area CA3 of the rabbit hippocampus were made in the basic experiments. Either the SFA or DG was stimulated to induce minimal EPSPs by the paired-pulse procedure (Chap. 4). In some cases, an additional movable stimulating microelectrode was inserted in the DG to search for the most effective site with minimal threshold currents (2–10 μA) to induce minimal EPSPs. Measurements of the EPSPs were made with analyzer TAS (Leitz, FRG) and laboratory computers. Areas were usually measured instead of amplitudes, to increase the accuracy of the measurements and signal-to-noise ratio (Sect. 2.2.6). The areas were converted into amplitudes by normalizing the measurements from the related average responses. Intervals for the measurements (from the beginning to a point near the peak of the EPSP) were chosen from the averages. "Quasi-stationary" plateau regions [475] without significant amplitude drifts were used for the quantal analysis. Each plateau region contained from 20 to 100 measurements with the mean number of measurements equal to 48±21 (±SD) for 41 regions. Data processing, which included averaging, calculation of quantal parameters, estimates of their errors and construction of theoretical distributions, was performed with computer algorithms written in collaboration with Drs. A.Baskys, V.I.Derevyagin, A.G.Gusev and D.R.Shidarev. The lower limits of errors of the estimates of the basic quantal (m, v) and binomial (p, n) parameters were approximately evaluated according to formulae similar to those given in several publications [670, 693, 833]. For the histogram method (Sect. 8.4) the error of v was tentatively taken as equal to the bin size. It should be stressed that these estimates are only rude approximations of the lower limit of possible errors.

Identifiable monosynaptic pathways mentioned in Section 2.2.1 were activated to evoke unitary EPSPs in the snail parietal ganglia (see [620, 621] for the details of the methods).

8.3 Consideration of Known Methods of Determination of Quantal Parameters

As already noted (Sect. 4.4; see Fig. 4.4C), minimal EPSPs of hippocampal neurones induced by testing stimuli during the paired pulse stimulation displayed amplitude fluctuations and response failures. Figure 8.1A–C illustrates the case with responses recorded from the same hippocampal neurone following different stimulus strengths. In spite of small N available for the stationary regions, the histograms show clear separate peaks with similar distances for different stimulus currents. Note a relatively small noise level on the background of the conditioning IPSP (Fig. 8.1D1) and a clear separation of failures (Fig. 8.1D1) from "single" (Fig. 8.1D3) and "double" (Fig. 8.1D2) responses corresponding to respective peaks in the histogram (Fig. 8.1A). Figure 8.2A illustrates EPSPs evoked by microstimulation with occasional response failures (Fig. 8.2A, upper sweep). Corresponding amplitude distributions (Figs. 8.1A–C, 8.2B) showed distinct peaks and troughs. The distances between peaks were approximately equal to the distance from zero to the first maximum (Fig. 8.1A, B, arrows; Fig. 8.2B) or the former distance was a multiple integer of the latter (Fig. 8.1C). Therefore, the behaviour of hippocampal minimal EPSPs formally satisfies the basic postulate of the quantum hypothesis [Eq. (7.2–1)], and therefore an attempt could be made to calculate quantal parameters assuming that the amplitude fluctuations (Figs. 8.1 and 8.2) depended on the different number of transmitter quanta released by the activated fibres.

Based on the known literary data (Chap. 7), we assumed the binomial model for the calculations of the quantal parameters. Most methods of determination of the binomial parameters used in the literature (7.6.1–7.6.4) were based on the knowledge of μ. Such methods seemed not to be suitable for CNS neurones with a background synaptic bombardment from many inputs. There were only initial attempts to use automatic optimization methods (Sect. 7.6.5) before we started our studies [308–310, 709], and the attempts did not seem to be quite satisfactory. For example, Edwards et al. [308] came to a conclusion about the non-quantal nature of the central synaptic transmission, which was rejected later. The problem of the uniqueness of the solutions was far from being clear when we started our studies [1012] and it is not completely resolved now [230]. The automatic analysis implied large samples, which were not available under our conditions with recordings from the in vivo preparation. Therefore, we did not use these methods until

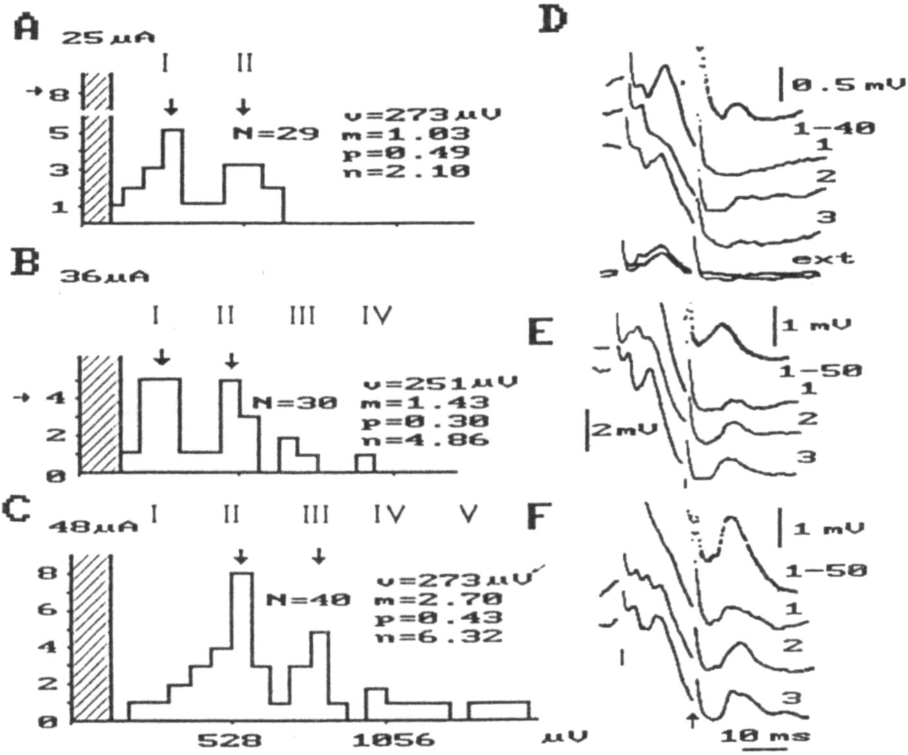

Fig. 8.1A–F. Multimodal distributions of amplitudes of EPSPs evoked in a hippocampal neurone by three different currents (**A–C**) and examples of corresponding responses (**D–F**). A–C *Ordinate* number of responses; *abscissa* EPSP amplitude, µV. *Arrows* denote peaks. *Roman numerals* denote values multiple to presumed v (see text). *N* is the sample size. *Shaded areas* indicate the noise level. *Horizontal arrows* near ordinate (**A**, **B**) show the number of responses, which did not exceed the noise level. **D–F** Upper (marked *1–40* or *1–50*) and lower (*1–3*) oscillograms show averaged and single responses, respectively. Calibrations: 0.5 and 1 mV for averaged records; 2 mV for single records. [1045]

more recently when they were modified and tested with computer simulations (see Chaps. 13–15).

A method described in Section 7.6.6 had been used in practice only to estimate the mean value of p for several neurones [475, 570, 578]. We tried to use it for single neurones, but it appeared to be inaccurate under our conditions. Theoretical distributions based on the parameters calculated by this method showed significant differences from observed (experimental) distributions ($P<0.05$; Chi-square test) in several pilot evaluations. An example is shown in Fig. 8.2D (dotted curve and numbers in parenthesis). Note the differences in the position of peaks in theoretical (dotted curve) and experimental (columns) histograms. In addition, the estimates of this method were usually uncertain: calculated errors were comparable with or even larger than the values of the parameters themselves: e.g., for the illustrated case m = 1.7Â0.8 and p = 0.2±0.67. The first method from Section 7.6.7 will be considered in Chapters 9 and 10, while the second one is suitable only for in vitro preparations.

In general, these considerations indicated the necessity of modifying the known methods and applying additional methods to perform quantal analysis of EPSPs recorded from central neurones in vivo.

8.4 The Histogram Method

Because of the presence of distinct peaks and troughs in the majority of experimental amplitude histograms, it was natural to try to employ the histogram method (Sect. 7.3.3). It had been used in practice mainly as an "initial peak method" or as a method to determine binomial parameters after the division of the histogram into classes according to the known amplitude of mPSPs [1085]. We had no possibility to reliably select respective mPSPs in the central neurones and it was noted already that the "initial peak method" was not reliable. However, we made use of the fact that in most cases, at least two peaks with appropriate distances from zero were visible. Therefore, we applied a version of the method 7.3.3 for calculations of the basic quantal parameters (v and m) and evaluation of approximate values of binomial parameters.

If k+1 successive peaks with numbers from 0 to k are distinguishable in the histogram, and $l_{i-1,i}$ is the distance between peaks with numbers i–1 and i,

$$v = (1/k) * \Sigma\, l_{i-1,i}, \qquad (8.4\text{–}1)$$

with the summation from i = 1 to i = k. When the distance for all peaks from zero (for k>2) is taken into account

$$v = [1/(2k-1)] * [\Sigma\, l_{i-1,i} + \Sigma\, l_{0,i}/i], \qquad (8.4\text{–}2)$$

with the summation from i = 1 to i = k. The amplitudes were grouped according to the v value, and p was calculated by Eq. (7.6.2–1) or (7.6.2–2). Then n and m were calculated from Eqs. (7.2–2) and (7.2–3).

An example of histogram analysis is given in Fig. 8.1A–C, where arrows denote centres of intervals selected for the grouping of amplitudes according to Eq. (8.4–2). The quantal parameters obtained in this experiment are given in insets of Fig. 8.1A–C and in Table 8.1 (numbers without parentheses). Values of m_0 were calculated according to Eq. (7.3.4–2) with parameter p determined either by the method described here (numbers without parentheses) or in the following section (numbers with parentheses). The number of responses not exceeding the noise level (shaded columns in Fig. 8.1A–C) was taken as N_0. Table 8.2 compares theoretical values of N_x [see Eq. (7.2–5)] calculated from parameters given in Table 8.1 with respective experimental values obtained from the division of the experi-

Table 8.1. Quantal parameters of EPSPs recorded in a hippocampal neurone at different strengths of septo-fimbrial stimulation

μA	N	v±S$_v$' (μV)	m±S$_m$'	p±S$_p$'	n±S$_n$'	m_0
25	29	273±33	1.03±0.13	0.49±0.12	2.10±0.35	0.93
		(266±108)	(1.12±0.43)	(0.45±0.06)	(2.48±1.01)	(0.96)
36	30	251±33	1.43±0.18	0.30±0.18	4.86±1.66	1.45
		(228±71)	(1.72±0.59)	(0.40±0.04)	(4.34±1.55)	(1.40)
44	49	264±33	2.02±0.16	0.37±0.13	5.46±1.19	1.83
		(239±72)	(2.34±0.67)	(0.47±0.04)	(4.92±1.48)	(1.67)
48	40	272±33	2.70±0.19	0.43±0.12	6.32±1.34	-
		(213±61)	(3.67±1.01)	(0.49±0.04)	(7.40±2.12)	-

Notes. Stimulation current is given in the first column (μA). For each column the top raw and the bottom raw (numbers in parenthesis) represent parameters calculated by method 8.4 and 8.5, respectively. N is the number of stimuli. S_v', S_m', S_p' and S_n' are approximate estimates of lower limits of the errors of parameters v, m, p and n, respectively (see Sect. 8.2).

mental histograms according to the estimated v. The similarity of the observed (Experimental) and predicted (Theoretical 8.4) values in Table 8.2 indicates that the experimental histograms can be fitted to the binomial distribution with the parameters found by the histogram method.

Figure 8.2B gives a more detailed comparison of an experimental histogram with a theoretical distribution based on parameters determined by the histogram method. The theoretical distribution was constructed (see Sect. 8.7) assuming the binomial model. A good agreement between experimental and theoretical distributions was confirmed by the Chi-square test (P>0.9). Similar insignificant differ-

Table 8.2. Comparison of experimental and theoretical amplitude distributions for EPSPs of the same neurone at different stimulus strengths

Current (μA)	Distribution	N_0	N_1	N_2	N_3	N_4	N_5	N_6	N
25	Experimental	8	12	9	0	0	0	0	29
	Theoretical 8.4	8	14	7	0	0	0	0	29
	Theoretical 8.5	9	14	6	0	0	0	0	29
36	Experimental	5	12	9	3	1	0	0	30
	Theoretical 8.4	5	12	9	3	0	0	0	29
	Theoretical 8.5	4	11	10	4	1	0	0	30
44	Experimental	6	9	17	12	5	0	0	49
	Theoretical 8.4	5	14	17	10	3	0	0	49
	Theoretical 8.5	2	9	16	15	7	1	0	50
48	Experimental	0	4	18	10	4	2	2	40
	Theoretical 8.4	1	5	11	12	8	3	0	40
	Theoretical 8.5	1	4	10	12	9	3	1	40

Notes. Theoretical values of N_x and N were taken as nearest integers. See Table 8.1 and text for further explanation.

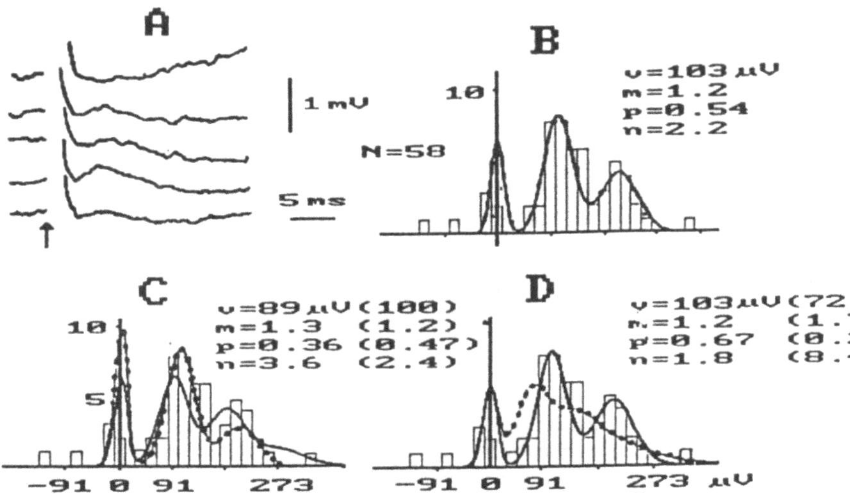

Fig. 8.2A–D. Quantal analysis of minimal CA3 EPSPs following microstimulation (2.5 µA, 0.3 ms) of the DG. **A** Responses to the second (testing) stimulus in a paired-pulse paradigm. **B–D** Experimental (*columns*) and theoretical (*curves*) amplitude distributions. Parameters were calculated by methods 8.4 (**B**), 8.5 (**C**) and by the variant of method 8.5 based on the estimates of number of failures (**D**). In addition, the method described in Section 7.6.6 was tested (*parameters in parentheses* and *dotted line* in **D**). In **C** the continuous line corresponds to p calculated from E_{max}, while the *dotted line* and parameters in parentheses were based on submaximal amplitude value instead of E_{max}. See Fig. 8.1 for additional explanations. [1012]

ences between experimental and predicted binomial distributions were found for 40 out of 41 tested amplitude histograms of hippocampal minimal EPSPs ($P>0.1$, Chi-square test, see examples of values in Table 8.3, cases 5–8).

8.5 Method Based on the Ratio of the Mean to Maximal Amplitude

To have another method, which would not depend on subjective evaluation of the histogram peaks and which less sensitive to N, we made use of a property of the binomial distribution, which relates mean (E) and maximal (E_{max}) values. The relationship $E_{max} = nv$ is obvious for the binomial distribution. After combining this relationship with basic equations of the quantum hypothesis [Eqs. (7.2–2) and (7.2–3)]

$$p = E/E_{max}. \qquad (8.5\text{–}1)$$

This equation was offered by Balanter [64] and independently by V.I.Derevyagin [88, 1045]. To our knowledge, it has not been practically used before our publications [88, 1045].

With a sufficient number of failures (e.g. $N_0>2$), it is possible to calculate m from Eq. (7.3.4–2) after determination of p from (8.5–1). When failures are not present, m can be calculated from the CV [Eqs. (7.3.5–3), (7.3.5–4) or (7.3.5–5)]. Afterwards, n and v can be calculated, as usual, from Eqs. (7.2–2) and (7.2–3), respectively. These two variants will be termed failures and variance methods, respectively.

The numbers in parenthesis in Table 8.1 represent parameters calculated by the variance (v, m, p, n) and the failures (m_0) methods from the experiment shown in Fig. 8.1. Table 8.2 compares theoretical N_x values calculated on the basis of the variance method with both the experimental values and with the values obtained by the histogram method. Figure 8.2C compares more detailed predicted distributions (continuous curve, see Sect. 8.7 for the formula) with the observed distribution (columns). The comparison shows a satisfactory agreement, though it is worse than for the histogram method (Fig. 8.2B). It can be suggested that in a given case, some discrepancies between theoretical and experimental distributions depended on the noise, which might distort the true value of E_{max}. In fact, the use of a "submaximal" E value instead of E_{max} gave better agreement of the theoretical distribution (dotted curve in Fig. 8.2C) with both the experimental histogram and parameters calculated by the histogram method (Fig. 8.2B). It is evident that with other conditions equal, a relative error in the determination of p due to noise influences should increase with a decrease in n. A more detailed analysis of Eq. (8.5–1) and further modifications of our methods will be given below (Chap. 13). The similarity between theoretical and experimental distributions ($P>0.1$, the Chi-square test) was found in about 90% of 50 cases tested (Table 8.2, Figs. 8.2 and 8.4, see also Fig. 11.1 and other data in Sect. 11.2).

Figure 8.3 compares parameters calculated by the two considered methods (Sects. 8.4 and 8.5) in our in vivo experiments with the use of either macro- (circles) or microstimulation (triangles). In Fig. 8.3 and all following cases, parameters calculated by methods 8.4 and 8.5 are marked by index 1 and 2, respectively. It is clear that both methods gave similar values, on average, with only minor discrepancies. For example, p_1 and p_2 were similar at $p_1> 0.4$, but when p_1 was small $p_2>p_1$ (Fig. 8.3C). It is easy to understand taking into account that Eq. (8.5–1) is valid for the binomial distribution and becomes inaccurate with approximation to the Poisson distribution. In addition to the similarity of the mean values of parameters calculated by the two methods, high and statistically significant correlations between m_1 and m_2 (Fig. 8.3B) and p_1 and p_2 (Fig. 8.3C) should be noted.

8.6 Quantal Parameters of Unitary EPSPs of Snail CNS

Calculation of quantal parameters of unitary EPSPs recorded in the snail central neurones (Fig. 8.4A, upper beam) in response to intracellular stimulation of presynaptic triggering neurone (Fig. 8.4A, lower beam) was used for additional

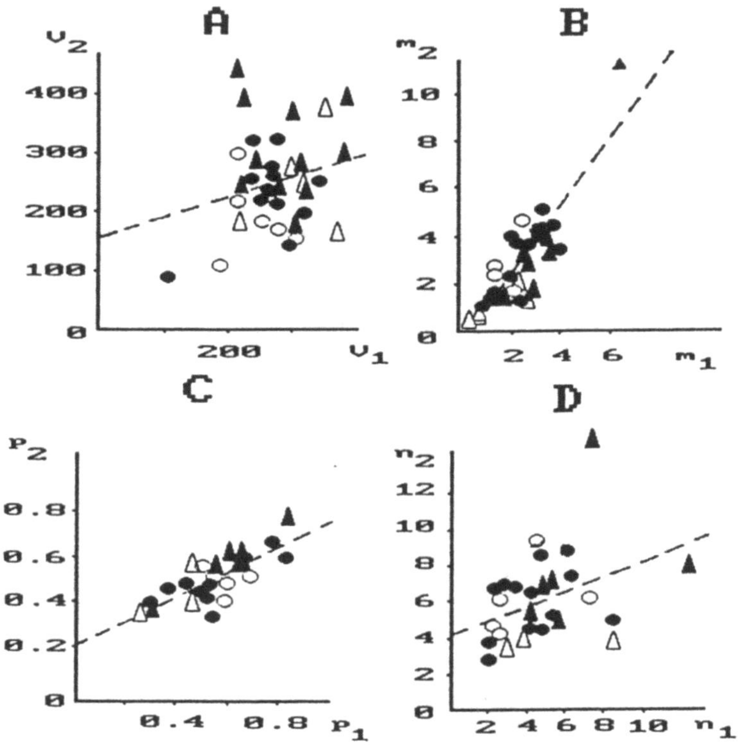

Fig. 8.3A–D. Comparison of quantal parameters calculated by two methods for minimal hippocampal EPSPs recorded in experiments with macrostimulation (circles) and microstimulation (triangles). **A–D** Parameters v, m, p and n, respectively. *Abscissa* parameters calculated by method 8.4; *ordinate* the same by method 8.5. *Interrupted lines* represent linear regressions. *Empty symbols* correspond to depressed (non-facilitated) responses recorded during late periods of the low-frequency testing stimulation. *Full symbols* correspond to non-depressed or potentiated responses recorded at initial stages of low-frequency stimulation or after tetanization in experiments with microstimulation (Chap. 12). For method 8.5, the variant based on the failures determination was used when $N_0>2$, otherwise the variance method was used. Mean values of parameters (±SD) and coefficients of correlation (r) are the following: $v_1 = 275 \pm 59\ \mu V$, $v_2 = 250 \pm 83\ \mu V$, $r = 0.24$ (A); $m_1 = 2.2 \pm 1.4$, $m_2 = 2.6 \pm 1.9$, $r = 0.86$ (B); $p_1 = 0.56 \pm 0.15$, $p_2 = 0.51 \pm 0.10$, $r = 0.79$ (C); $n_1 = 4.7 \pm 2.4$, $n_2 = 6.0 \pm 2.5$, $r = 0.37$. The number of measurements for **A** and **B** was equal to 35, for C and D it was 27. The difference is explained by exclusion of p and n, which were not significantly different from zero. [1050]

verifications of the two considered methods. A noticeable LFD similar to that described for the CNS of *Aplysia* (Sect. 2.5) was found in most cases during repeated testing stimulation with intervals of about 10 s. The mean amplitudes from five or ten adjacent responses were plotted (Fig. 8.4B) in order to choose "quasi-stationary" regions suitable for statistical analysis (Sect. 7.4.2). EPSP amplitudes decreased considerably during the initial part of the experiment. This fact is illus-

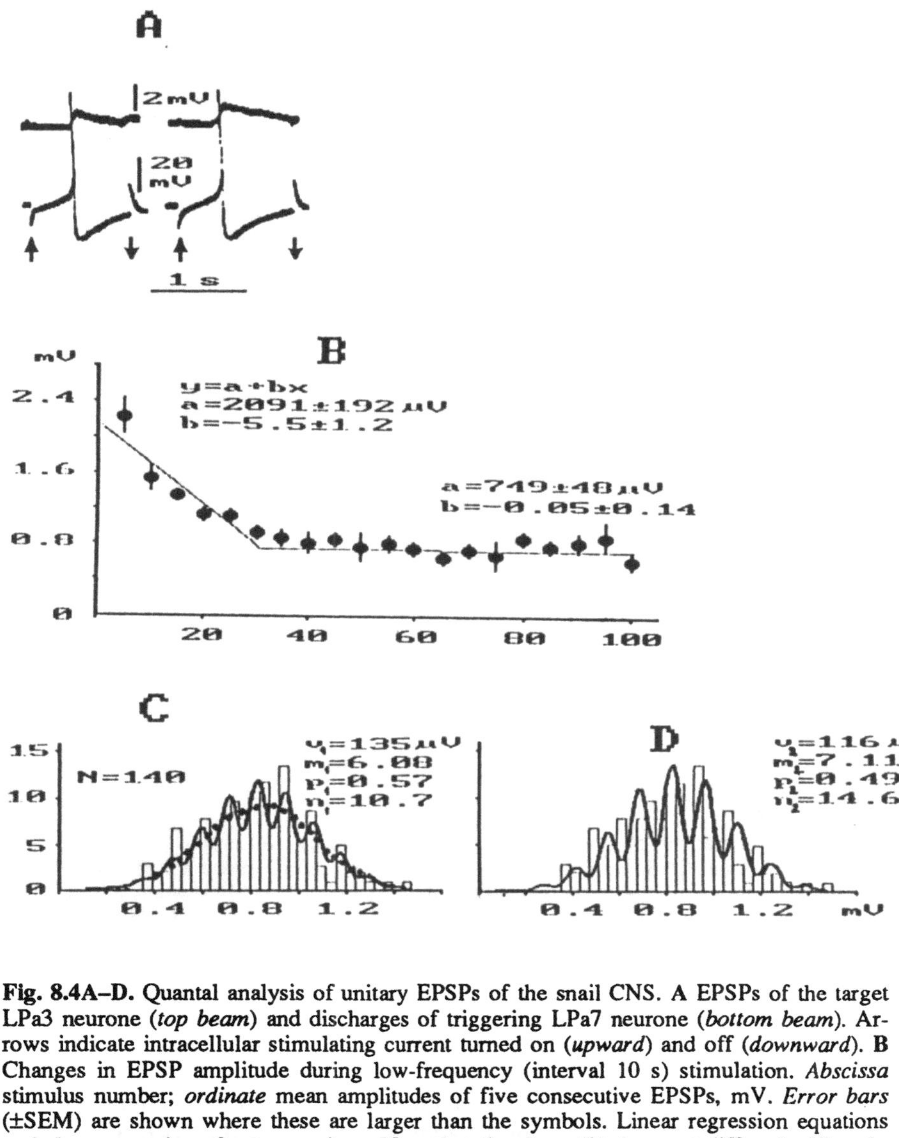

Fig. 8.4A–D. Quantal analysis of unitary EPSPs of the snail CNS. **A** EPSPs of the target LPa3 neurone (*top beam*) and discharges of triggering LPa7 neurone (*bottom beam*). Arrows indicate intracellular stimulating current turned on (*upward*) and off (*downward*). **B** Changes in EPSP amplitude during low-frequency (interval 10 s) stimulation. *Abscissa* stimulus number; *ordinate* mean amplitudes of five consecutive EPSPs, mV. *Error bars* (±SEM) are shown where these are larger than the symbols. Linear regression equations and slopes are given for two regions. Note that the slope (*b*) does not differ significantly from zero for the "quasi-stationary" region (stimulus numbers 35 to 100). **C**, **D** Comparison of experimental histograms (*columns*) with theoretical distributions (*curves*) calculated by methods 8.4 (**C**) and 8.5 (**D**) from data of several "quasi-stationary" regions. The *dotted* theoretical distribution in **C** was constructed according to Eq. (8.7–1) with $S_v = 0.25v$; the *continuous* distributions in **C** and **D** were constructed according to Eq. (8.7–3) with $S_v = 0$ and $S_n = 37\ \mu V$. [1012]

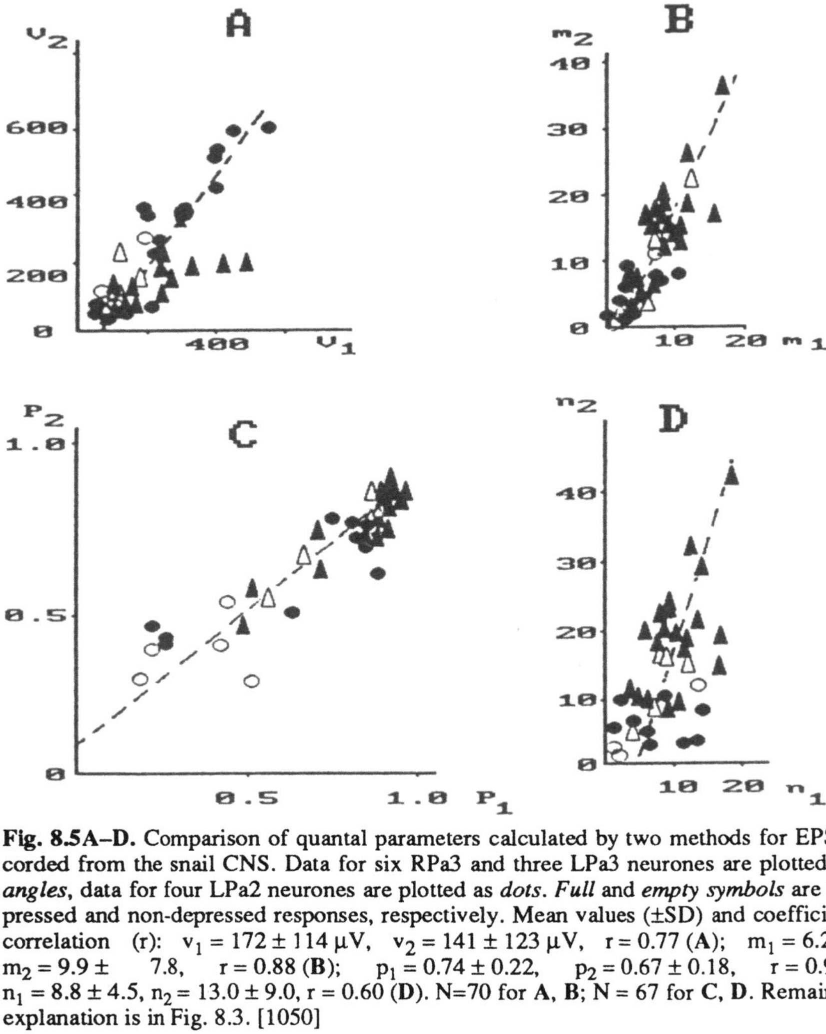

Fig. 8.5A–D. Comparison of quantal parameters calculated by two methods for EPSPs recorded from the snail CNS. Data for six RPa3 and three LPa3 neurones are plotted as *triangles*, data for four LPa2 neurones are plotted as *dots*. *Full* and *empty symbols* are for depressed and non-depressed responses, respectively. Mean values (±SD) and coefficients of correlation (r): $v_1 = 172 \pm 114$ µV, $v_2 = 141 \pm 123$ µV, $r = 0.77$ (**A**); $m_1 = 6.2 \pm 4.1$, $m_2 = 9.9 \pm 7.8$, $r = 0.88$ (**B**); $p_1 = 0.74 \pm 0.22$, $p_2 = 0.67 \pm 0.18$, $r = 0.91$ (**C**); $n_1 = 8.8 \pm 4.5$, $n_2 = 13.0 \pm 9.0$, $r = 0.60$ (**D**). N=70 for **A**, **B**; N = 67 for **C**, **D**. Remainder of explanation is in Fig. 8.3. [1050]

trated by the first regression line drawn through the initial six points (30 responses). In contrast, the later part of the amplitude course in Fig. 8.4B was fitted by a regression line with the slope, which did not differ significantly from zero ("quasi-stationary" region). Amplitude distributions and the quantal parameters for this "quasi-stationary" region are given in Figure 8.4C and D. The parameters calculated by the two methods were similar within the errors of calculations. A good agreement between experimental (columns) and theoretical distributions (continuous curves) should be also noted (P>0.9, Chi-square test for both Fig. 8.4C and D).

Figure 8.5 summarizes data obtained with analysis of 70 "quasi-stationary" regions based on recordings of responses of 13 neurones. One can see that, despite some differences between parameters calculated by the two methods, the mean values were of the same order. Parameters calculated by the two methods show highly significant correlations with the coefficient of correlation in the range from 0.55 to 0.89 (see Fig. 8.5, legend). Similarly to the case with the hippocampal EPSPs (Fig. 8.3), p_1 and p_2 showed especially close mean values and the highest correlation.

8.7 Theoretical Distributions of PSP Amplitudes

To build a predicted (theoretical) distribution, the following formula was used in the literature [106, 158, 475, 485, 709, 1084]:

$$N(E) = N \sum_{x=0}^{n} \left[(nx)\, p^x (1-p)^{n-x} \right] \frac{1}{S_v \sqrt{2\pi x}} \exp - \left\{ \frac{(E - xv)^2}{2\, x\, S_v^2} \right\} \qquad (8.7\text{–}1)$$

or a formula analogous to Eq. (8.7–1) with expression in the brackets corresponding not to the binomial [Eq. (7.2–5)] but to the Poisson distribution [Eq. (7.2–7)]. Eq. (8.7–1) was deduced from the hypothesis that the amplitude distribution N(E) are determined by the sum of normal distributions with weight coefficients obeying the binomial (or Poisson) law. The value of S_v was either determined from distributions of amplitudes of mPSPs [106, 158, 485] or taken to be 0.2–0.3 [475]. Another variant is the optimization of S_v by the trial and error method [570, 709]. If the amplitude histogram of mPSPs is better described by the gamma-distribution [692, 693], Eq. (8.7–1) must be modified correspondingly.

However, we noticed that theoretical distributions (Fig. 8.4C, dotted curve) calculated according to Eq. (8.7–1) were usually "smoother" than the respective experimental ones (Fig. 8.4C, columns) and did not contain peaks, especially around the distant modes, which might be clearly distinguishable in experimental distributions. We noted that similar phenomena were seen in published distributions of other authors [106, 158, 475, 570, 571]. The reason for this discrepancy might be that only S_v (SD of one quantal event) is taken into account by Eq. (8.7–1), while both S_v and S_n should be considered [89]. In fact, if a noise is present with a variance S_n^2, then the variance of a single quantum (i.e. the variance of PSP amplitudes for x = 1) will be $S_1^2 = S_v^2 + S_n^2$. With $x>1$, only this part of the variance will increase, which is related to the intrinsic variance of v, i.e.

$$S_x^2 = xS_v^2 + S_n^2. \qquad (8.7\text{–}2)$$

Therefore, Eq. (8.7–1) assumes the form

$$N(E) = N \sum_{x=0}^{n} [(nx)\, p^x\, (1-p)^{n-x}] \frac{1}{\sqrt{2\pi(S_n{}^2 + xS^2)}} * \exp\left\{ \frac{(E - xv)^2}{2(S_n{}^2 + S_v{}^2)} \right\} \qquad (8.7\text{–}3)$$

To test these considerations, distributions calculated from Eqs. (8.7–1) and (8.7–3) were compared with experimental histograms of both hippocampal minimal EPSPs and unitary EPSPs of the snail CNS. Quantal parameters were calculated according to the histogram method. Values of S_v and S_n, which gave the best fit of distributions N(E) and N'(E) to the experimental histograms were found by a computer algorithm. The Chi-square test was used as the optimization criterion. It was originally planned to determine S_n directly from the noise measurements. However, the values obtained in this way were in many cases close to the optimum S_1 found by the algorithm, or even they exceeded it. In these cases, theoretical distributions contained no peaks, and the use of Eq. (8.7–2) led to a clearly underestimated (or even negative) S_v. The reason for this fact, which had already been noted by other authors [709] was not clear. Superficially, it looks like a decrease in S_n during generation of minimal and unitary EPSPs (see also Sect. 13.4.6 for later observations and additional discussion) but mechanisms for such a decrease are not understood, and this problem requires further analysis.

The main results of our determinations of the S_n and S_v are presented in Table 8.3. In addition to our data obtained for the two types of neuronal synapses, Table 8.3 presents results of the similar analysis of the "classical" histogram of EPP amplitudes [158] reproduced in several reviews [55, 475, 485].

We found that with the proper selection of S_v and S_n, Eq. (8.7–3) gave good agreement with the experimental data (Figs. 8.2B). When only intrinsic variations of v were taken into account, i.e. S_n was ignored ($S_n = 0$ in Table 8.3), Eq. (8.7–3) transformed into Eq. (8.7–1). In this case, the best fit was always inferior to that found with Eq. (8.7–3) (see values of P in Table 8.3). Our subjective impression was that N'(E) gave essentially better agreement with experimental histograms than N(E), even when P values (Table 8.3) showed only minor differences. This might point to a low sensitivity of the Chi-square test under the present conditions, probably because of small N. Distribution N(E) was particularly smooth compared with the experimental histograms when S_v was taken to be equal to 0.2v–0.3v (Fig. 8.4, dotted curve). Good fits between theoretical distributions N'(E) and experimental histograms and reasonable S_n values were also obtained when the intrinsic quantal variations were taken to be negligible compared to the noise variations (Fig. 8.4C, D, continuous curves).

Table 8.3. Standard deviations of the quantal size (S_v) and noise (S_n) and the coefficient of variation of the quantum (cv), which gave the best fit between theoretical and experimental distributions

##	S_v(μV)	S_n(μV)	cv	P
1	70	50	0.18	0.97
	86	0	0.22	0.96
	0	86	–	0.20
2	0	37	0.00	0.98
	12	0	0.10	0.90
	0	37	–	0.98
3	25	50	0.08	0.39
	70	0	0.23	0.22
	0	70	–	0.24
4	30	33	0.13	0.13
	33	0	0.14	0.12
	0	70	–	0.06
5	23	75	0.08	0.92
	44	0	0.16	0.87
	0	89	–	0.72
6	32	50	0.14	0.72
	54	0	0.23	0.62
	0	69	–	0.71
7	22	11	0.21	0.69
	29	0	0.28	0.60
	0	28	–	0.65
8	17	83	0.06	0.26
	66	0	0.24	0.14
	0	116	–	0.21

Notes. For each case the top row was calculated by Eq. (8.8–3), the middle row by Eq. (8.7–1), and the bottom row by Eq. (8.7–3) at $S_v^2 << S_n^2$. 1 = EPP of a mammalian neuromuscular junction [157]; 2–4 = EPSPs of the snail CNS; 5–8 = EPSPs of the rabbit hippocampus. P is the significance level for comparison of the experimental and theoretical distributions by the Chi-square test.

8.8 Discussion

Experiments described in the present chapter allowed us to select methods suitable for statistical analysis of synaptic responses of central neurones. The methods known in the literature were considered, tested and modified. We selected two basic methods, which did not require knowledge of the amplitude of mPSPs. These methods gave consistent estimates of statistical parameters of minimal and unitary PSPs recorded from the central synapses. The presence of regular peaks and troughs under our conditions allowed us to use the histogram method with determination of v as an average distance between the histogram modes. Despite small N, we believe that the peaks did not result from occasional statistical variations (see below). Several factors contributed to the appearance of the regular clear peaks in the histograms. Especially important seems to be the procedure of the detection of the EPSP on the background of the strong conditioning IPSP, which suppressed synaptic noise (Sect. 4.7). The other favourable factors include the measurement of EPSP areas instead of amplitudes and comparatively small values of n and m for the hippocampal synapses. Both methods suffer from shortcomings, but they are partially complimentary. For example, method 8.4 is universal in the sense that the determination of v and m does not depend on the type of the distribution. It can be applied to the Poisson, simple binomial and compound binomial distributions. Method 8.5 may give errors when the histogram essentially deviates from the simple binomial distribution, i.e. when p is small (Poisson distribution) or non-uniform (compound binomial distribution). The agreements between the basic quantal parameters determined by the two methods is consistent with the assumption that the simple binomial model is adequate for the majority of the cases under our conditions. Method 8.4 might be subjective, especially with small samples used in the present study, when random amplitude variations with superimposed noise might be taken for the histogram peaks and troughs. However, similar intervals between peaks in successive samples from the same neurone (Fig. 8.1), reasonable values of the parameters and the agreement with the estimates of two variants of the other method are against this objection. Method 8.5 based on the CV is objective and does not demand large N. Approximate errors were usually <30% of the estimated v_2 and m_2 even for N = 30 or 40. One can conclude that the best way is to use both methods, which were shown to be consistent and suitable for statistical analysis of both invertebrate and vertebrate central synapses.

Our estimates for the snail central neurones (Fig. 8.5) gave comparatively broad ranges of both m (most often 1–12) and v (most often 50–400 μV). A more detailed analysis made in Chapter 11 showed that these large variations can be explained by different values for two different identified neurones and under different experimental conditions, i.e. for initial and final stages of the low-frequency stimulation (Sect. 11.4). On the whole, the m and v of the snail central synapses were found to be appreciably higher than corresponding values for the *Aplysia* CNS (Sects. 7.8.2, 7.9.2). The difference is especially large if initial stages of low-frequency stimulation are taken (Sect. 11.4). One of the reasons for these discrepancies might be a depressed state of the synaptic connections in *Aplysia* due to modified ionic composition of the extracellular medium. A lower R_{in} of *Aplysia* neurones may explain comparatively small v.

The values of v estimated for minimal hippocampal EPSPs (100–300 μV) lie within the limits obtained for the cat spinal motoneurones (Sect. 7.8.3, Table 7.1). They are in agreement with more recent data obtained for hippocampal neurones in vitro (Sect. 7.8.4). The values of m obtained for the hippocampal neurones with stimulation at the lowest currents (Table 8.1, Fig. 8.3) in non-potentiated state (Table 12.2) are also close to m given in Sections 7.8.3 and 7.8.4. Binomial parameter p was in the range of 0.4–0.6 and n was equal to about 2 for the lowest currents. However, it should be kept in mind that the EPSPs were usually recorded under conditions of paired-pulse stimulation, hence p and n might increase due to FF. It will be shown below in experiments with microstimulation (see Chap. 12) that p close to zero and $m<1$ might be more typical values for the non-facilitated synapses of the CA3 area in the unanaesthetized mammalian brain.

Some reservations should be made regarding the interpretation of statistical parameters of amplitude fluctuations when MPSPs rather than UPSPs are analyzed. The nature of minimal EPSPs recorded in our experiments has been already discussed (Sect. 4.7) and the EPSPs were considered to be mainly monosynaptic. Nevertheless, a "non-quantal" interpretation might be valid, in principle. In terms of the non-quantal interpretation, the amplitude fluctuations of minimal EPSPs are exclusively due to intermittent excitation of several afferent fibres or collaterals. In this case, the meaning of estimated statistical parameters is different from that implied in the present study, i.e. m and n will represent the mean and maximal number of excited fibres, respectively, p will be the mean probability of excitation of a single fibre, and v will be the average amplitude of a unitary EPSP for a given group of fibres (see also Sect. 3.6). On the basis of arguments given below (see also Sect. 7.7 and review [823]), we favour the "quantal" (or more exactly "mixed") interpretation of the statistical parameters calculated here. This interpretation is similar to that implicated by other authors who recorded minimal PSPs of central neurones (Sects. 7.8, 7.9). Within this interpretation, m and n reflect the sum of the corresponding quantal parameters for all "effective" fibres, i.e. fibres excited with a sufficiently high response probability. The calculated p will be close to the "true" p value (i.e. quantal release probability) only when the probability of activation of a given fibre is close to 1. The interpretation based on the quantum hypothesis is supported by the similarity of the parameters of the minimal EPSPs with quantal parameters of UPSPs obtained both in our experiments and in those of other authors. In the frame of the alternative (non-quantal) interpretation, the true v would be considerably smaller than the values obtained in our studies. This seems to be unlikely in view of the results for other central synapses (Chap. 7) including our results (Sect. 8.6). A detailed analysis of statistical parameters for different stimulus currents in one experiment (Table 8.1, Fig. 8.1) also supports the quantal interpretation: note a relative independence of p on stimulus current. This observation not only supports the quantal interpretation, but is also in agreement with a suggestion that the effective probability of fibre activations might be close to 1 under our conditions. Another support for the quantal interpretation of amplitude fluctuations of minimal hippocampal EPSPs was obtained more recently by McNaughton et al. [700]. They found estimated v to be relatively independent of changes in EPSP amplitudes either after manipulations with Ca^{2+} and Mg^{2+} concentrations in the external media or after increasing the stimulus frequency (see also Sect. 7.10.3).

Our analysis (Sect. 8.7) confirmed the necessity of allowing for noise variations when theoretical distributions are calculated. A similar conclusion was reached independently by Matteson et al. in publications [679, 680] which studied multicomponent quanta of neuromuscular preparations (Sect. 7.4.4) and which appeared nearly simultaneously with our first related publications [88, 89]. The modified Eq. (8.7–3) was proposed and was used to calculate theoretical distributions. It should be noted that the construction of theoretical distributions and their comparisons with experimental histograms had been used before our publications only in rare cases (mainly in studies using automatic analysis of experimental histograms, Sect. 7.3.7). Our analysis supports the view that the true value of cv should usually be smaller than the 0.2–0.3 taken in literature on the basis of studies of amplitudes of mPSPs. It should also be noted that in the original publication [331], the value 0.3 for the cv actually included S_n. Therefore, the overestimate of the cv might be explained by disregarding the variations of noise. The procedure used in the present chapter can be considered as an indirect method to estimate S_v under conditions when a detection of mPSPs encounters difficulties. On the basis of our data, one can conclude that the cv for the neuro-neuronal synapses lies within the range of 0.05–0.20. The mean value was equal to 0.1 ($N = 7$) with 0.07 ($N = 3$) and 0.12 ($N = 4$) for the unitary EPSPs of the snail parietal ganglia and minimal hippocampal EPSPs, respectively. Similar estimates of cv were obtained more recently by several authors for other central synapses. The conclusion was that the intrinsic variations of v were either negligible [464, 823] or about 0.05 [1068] or below 0.15 [307, 526, 530]. I believe that the S_v values of 0.4–0.5v, given by Bekkers et al. [101], are overestimates, probably due to occasional release of multiple quanta. In fact, these authors noted regular peaks in the amplitude distributions of the presumed mPSPs. Based on the present data, cv equal to 0.05 or 0.1 were used in our further studies for calculations of theoretical distributions.

8.9 Summary

The applicability of known methods of quantal analysis to central synapses is considered. Two basic independent methods of determination of the mean quantal content (m), quantal size (v) and binomial parameters (n and p) were adopted after modifications of the known methods. The first (histogram) method is based on the subdivision of experimental amplitude histograms according to visible peaks and troughs with a subsequent calculation of m, p and n. The second method was based on the calculation of parameter p as a ratio of the mean amplitude to the maximal amplitude with a subsequent calculation of m from the amplitude variance or from the number of failures. These two methods were used to calculate parameters of both minimal excitatory postsynaptic potentials (EPSPs) recorded in the rabbit hippocampus and unitary EPSPs recorded in snail parietal ganglia. In most cases, both methods gave similar values highly correlated with each other. Experimental distributions of the EPSP amplitudes were not significantly different from theoretical distributions based on the binomial model. Similar v values (usually 100–300 μV) were obtained for both neuro-neuronal junctions studied. For the hippocampal EPSPs m was about 1–2 in experiments with the weakest stimulus currents. Quantal interpretation of the amplitude fluctuations and polymodal amplitude distributions of the hippocampal minimal EPSPs was supported by reasonable values of the statistical parameters and their similarity with the quantal parameters of unitary EPSPs recorded in our experiments on the snail CNS and in experiments of other authors in other structures. A modified equation for the theoretical distribution of PSP amplitudes was proposed, which took into consideration both intrinsic variations of v and noise variations. The modified equations gave a better fit to the experimental amplitude histograms of recorded minimal and unitary EPSPs in comparison with the equation traditionally used. It is concluded that the coefficient of variation of v was often overestimated in the literature, and it might be no higher than 0.15 for central synapses.

9 Models of Transmitter Depletion and Their Application to Analysis of Synaptic Plasticity (Literature Review)

9.1 Introduction

Methods of the quantum hypothesis gave consistent results with analysis of MP-SPs and UPSPs (Chap. 8). However, the two methods used suffer from shortcomings (Sect. 8.8) and make high demands on both sample sizes and their stationarity (Sects. 7.4.2, 8.6). In many cases, these conditions are difficult to satisfy because of technical difficulties and the plasticity of central neurones (e.g. Chap. 4). Therefore, a detailed analysis of other quantitative methods seems to be justified. One approach mentioned in Section 7.6.7 is based on the analysis of transmitter depletion from a precursor pool (or from a "mobilization store"). Initial formulations of this model had been given by Rall in the Appendix to the paper of Liley and North [608]. It is interesting to note that a similar approach had been formulated by Turpaev [985] a year before to describe ACh release and destruction in the frog myocardium (see also [986]). The model of Rall-Liley-North and its modifications had been widely used for analysis of neuromuscular transmission under various names: the model of transmitter turnover [686], transmitter mobilization [596], transmitter flow [1115], transmitter depletion [1146]. The latter designation will be used here. More recently, applications of this model were initiated for neuro-neuronal synapses at peripheral nervous system [356, 357, 954], invertebrate CNS [226, 583, 762, 974, 1115–1117] and vertebrate CNS [205, 316, 829]. The aim of the present chapter is to describe methods based on the model of transmitter depletion and to review applications of this model to analysis of synaptic plasticity.

9.2 Basic Model Parameters and Recording Conditions

The model [608] is based on the assumption that transmitter available for release is in an equilibrium with its precursor or with a large depot store (Fig. 9.1). The amount of the available transmitter and its precursor will be designated C and T, respectively (Fig. 9.1). The amount of the transmitter released by a presynaptic spike is

$$R = FC, \qquad (9.2\text{–}1)$$

where F is a fraction of the available transmitter released by the presynaptic spike. The release produces a PSP with the mean amplitude

$$E = qR = qFC, \qquad (9.2\text{–}2)$$

where q is a proportionality factors which converts the amount of transmitter (expressed in mM) into PSP amplitude (expressed in mV). Non-linear summation may took place at sufficiently large amplitudes and appropriate corrections should be introduced (see Sect. 7.4.1). Factor q is a constant for a given preparation, i.e.

$$q(t) = \text{constant}. \qquad (9.2\text{–}3)$$

This postulate of the initial model seems to be approximately satisfied for most cases studied (see below).

Strictly speaking, UPSPs should be recorded to use the model for determination of synaptic parameters (see Sect. 2.2). Extracellular recording from local sites should be preferred in case of multiple and large junctions typical for many neuromuscular preparations (Sect. 7.3.2). Measurements of aggregate FPs were used in some publications including all known studies on vertebrate CNS [205, 316, 695, 829]. Parameters obtained for these cases can be regarded only as rough approximations of mean parameters of corresponding synapses.

9.3 Model of Transmitter Depletion Without Mobilization

9.3.1 Method of Paired Stimulation

In the present and the following two Sections, the simplest situation will be considered when a replenishment of the transmitter from the store pool (T in Fig. 9.1) into the pool of the transmitter available for release (C in Fig. 9.1) can be ignored. The method of conditioning-testing can be used. The simplest case will be examined here when single pulses are delivered as both conditioning and testing stim-

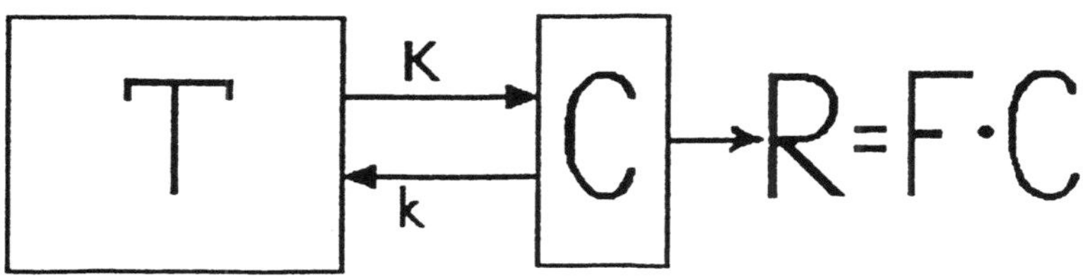

Fig. 9.1. Schematic representation of the model of transmitter depletion of Liley and North [608]. *T* Amount of transmitter precursor or transmitter in a depot pool; *C* amount of transmitter available for release; *K* and *k* reaction velocity constants governing the equilibrium of pools *T* and *C*; *R* amount of transmitter released by a single presynaptic spike; *F* fractional release from the available pool per one presynaptic spike

uli. Trains were also used as conditioning stimuli [122, 316, 583] but this does not introduce principle changes in the considerations. If a paired-pulse depression (PPD, see Sect. 9.7.1) is observed and if it is assumed that no significant mobilization from the store pool occurs during the interpulse interval, the following reasoning is valid. The amount of transmitter remaining in the pool of the available transmitter after the first (conditioning) stimulus is $C_1' = C_1 - R_1$. Here, C_1 is the initial amount of the transmitter and R_1 is the amount of the transmitter released by the first pulse. Substituting Eq. (9.2–1) for R_1,

$$C_1' = C_1 - FC_1 = C_1(1 - F). \qquad (9.3.1\text{–}1)$$

Since no replenishment of the transmitter occurs between stimuli, its amount before the second stimulus is $C_2 = C'_2$ and $F = 1 - (C_2/C_1)$. If $F_1 = F_2$ or, more generally,

$$F_i = F = \text{constant}, \qquad (9.3.1\text{–}2)$$

and condition (9.2–3) is satisfied, then considering (9.2–2), one can obtain

$$F = 1 - (E_2/E_1), \qquad (9.3.1\text{–}3)$$

$$C_1 = E_1/F. \qquad (9.3.1\text{–}4)$$

These equations permit the determination of basic parameters of the depletion model from experimental values (E_1 and E_2).

It is necessary to remember that this method is suitable only for junctions with a depression during paired-pulse stimulation. The method is based on conditions (9.2–3) and (9.3.1–2) and on the postulate that the sole mechanism of the PPD is the transmitter depletion caused by the conditioning pulse.

9.3.2 Method of Amplitude Dependence on Serial Stimulus Number During Frequency Stimulation

If postulates (9.2–3) and (9.3–2) are satisfied and there is no transmitter mobilization during continuous stimulation, the amplitude approaches zero as a geometrical progression. Indeed, in the absence of mobilization, the amount of transmitter available for release preceding the stimulus number i can be expressed as $C_i = C_i - 1 = (1 - F)C_{i-1}$. From this equation and Eq. (9.2–2), one has $E_i = E_{i-1}(1 - F)$. Properties of the geometrical progression give

$$E_i = E_1(1 - F)^{i-1}. \qquad (9.3.2\text{–}1)$$

On a semilog scale, the relationship (9.3.2–1) will be linear with a slope equal to $\ln(1 - F)$:

$$\ln E_i = \ln E_1 + (i-1)\ln(1-F). \qquad (9.3.2\text{–}2)$$

After determination of F from the slope of the line, qC_1 can be calculated from Eq. (9.3.1–4).

9.3.3 Method of Relationship Between Amplitude and Sum of Amplitudes of Previous Responses

Under conditions considered in the previous section, it is convenient to use a mathematically equivalent method [312]. Both might be termed methods of frequency depression (FD) without mobilization. The amount of transmitter which remained before a stimulus number i is given by $C_i = C_1 - F * \Sigma C_l$ with summation from $l = 1$ to $l = i{-}1$. Combining Eqs. (9.2–2), (9.2–3) and (9.3.1–2),

$$E_i = E_1 - F * \Sigma E_l \qquad (9.3.3\text{–}1)$$

also with l from 1 to i–1. The slope of the relationship of E_i and the sum of amplitudes of previous responses gives F. The value of qC_1 can be calculated from Eq. (9.3.1–4). In practice, only an initial part of the relationship (9.3.3–1) proves to be linear and when methods 9.3.2 and 9.3.3 are used, a straight line is usually drawn only through the initial two to five points. Note that when only two initial points are taken, methods 9.3.2 and 9.3.3 are identical to method 9.3.1.

Significant deviations from the linear relationship predicted by Eqs. (9.3.2–2) and (9.3.3–1) are usually observed after several stimuli. This can be explained by transmitter mobilization (Fig. 9.1). The methods given in this section should be used mainly for preliminary estimates, since methods allowing for the mobilization are known (see below). Nevertheless, in some cases, simple estimates by the methods presented in this section can prove to be sufficient. Comparison of the behaviour of response amplitudes at different stimulus frequencies (f) is a useful control procedure [312]. If $E_i(f)$ = constant, it can be taken that mobilization is negligible.

9.4 Model of Transmitter Depletion with Mobilization

9.4.1 Restoration After Stimulation. Method of Paired Stimulation [608, 1115, 1146]

Provided T (Fig. 9.1) and the volume of the presynaptic terminal are constant, the expression $dC/dt = KT - kC$ for the rate of replenishment of available transmitter after a single pulse can be written on the basis of mass action law and the kinetics of reversible reactions [508]. Substituting M ("mobilization") for KT in this equation,

$$dC/dt = M - kC. \qquad (9.4.1\text{–}1)$$

In the equilibrium state before the first pulse, $dC/dt = 0$ and $C = C_1$, therefore

$$M = kC_1. \qquad (9.4.1\text{–}2)$$

Solving Eq. (9.4.1–1) under the condition

$$M(t) = \text{constant};\ k(t) = \text{constant} \qquad (9.4.1\text{–}3)$$

and under the initial condition $t = 0$ (i.e. immediately after the first pulse), $C = C_1'$, one obtains

$$C = C_1[1 - F * \exp(-kt)]. \qquad (9.4.1\text{–}4)$$

When conditions (9.2–3) and (9.3.1–2) are satisfied, it follows from Eqs. (9.2–2) and (9.4.1–4)

$$(E_1 - E_2)/E_1 = F * \exp(-kt), \qquad (9.4.1\text{–}5)$$

where t is the interval between responses with amplitudes E_1 and E_2 evoked by conditioning and testing stimuli, respectively. Equation (9.4.1–5) converts into Eq. (9.3.1–3) when kt approaches 0.

It is convenient to represent Eq. (9.4.1–5) in the form

$$\ln[(E_1 - E_2)/E_1] = \ln F - kt, \qquad (9.4.1\text{–}6)$$

and to plot the logarithm of the "depression" (i.e. the left side of the equation) versus t. The graph often gives a nearly straight line at sufficiently large t [122, 608, 1115]. In this case, the following four parameters of the model can be determined: F, from the point of the intersection of the line with abscissa; k, from the slope of this line; qC_1 and qM, from Eqs. (9.3.1–4) and (9.4.1–2), respectively. In practice, the initial part of the relationship (9.4.1–6) is often influenced by the processes related to the frequency facilitation (FF, see Sect. 9.7.1) at short interstimulus intervals. Recently, a computer algorithm was used [848] to fit parameters of Eq. (9.4.1–5) to experimental data.

9.4.2 Frequency Stimulation. "Continuous" Model [608]

To describe processes related to transmitter depletion during stimulation at frequency f, an additional term should be introduced in Eq. (9.4.1–1). Then

$$dC/dt = M - kC - fFC. \qquad (9.4.2\text{–}1)$$

Assuming Eqs. (9.3.1–2) and (9.4.1–3) and integrating under the initial condition $t = 0, C = C_1$ one obtains

$$C = M/(k+fF) + \{C_1 - [M/(k+fF)]\} * \exp[-(k+fF)t]. \quad (9.4.2\text{–}2)$$

As t approaches infinity, C approaches a steady state

$$C_s = M/(k + fF), \quad (9.4.2\text{–}3)$$

so that the depletion of the transmitter is in equilibrium with its mobilization. From Eqs. (9.4.2–3) and (9.4.2–2), $C = C_s + (C_1 - C_s) * \exp-[(k + fF)t]$. If condition (9.2–3) is satisfied, the latter can be rewritten as

$$E(t) = E_s + (E_1 - Es) * \exp[-(k + fF)t], \quad (9.4.2\text{–}4)$$

where E(t) is the amplitude at time t. After taking the natural logarithm on both sides,

$$\ln[E(t) - E_s] = \ln(E_1 - E_s) - (k + fF)t. \quad (9.4.2\text{–}5)$$

Here, E_1 and E_s are amplitudes of the first response and response recorded in the equilibrium state, respectively. The value of k + fF can be determined from the plot of $\ln[E(t) - E_s]$ versus t provided that this relationship is linear. However, the net parameters of the model can be obtained only if one of them was determined by an independent method. For example, if F is calculated by method 9.3.1, k can be determined from k + fF. As usually, qC_1 is calculated from Eq. (9.3.1–4). To calculate qM, the following relationship can be used:

$$qM = E_s(k + fF)/F. \quad (9.4.2\text{–}6)$$

It is derived from Eqs. (9.4.2–3) and (9.2–2). When F is known, k can be also calculated with less accuracy but more easly from values of E_1, E_2 and E_s. Because $E(t) = E_2$ at $t = 1/f$, one can rearrange Eq. (9.4.2–5) into $k + fF = f * \ln[(E_1 - E_s)/(E_2 - E_s)]$. Then,

$$k = f\{\ln[(E_1 - E_s)/E_2 - E_s] - F\}. \quad (9.4.2\text{–}7)$$

In this case, $F \leq \ln[(E_1 - E_s)/(E_2 - E_s)]$ must be satisfied. According to our observations (Chap. 10), F calculated by Eq. (9.4.1–5) does not always satisfy this condition. From Eq. (9.4.2–7), this leads to $k \leq 0$, which contradicts the model (Fig. 9.1).

Equation (9.4.2–2) can be easily solved for the case without mobilization under condition (9.3.1–2). Since, in this case, M = 0 and k = 0, one can obtain $E(t) = E_1 * \exp(-fFt)$ or

$$\ln E(t) = \ln E_1 - fFt. \qquad (9.4.2\text{–}8)$$

To compare with Eq. (9.3.2–2) obtained earlier, an account should be taken that $t = (i - 1)/f$ for $E_i = E(t)$. Here, $1/f$ is the period of the stimulation. Then, $ft = i - 1$ and Eq. (9.4.2–8) assumes the form $\ln E_i = \ln E_2 - (i - 1)F$. This corresponds to Eq. (9.3.2–2) when F is small (below 0.1–0.2). Deviations for higher values of F can be explained by the fact that the discrete case (i.e. stimulation with intervals equal to 1/f) is approximated by a continuous model. Therefore, it is better to estimate F from Eq. (9.3.2–2) and not from Eq. (9.4.2–8) used sometimes [206].

9.4.3 Approximate Estimates of Parameters of the Continuous Model

As has been seen, the depletion model with transmitter mobilization does not give independent methods of calculation of all parameters under conditions of frequency stimulation. However, F and k can be determined from $k + fF$ provided some additional assumptions are made.

a) $k << fF$ [686]. This is the case of sufficiently large F and f so that one can ignore transmitter depletion which is not related to the stimulation. One can determine F from the slope of the linear relationship (9.4.2–5). A simple relation $F = \ln[(E_1 - E_s)/(E_2 - E_s)]$ can be obtained for this case from Eq. (9.4.2–7) at $t = 1/f$. From Eq. (9.4.2–6) at $k << fF$, it follows

$$qM = E_s f. \qquad (9.4.3\text{–}1)$$

This equation suggests that at equilibrium state, the rate of transmitter release is equal to the rate of mobilization. It was often used without special references to the condition $k << fF$ [313, 1104]. It should be noted (see below) that this condition is unlikely to be satisfied at low f, for example, at 1/min [313] when $fF = 0.004$ for $F = 0.27$ (Table 9.1, #3).

b) $F << k$ [653, 654]. In this case, Eq. (9.4.2–5) assumes the form:

$$\ln[E(t) - E_s] = \ln(E_1 - E_s) - kt. \qquad (9.4.3\text{–}2)$$

It gives a linear relation between $\ln(E_1 - E_s)$ and t with the slope equal to k. Equation (9.4.3–2) can be rewritten as follows:

$$kt = \ln\{(E_1 - E_s)/[E(t) - E_s]\}. \qquad (9.4.3\text{–}3)$$

Sometimes, it is convenient to use the equation transformed into $E(t)/E_s = 1 + [(E_1/E_s) - 1] * \exp(-kt)$. This equation was deduced by Maeno [652] in a more complicated way. An equation similar to Eq. (9.4.3–3) was used [652] to determine k (Table 9.1, #4). He made use of changes in E(t) with a shift from a higher (5 Hz) to a lower f (0.5 Hz), which produced a lower FF. Limitations of the meth-

Table 9.1. Parameters of the transmitter depletion model calculated for responses from neuromuscular preparations (## 1–5), neuronal synapses (## 6–8) and for aggregate responses of the spinal cord (#8)

#	Preparation	Method	F	k(1/s)	qC1(mV)	qM(mV/s)	Reference
1	Rat muscle	9.4.1	0.45	0.53	–	–	[608]
2	Rat muscle	9.4.1	0.39±0.13 0.16–0.60	0.25	4.9	1.23	[225]
3	Human muscle	9.3.3	0.27	–	37.5	–	[312]
4	Frog muscle	9.4.3	–	0.065 0.04–0.08	–	–	[652]
		9.4.4	0.09	0.43	32.5	1.41	
5	Crayfish muscle	9.4.1	0.60	0.035	15.8	0.06	[1146]
6	*Aplysia* CNS	9.4.1	0.21±0.05 0.13–0.31	0.06±0.02 0.16	54±24 19–148	3.5±1.5 1.2–9.5	[1115]
7	Snail CNS	9.4.4	0.48±0.03	0.022	52	1.2	[226]
8	Toad sympathetic ganglion	9.4.1	0.13	0.22	–	–	[954]
9	Cat spinal cord	9.4.4	0.45 0.38–0.47	0.62 0.5–1	1.1	0.68	[205]

Notes. Either means ±SD and ranges or the mean ±SEM and maximal estimates (#7) are given. Some values were calculated by the present author according to data published.

ods described in this section should be emphasized: it is clear from Table 9.1 that, in practice, values of fF and k are often of the same order.

9.4.4 Frequency Stimulation". Discrete" Model [205]

The basic Eq. (9.4.2–1) assumes continuous processes of transmitter depletion and mobilization. It gives a reliable description of real events only at sufficiently high f. Actually, the depletion of transmitter [term fFC in Eq. (9.4.2–1)] is not continuous. It occurs at discrete moments after each presynaptic spike, whereas the replenishment takes place continuously. The replenishment time course is determined by a solution of Eq. (9.4.1–1) under initial conditions which are different for different stimulus numbers. The discrete model [205] was an important advance, since it permitted independent determination of all basic parameters under conditions of frequency stimulation. I shall present a deduction of the corresponding equations which is similar to the original deduction [205] but formally more strict [1025]. In terms of the discrete model, the amount of the available transmitter is equal to C_1 at $t = 0$ (before the first stimulus). Also at $t = 0$, but immediately after the first stimulus, the amount of the available transmitter C_1' is determined by Eq. (9.3.1–1). The size of the available pool before a stimulus with the sequential number i (i>1) is given by a solution of Eq.(9.4.1–1) under the initial condition $C_{i-1} = C_{i-1}(1 - F)$, i.e.

$$C_i = C_1 - [C_1 - C_{i-1}(1 - F)] * \exp(-k/f), \qquad (9.4.4\text{–}1)$$

where C_{i-1} represents the amount of the available transmitter before the stimulus number i–1. At the equilibrium, $C_i = C_{i-1} = C_s$ and from Eq. (9.4.4–1),

$$\exp(-k/f) = (C_1 - C_s)/[C_1 - C_s(1 - F)]. \qquad (9.4.4\text{–}2)$$

Inserting Eq. (9.4.4–2) into (9.4.4–1) and rewriting,

$$F = (C_1 - C_2)(C_1 - C_s)/[C_1(C_1 - C_s) - C_s(C_1 - C_2)].$$

Conversion of values of C into corresponding values of E(t) under conditions (9.2–3) and (9.3.1–2) gives:

$$F = (E_1 - E_2)(E_1 - E_s)/\{E_1[E(t) - E_s] - E_s(E_1 - E_2)\}. \qquad (9.4.4\text{–}3)$$

From Eq. (9.4.4–2), it is easy to obtain another basic parameter

$$k = f * \ln\{[E_1 - E_s(1 - F)]/[E(t) - E_s]\}. \qquad (9.4.4\text{–}4)$$

Parameters qC_1 and qM are calculated from Eqs. (9.3.1–4) and (9.4.1–2).

As was mentioned, the advantage of the discrete model is the possibility of determining all parameters from one stimulus series. An attempt was made to get rid of the postulate [Eq. (9.4.1–3)] and to approximate the time-course of the replenishment by several exponents [226]. However, this modification of the discrete model for snail CNS led to the introduction of three different constants of mobilization even with a short series consisting of ten stimuli. In the next section, other approaches will be discussed which avoid the restriction [Eq. (9.3.1–2)] and suggest changes in F during stimulation.

9.5 Modifications of the Depletion Model for Variable F

9.5.1 Model of "Partial Depletion" [122]

A great number of quantitative and semiquantitative studies at different neuromuscular junctions [312, 443, 608, 652–654, 785, 950, 966] and neuro-neuronal synapses [583, 762, 866, 867] supported the adequacy of the models described above. On the other hand, some experimental results deviated from the predictions of these models. For example, the basic model assumes a linear relation between E_1 and the "magnitude of depression" $(E_1 - E_2)/E_1$. In fact after combination of Eqs. (9.2–2) and (9.4.1–5), $(E_1 - E_2)/E_1 = E_1 * \exp(k/f)/qC_1$. First Betz [122] and later others [224, 1146] noted deviations from the linear relation when response amplitudes were changed due to alterations either in the number of conditioning stimuli or in the external Ca^{2+} and Mg^{2+} concentrations. It should be noted here that linearity is preserved only when C_1 is constant, which is hardly satisfied with the procedures used by the authors cited above.

They found [122, 224] that response changes during PPD and LFD were best fitted if F was accepted to be not a constant but to vary in proportion to C, i.e.

$$F_1/F_2 = C_1/C_2. \qquad (9.5.1\text{–}1)$$

9.5.2 "Partial Depletion" with Paired Stimulation

Combining Eqs. (9.5.1–1), (9.2–2), (9.4.1–4) and rewriting, one has $E_2/E_1 = [(1 - F_1 * \exp(-k/f)]2$ or

$$\ln(1 - \sqrt{E_2/E_1}) = \ln F_1 - k/f \qquad (9.5.2\text{–}1)$$

Similarly to the case with Eq. (9.4.1–6), it is possible to obtain F_1 and k by plotting the left side of Eq. (9.5.2–1) versus 1/f at different f. Parameters qC_1 and qM are given by Eqs. (9.3.1–4) and (9.4.1–2) as in the other cases.

If mobilization is ignored, one has from (9.5.2–1)

$$F_1 = 1 - \sqrt{E_2/E_1} \qquad (9.5.2\text{–}2)$$

[compare with Eq. (9.3.1–3)]. F calculated on the basis of the "simple" depletion model (Sects. 9.3 and 9.4) might be twice as high as F_1 obtained under condition (9.5.1–1) [224].

Zucker and Bruner [1146] attempted to represent the relation (9.5.1–1) in a somewhat more general form

$$F_1/F_2 = aC_1/C_2, \qquad (9.5.2\text{–}3)$$

where a is an arbitrary constant which is independent on time. They deduced equations for calculations of a and F_1 from experiments with paired-pulse stimulation. In fact, under condition (9.5.2–3), one can have from Eq. (9.4.1–4)

$$E_2/E_1 = [(1 - F_1 * \exp(-k/f)]^2/a. \qquad (9.5.2\text{–}4)$$

By extrapolation of this equation to t = 0, it is easy to obtain [1146] $a = [E_1(1 - F_1)^2]/F_2$. However, it is clear from Eq. (9.5.2–4) that at large t, the ratio E_1/E_2 is approaching a. In this case, the factor a is identical to 1, since sooner or later, the recovery of the depressed response will occur and E_2 will become equal to E_1. This shows that the attempt to derive a more general relation between F and C [1146] fails since at a = 1, Eq. (9.5.2–3) is identical to (9.5.1–1).

9.6 Transmitter Mobilization and Parameters of the Store Pool

9.6.1 Relationship Between Mobilization and Stimulus Frequency

Under conditions of frequency stimulation, qM is calculated from Eq. (9.4.2–6). It was noted that this equation had often been used without appropriate reservations. In fact, the dependence of M upon f was often found to be stronger than that predicted by Eq. (9.4.2–6) [583, 867, 1146], e.g. M stopped falling when t was increased above a certain limit [1146].

Equation (9.4.2–6) can be easily modified into

$$E_s = qMF/(k + fF) = qM/[(k/F) + f]. \qquad (9.6.1\text{–}1)$$

This relation suggests a monotonous diminution in E(t) when f increases and other parameters are constant. Therefore, it has been suggested that frequency stimulation favours an increase in transmitter mobilization. Some difficulties for this hypothesis were created by the data [583, 1146] which indicated that the period of the poststimulus increase in mobilization had to be very short, for example less than 20 ms [1146].

Attempts have been made to determine a relationship between M and f [64, 654]. A general relationship M = z(f)M' can be considered, where M' is a mobilization without stimulation and z(f) is a non-linear function which describes an increase in M with increased f. The function is equal to 1 at f = 0. It follows from Eq. (9.4.2–3) that when k>>fF, C_s = z(F)M'/k. From Eq. (9.4.1–2), C = M/k, which gives: $C_s = z(f)C_1$ or $E_s = z(f)E_1$ under conditions (9.2–3) and (9.3.1–2). Experimentally, the relation $E(t) = E_1 * \exp(hf)$ was obtained with variation f from 0.3 to 10 Hz [653]. It gave z(f) = exp(hf) with h from 0.21 to 0.55 s in 10 neuromuscular preparations.

9.6.2 Dependence of Mobilization on Time

The assumption (9.4.1–3) is not necessarily satisfied in real experimental situations [226]. Changes in M at different f (see Sect. 9.6.1) presumably do not arise instantly and should develop gradually during stimulation. The data obtained from analysis of UPSPs at CNS of *Aplysia* [1115] suggested an increase in M after a single stimulus ("M-wave"). However, the precise time course of the M-wave cannot be described on the basis of the available data because of simultaneous changes in F and k. Based on the observations that the dependence of the amplitude on time satisfied condition (9.4.1–6) within the initial 10 or 20 s after a single stimulus, a conclusion was reached about relatively short-term changes in model parameters [1115]. It was suggested that the parameters took their initial (prestimulus) level within 10 or 20 s. This suggestion allowed calculating parameters on the basis of the later part of the amplitude dependence on time interval between conditioning and testing stimuli. The same authors gave a semiquantitative description of the time course of M on the basis of experiments with paired-pulse testing during and after frequency stimulation [1115]. According to these data, M increased during stimulation and reached an approximately stable level after 10 to 15 stimuli at $f = 1$ Hz. After cessation of the stimulation, M returned to the initial value approximately in 30–120 s. It should be emphasized that such estimates were based on some additional assumptions, like a slow change in F after frequency stimulation (see Sect. 9.7.5). Strictly speaking, the models considered in this chapter are unable to describe the precise time course of M.

9.6.3 Total Transmitter Store

Attempts have been made to estimate a total depot store (T in Fig. 9.1) using prolong frequency stimulation. For example, the stimulation was continued for 7 min which led to a gradual decrease in E [313]. Some difficulties of this approach, for example a development of presynaptic block during prolonged stimulation, were indicated [313]. On the basis of method 9.3.3, T = 23,000 quanta was estimated. This value is 60 times as high as C estimated for the same preparation (Table 9.1, #3). The rate of mobilization was estimated to be 0.014/s expressed as a fraction of T. The authors indicated the agreement of this estimate with measurements of ACh metabolism in the sympathetic ganglion [130]. Capek et al. [206] used a prolonged (20 min) frequency activation of the frog neuromuscular junction and found that the T/C ratio was equal to 50. Their estimate of T was about 50,000 quanta.

9.7 Depletion Model in Studies of Synaptic Plasticity

9.7.1 Applications of the Depletion Model and Simple Synaptic Plasticities

Examples of initial estimates of parameters of various junctions are given in Table 9.1. Similar to the quantum hypothesis (Sect. 7.10.1), the depletion model can be used to study mechanisms of action of different physical and chemical agents. It is possible to decide which of the parameters are changed and to draw a conclusion about influences on available store of transmitter (qC), mechanisms of its release (F) and mobilization (qM). Such an approach was used to investigate effects of ionic changes [312, 471, 608, 794], various drugs [312, 974], temperature [226, 471] and other factors [238].

Even more publications studied mechanisms of synaptic plasticity such as FF, PPF, FD, LFD, PTP and PTD defined above (Introduction, Sects. 2.5, 7.10). Similarly to the PPF, paired-pulse depression (PPD) can be defined as a decrease in response amplitude for a testing stimulus delivered after a previous (conditioning) stimulus. It is convenient to define the following two terms: "postactivation depression" (PAD) which includes PTD and PPD, and "postactivation facilitation"

(PAF) which includes PTP and PPF. It has been already mentioned that the manifestation of a given plasticity phenomenon depends on several conditions, including the type of a synaptic junction, its state and stimulus pattern. For example, many neuromuscular junctions display mainly PAD and LFD under normal conditions or after a partial suppression of the synaptic transmission by *d*-tubocurarine, while they show mainly PAF and FF under Mg^{2+} suppression [443, 474, 653, 785]. Depression and facilitation can follow each other both during frequency stimulation and after a tetanus or single pulse. A rather typical sequence after a single stimulus is PAF, which has a duration of several scores or hundreds of ms followed by a PAD with a more gradual recovery during several s [122, 829, 954, 1115]. For vertebrate neuromuscular junctions, the recovery usually has an exponential course. Its time constant (which is the inverse value of k given in Table 9.1) varied from 2 [606, 966] to about 5 s [122, 225, 632]. A slower recovery was observed in neuro-neuronal synapses (Table 9.1, #6 and #7) and especially (Table 9.1, #5) in crayfish neuromuscular junctions [181, 182, 502, 1146]. Time courses of PAF and PTP are usually even more complicated and variable than that of PAD. For example, PAF following a brief conditioning train may vary from 20 ms [611] to several s [654] and may be described by double [611, 664] or even triple exponents [654]. PTP time course depends on the duration of the conditioning train and it also depends on the preparation (see references in Sects. 2.5 and 7.10.4). PTP, lasting from several s [352] to several min [654], is characteristic of most synapses studied in the context of the depletion model. However, at central neuro-neuronal synapses of *Aplysia*, PTP lasted up to several scores of minutes [867] or even longer [155, 199] see also Section 2.5.3).

9.7.2 Frequency Depression and Postactivation Depression

PAD of several seconds duration was initially described in experiments with paired stimulation at mammalian [303, 608] and amphibian neuromuscular junctions [632]. Sensitivity of the postsynaptic membrane to ACh was not changed during PAD [785]. This finding justifies the suggestion that a decrease in R underlies the depression. It was the phenomena of PAD and FD that permitted the formulation of the depletion model on the basis of the assumption that a decrease in R is due to an exhaustion of C at a constant F [608]. This suggestion was consistent with the exponential course of amplitude recovery [Eq. (9.4.1–5)] and approximately linear increase in the magnitude of depression (Sect. 9.5.1) with an increase in E_1 in some cases [608, 950, 966]. Therefore, in the context of the depletion model, the time-course of the PAD reflects a gradual recovery of C diminished by the conditioning stimulation. Similar processes were suggested for the case of frequency stimulation with a FD. In addition, it was suggested that the frequency stimulation influenced M. According to this interpretation, the decrease in response amplitude during frequency stimulation is explained by the fact that transmitter depletion failed to be compensated by its replenishment. A steady state is reached when both processes are in equilibrium. A further amplitude diminution after very prolonged tetanizations (scores of seconds, minutes) may be due to a decrease in T (Sect. 9.6.3). When M is low, for example after suppression of transmitter synthesis [312, 376], the response amplitude may fall to 0.

Naturally, the agreement of experimental data with predictions of the depletion model does not prove the uniqueness of the model. Moreover, disagreements of experimental data with the original depletion model were found (Sect. 9.5.1). It seems likely that the model of "partial depletion" (see Sect. 9.5) gives a somewhat better description of real situations. Therefore, decreases in both C and F can underlie PAD and FD. In addition, available experimental data indicate an involvement of changes in k and M (Sect. 9.6). It makes the interpretation of PAD and FD more complicated in comparison with the original interpretation upon which the depletion model was based [608]. It should be noted that Eq. (9.4.1–1) with time-dependent parameters k and M has no simple solution.

9.7.3 Postactivation Facilitation

In the previous section, difficulties in explaining the PAD in the context of the depletion model were considered. The simple variant based on Eq. (9.4.1–1) naturally is even less suitable for the description of complicated amplitude time courses which include PAF (Sect. 9.7.1). One possible approach to explain the PAF is the implication of a so-called "two-compartment" model [130, 352, 829, 376]. This model postulates that a small part of T becomes available for release after conditioning stimulus ("conditionally available transmitter"). A solution of the corresponding equation [829] predicted a simple exponential decay of the amplitude from a potentiated value. However, the time-course of the PAF obtained in experiments with recordings of FPs from the olfactory cortex was better described by two exponents [829]. Superposition of two processes was postulated: a decrease in the amount of "conditionally available transmitter" and a change in F.

Attempts have been made to derive more general equations which could describe postactivation changes both in C and F [705, 706, 954, 1115]. If condition (9.3.1–2) is not satisfied, the transformation of Eq. (9.4.1–1) into the corresponding equation for amplitudes becomes more complicated. For example, Tashiro et al. [954] examined such a transformation for paired stimulation assuming a certain dependence of F on t. In fact, assuming explanation of the PAF on the basis of the "residual calcium" hypothesis [232, 364, 491, 792, 814],

$$F(t) = F_1[1 + \exp(-bt)]^d, \tag{9.7.3–1}$$

where F_1 is a fraction of transmitter released by the first impulse, b and d are constants related, respectively, to time of removal of the Ca^{2+} ions and to the number of Ca^{2+} ions necessary for activation of one transmitter molecule. A solution of Eq. (9.4.1–1), similar to that considered in Section 9.4.1 but under condition (9.7.3–1), gives

$$E(t) = E_1[1 - F_1 * \exp(-kt)][1 + \exp(-bt)]^d. \qquad (9.7.3\text{–}2)$$

This equation described satisfactorily the amplitude time-course at amphibian sympathetic ganglia [954] after a single presynaptic stimulus provided that $d = 1$, $b = 6.3/s$ (F_1 and k are given in Table 9.1, #8). F_1 increased and b decreased with an increase in Ca^{2+} concentration which supported "residual calcium" hypothesis [954]. However, it should be noted that the introduction of too many parameters argues against the uniqueness of this description of PAF. In addition, an equation similar to (9.7.3–2) may be proposed with the assumption that the residual Ca^{2+} ions induce an increase in C rather than F. In fact, Eq. (9.7.3–2) can be rewritten as $E(t) = F_1C_1[1 - F_1 * \exp(-kt)][1 + \exp(-bt)]^d$, where F_1 = constant and the other terms describe a time-course of C. As a support of this alternative interpretation of PAF, a hypothesis of Bennett and Fisher [105] might be mentioned. They suggested that residual Ca^{2+} ions produced changes in C and on the basis of considerations independent on the models reviewed in this chapter, they derived equations which satisfactorily described the phenomena of FD and FF at neuromuscular junction (see also [792]). The situation will be even more uncertain when the time-course of PAF is described by more than two exponents. Finally, the data which indicate postactivation changes in M (see Sect. 9.6.2) should be recalled.

Melkonyan et al. [705, 706] introduced a time dependent F and described time-courses of FD in spinal motoneurones at different f [247] and a facilitation related to learning. However, a detailed experimental testing of this interesting model, as well as convincing physiological justification of the introduced parameters, are still lacking.

9.7.4 Frequency Facilitation

When $E(t) < E_1$, Eq. (9.4.2–4) describes FD. When mobilization is negligible, E_s from Eq. (9.6.1–1) approaches 0. On the contrary, when $E_s > E_1$, FF takes place, which was attributed to an increased M [652]. The FF was not sufficiently well investigated in the framework of the depletion model. In many cases, complicated changes in response amplitudes were observed in the course of stimulation (Sect. 9.7.1), for example, an initial increase (FF) with a subsequent diminution (FD) below E_1 level. This may signify a time-dependence of M, although Eq. (9.4.2–4) assumes its constancy. It is clear that F can also be modified during the FF (see Sect. 9.7.3). Woodson et al. [1115] estimated a relative change in F (A_F) during FF at *Aplysia* central synapses. They postulated that PTP (see Sect. 9.7.5) is explained mainly by changes in F, i.e.

$$A_F = F_s/F_1, \qquad (9.7.4\text{–}1)$$

where F_s and F_1 are potentiated and initial values of F, respectively. If a decrease in F during the period from the cessation of the tetanization to the PTP maximum is ignored, $A_F = E_{PTP}/E_1$. Here, E_{PTP} is the response amplitude at the maximum of the PTP. This expression was derived from Eqs. (9.7.4–1) and (9.3.1–4) under condition (9.2–3) assuming additionally $C = C_1$ at the PTP maximum. It should be emphasized once again that since the available data indicate changes in F during stimulation, solutions of Eq. (9.4.2–1) are not correct and additional information is necessary to obtain related to F dependence on t.

9.7.5 Post-Tetanic Potentiation

The depletion model was used for analysis of PTP in several qualitative or semi-quantitative studies [608, 842, 867, 1104, 1115]. Pretetanic parameters of the model were compared with those calculated for the post-tetanic period. Amplitudes were measured during presentation of either paired stimuli or stimulus trains. The methods described in Sections 9.3–9.4 was used. On the whole, available data provided a support to the suggestion that an increase in F is the basic mechanism of PTP. According to data of Wilson and Skirboll [1104], C determined by method 9.3.3 and M determined from Eq. (9.4.3–1) decreased after tetanization, and then approached their initial level exponentially. Therefore, PTP at neuromuscular junctions may be explained exclusively by an increase in F [842, 1104], while a delay in PTP development is related to a recovery of C and M from diminished post-tetanic values. Wilson and Skirboll [1104] support the explanation of PTP based on the residual calcium hypothesis, showing a maximal increase in F immediately after tetanic stimulation and a decrease in F (together with a decrease in PTP itself) at low Ca^{2+} concentrations. A similar interpretation was offered to explain a more prolonged (scores of minutes) PTP of UPSP at the *Aplysia* CNS [867, 1115]. However, a FF was observed following a short period of a FD during stimulation at 0.5–10 Hz frequency. The FF was explained by increases in both F and M. It was suggested that these parameters remained large after the cessation of the conditioning stimulation, and that the gradual PTP development is due to an increase in C. As for neuromuscular junctions, a later PTP decline was explained by exponential restoration of F.

9.8 Conclusion

This chapter shows that the original depletion model [608] has undergone a great number of refinements and modifications. Naturally, the appearance of these modifications was dictated not by the idea of complicating the elegant original formulations but rather by numerous disagreements between the model predictions and experimental observations. It should be stressed that detailed quantitative comparisons of the model predictions with precise experimental measurements were made very infrequently. When it was made, significant discrepancies were often discovered [1146]. Nevertheless, despite all imperfections of the depletion model, it should be noted that it played a positive role in the development of the physiology of synaptic transmission. It stimulated precise experiments, gave methods of compressed presentation of experimental results and methods of quantitative verification of theoretical predictions. Similarly to the quantum hypothesis, the depletion model stimulated a search for biochemical (e.g. [130]) and morphological (e.g. [216]) correlates of physiological parameters. For example, bimodal distributions of the distances from vesicles to the active zones in spinal synapses were shown [847] which supported the two-compartment model (see Sect. 9.7.3).

An unresolved problem is the relationship between the depletion model and the quantum hypothesis. Many authors actually postulated that parameters F and C of the depletion model were identical to binomial parameters p and n calculated on the basis of the quantum hypothesis [122, 205, 226, 445, 471, 474, 475, 670, 684, 695, 848, 842]. Such an identification is understandable in view of the fact that the basic Eqs. (9.2–1) and (9.2–2) are formally identical to Eqs. (7.2–2) and (7.2–3), respectively. On the other hand, it should be emphasized that the quantal parameters are estimated from statistical considerations (Chap. 7) quite different from those used to estimate parameters of the depletion model F and C. Parameter q (factor of proportionality between amplitude and R) cannot be determined independently in the context of the depletion model. When v is known, the factor q can be accepted to be equal to v [122, 312, 313, 445]. In that case, C can be measured not in mV but in the number of quanta. In contrast, v, which corresponds to parameter q, is one of the basic parameters of the quantal hypothesis. For the depletion model, the equation $q = v$ signifies only a conversion of C to another dimension. When $q = v$, a combination of Eqs. (9.2–2) and (7.2–3) gives $m = FC$, i.e. an equation similar to Eq. (7.2–2). At first sight, this obliterates distinctions between the two models and indicates that binomial parameters p and n are identical to parameters of the depletion model. In fact, according to the original formulation of the quantum hypothesis [260, 261], n and p represent the number of transmitter quanta available for release and the mean probability of the release of one quantum, respectively (Sect. 7.2). Therefore, physical (physiological) correlates of these parameters are analogous to those of the depletion model parameters (C and F). However, the present day interpretation of parameter n as a number of active release sites (Sect. 7.11) is distinct from the original interpretation. Possible spatial non-uniformity of release probabilities introduces additional complications into the relationship between parameters of the depletion and quantal models.

The first attempt at experimental comparison of parameters p and F was made by Christensen and Martin [225]. In all nine experiments, the values of p (mean 0.12 ± 0.05) were found to be lower than F (mean 0.39 ± 0.14) and the correlation between F and p, calculated by the present author from the published data, was not significant ($r = 0.30$, $P > 0.1$). However, the determination of p was based on comparison of the CV at different Ca^{2+} concentrations. This method presumes that n is independent of Ca^{2+} concentration and, generally speaking, it is not correct (see Sect. 7.3.6). Since n increases with the increase in Ca^{2+} concentration, the values of p should be underestimated by this method. Attempts to compare p calculated by more precise methods and F calculated by method 9.3.3 were made at a peripheral neuro-neuronal synapse [357]. Significant differences between F (0.44, 0.38 and 0.34) and p (0.61–0.71, 0.16–0.35 and 0.60–0.92 for the same three cases, respectively) were obtained without any regular prevalence of one or another parameter. However, the small number of measurements (N = 3) and the somewhat primitive method of F determination (Sect. 9.3.3) used in this study should be noted. In general, the literature review shows that in many publications, the equations of the depletion model were used without appropriate analysis and reservations. The terminology of the model was often applied to semiquantitative considerations or purely qualitative speculations [184, 218a]. Meanwhile, when certain limitations are satisfied, the model of transmitter depletion gives accurate relations which permit calculations of both steady state parameters of synaptic transmission and their changes under various challenges. On the other hand, the model has a number of imperfections related in particular to an ambiguity in the parameters estimated. Therefore, the model requires further theoretical and experimental refinements. The relationship between the parameters of the depletion model and quantal parameters needs clarification.

9.9 Summary

This chapter presents a detailed description of the model of transmitter depletion based on the consideration of transmitter release and its replenishment from the pool of a precursor (or a large mobilization store). Derivations of basic equations of the original model of Rall-Liley-North and some of its known modifications are given. Methods which were used in the literature to estimate the following parameters are reviewed: the amount of available transmitter (C), fractional release per presynaptic pulse from the available pool (F) and also mobilization (M) and demobilization (k) constants. Publications are discussed which exemplify analysis of simple phenomena of synaptic plasticity in the context of the transmitter depletion model. It is noted that the relationship between parameters of the depletion model and parameters calculated in the context of the quantum hypothesis remains to be understood. It is concluded that the depletion model needs further refinements. Nevertheless, when certain limitations are satisfied, the model permits both determination of quantitative parameters of synaptic junctions and their changes under various challenges.

10 Parameters of Transmitter Depletion Model for Snail Central Synapses and Comparison of Depletion and Quantal Models

10.1 Introduction

Chapter 9 shows that the model of transmitter depletion was applied to study synaptic mechanisms. However, many authors used simple variants ignoring mobilization (Sect. 9.3). The more realistic "continuous" model (Sect. 9.4) also often gave deviations from experimental observations even when the "partial depletion" was accepted (Sect. 9.5). Moreover, this variant does not determine independently all model parameters. The "discrete" model (Sect. 9.4.4) seems to be more attractive but it was used only for analysis of aggregate spinal root responses [205]. The depletion model has been applied to EPSPs of neuro-neuronal synapses only occasionally [226, 954, 1115]. Parameters calculated by different methods were not compared for the same preparation. Many authors regarded parameters of the depletion model as being identical to parameters of the quantum hypothesis. However direct comparisons are scarce and they suggest rather a difference between the related parameters (Sect. 9.8). Therefore, the objectives of our following work were: (1) to modify known methods of the transmitter depletion model; (2) to compare the estimates from different methods with one another and with experimental data; (3) to compare parameters of the depletion model with statistical parameters calculated on the basis of the binomial model of the quantal release. Physiological experiments [1049] were performed on the above described identified monosynaptic connections between LPa7 neurone and giant LPa2 and LPa3 neurones of *Helix pomatia* (Sects. 2.2.1, 8.6). The related EPSPs are particularly convenient for analysis in the context of the depletion model because they show clear PPD without facilitation even at short interpulse intervals of less then 0.3 s. An interesting additional question was whether the parameters of the synapses formed by the same presynaptic neurone on different postsynaptic neurones were similar. The theoretical part (Sects. 10.2, 10.3) was elaborated together with Dr. V.I.Derevyagin [1025].

10.2 Discrete Model of "Partial" Transmitter Depletion

It is possible to generalize the assumption of the "partial" depletion (Sect. 9.5.1) for the case of frequency stimulation and to modify the "discrete" model (Sect. 9.4.4) in the following way. Equation (9.5.1–1) can be transformed into $F_{j-1}/F_j =$

C_{j-1}/C_j. By multiplying these equations from j = 2 to j = i, one has: $F_1/F_i = C_1/C_i$ or, taking into account Eq. (9.2–2),

$$F_1/F_i = \sqrt{E_1/E_i}. \qquad (10.2\text{–}1)$$

For the equilibrium state after stimulus number s is

$$F_1/F_s = \sqrt{E_1/E_s}. \qquad (10.2\text{–}2)$$

From Eqs. (9.2–2), (9.4.4–3) and (10.2–2) one obtains

$$F_1=(\sqrt{E_1} - \sqrt{E_2})(\sqrt{E_1} - \sqrt{E_s})/[(E_1 - \sqrt{E_1E_s}) - E_s(1 - \sqrt{E_2/E_1}). \qquad (10.2\text{–}3)$$

The value of k is calculated by Eq. (9.4.4–4), in which F is replaced by F_s and E_1 by $\sqrt{E_1E_s}$; qC_1 and qM are calculated from Eqs. (9.3.1–4) and (9.4.1–2).

10.3 Discrete Model with Arbitrary Changes in F

We made an attempt to find a method which imposes no limitations on the type of changes in F during stimulation. An equation similar to (9.4.4–1) can be transformed into a system of equations

$$qC_1 - qC_i = (qC_1 - qC_{i-1} + E_{j-1})\exp(-k/f);\ 2 \le i \le s+1. \qquad (10.3\text{–}1)$$

The system (10.3–1) consists of s equations with s+1 unknowns. By substitution of equation number i for equation number i+1, it is easy to transform Eq. (10.3–1) into a system $qC_i = qC_1 - \Sigma E_j[\exp(-k/f)]^{i-j}$ with summation from j = 1 to j = i–1. After subtracting this equation for j = s from the same one for i = s+1 and assuming that $qC_s = qC_{s+1}$, one obtains an equation of s–1 power relative to exp (-k/f): $E_1[\exp(-k/f)]^{s-1} + \Sigma(E_j - E_{j-1})[\exp(-k/f)]^{s-j} = 0$ with summation from j = 2 to j = s. When k/f>0, 0<exp(–k/f)<1. Then the value of exp(–k/f) can be found by the method of successive approximations in the interval [0, 1] with any assigned degree of accuracy. Afterwards, the value of k can be easily calculated.

Considering Eq. (9.2–2), Eq. (10.3–1) can be rewritten in the form $E_i/F_i = E_1/F_1 - \Sigma E_j[\exp(-k/f)]^{i-j}$ where the summation is made from j = 1 to j = i–1. Hence

$$F_i = E_1/\{E_1/F_1 - \Sigma E_j[\exp(-k/f)]^{i-j}\} \qquad (10.3\text{–}2)$$

with the same summation interval. From this equation, all values of F_i can be calculated if F_1 is known. The condition $0 \le F_i \le 1$, which follows from the physical meaning of F, limits the range of possible solutions for F_1. In practice, we determined the following three values of F_1 together with the corresponding values of

F_i: (1) F_1', the maximal value of F_1 at which the condition $0 \leq F_i \leq 1$ is satisfied for all F_i calculated from Eq. (10.3–3); (2) F_1'', for which an additional condition $F_i = F_1$, was imposed; (3) F_1''', for which condition (10.2–1) was imposed. The choice of the latter condition is based on the suggestion that F_i has to decrease during FD. In principle, the method allows adopting any other law of changes in F in the course of stimulation. Parameters qC_1 and qM were calculated from Eqs. (9.3.1–4) and (9.4.1–2) with $F=F_1$.

To analyze different variants of the depletion model, parameters were calculated by several methods. Five basic methods were considered. These included three known methods described in Chapter 9 (9.3.2, 9.3.3, 9.4.4) and two new methods (Sects. 10.2, 10.3). Parameters F estimated by method 9.3.2 and by mathematically identical method 9.3.3 will be indexed by the Roman numeral I. Parameters F estimated by methods 9.4.4 and 10.2 will be indexed by II and III, respectively. The designations of parameters F estimated by method 10.3 (F_i, F_1', F_1'' and F_1''') were given above.

10.4 EPSP Changes During Stimulation

EPSPs of 20 target neurones were recorded. The triggering neurone was stimulated by current pulses delivered through an intracellular microelectrode included into a bridge circuit. Series of 10–60 stimuli at 0.1–1.5 Hz were used with inter-

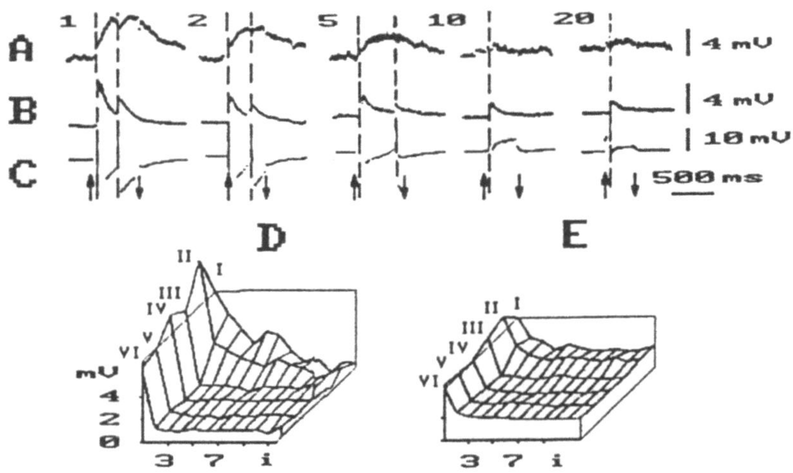

Fig. 10.1A–E. Unitary EPSPs of identified neurones of the snail parietal ganglia. **A–C** Simultaneous recordings from three neurones: LPa2 (**A**), RPa3 (**B**) and LPa7 (**C**). *Arrows* mark intracellular stimulation of the triggering neurone LPa7 as in Fig. 8.4A. *Broken lines* coincide with the triggering spikes in **C**. Sequential stimulus numbers are given. **D, E** Changes in amplitudes of 12 initial EPSPs of neurones LPa2 (**D**) and RPa3 (**E**) in six series (I–VI) from the same experiment. *Abscissa* sequential stimulus number (i); *ordinate* EPSP amplitude (mV). [1049]

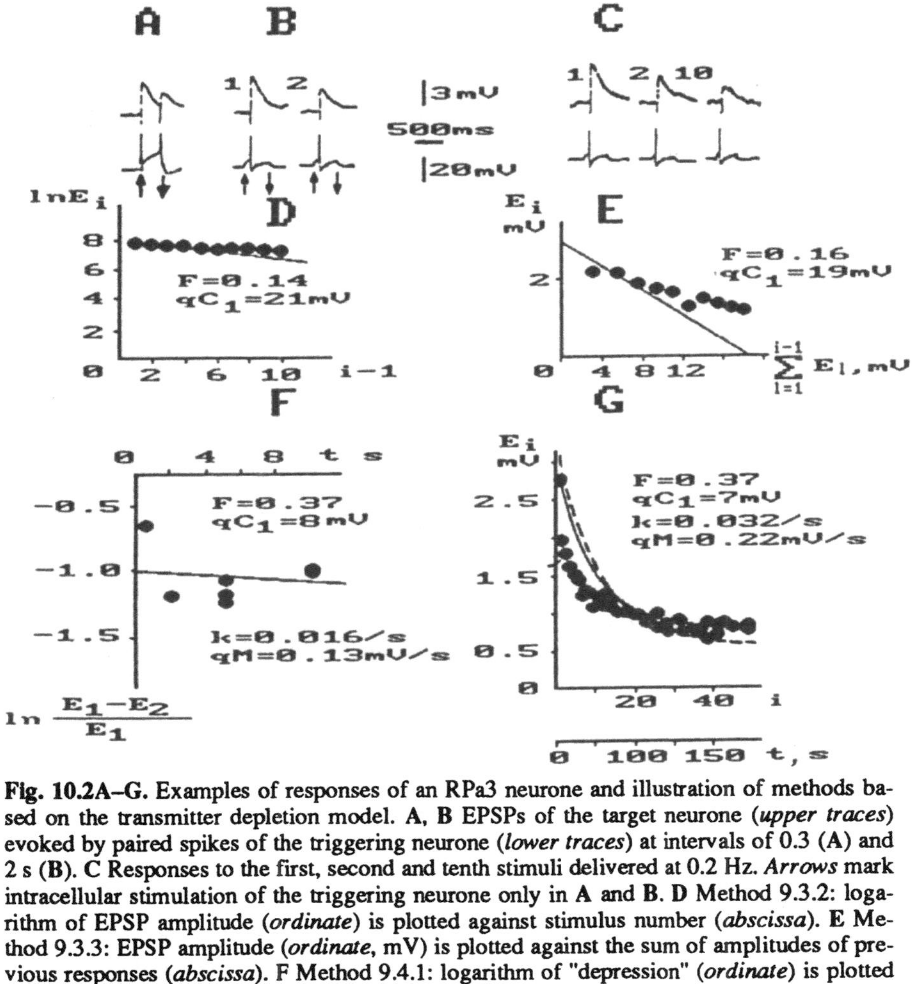

Fig. 10.2A–G. Examples of responses of an RPa3 neurone and illustration of methods based on the transmitter depletion model. **A, B** EPSPs of the target neurone (*upper traces*) evoked by paired spikes of the triggering neurone (*lower traces*) at intervals of 0.3 (**A**) and 2 s (**B**). **C** Responses to the first, second and tenth stimuli delivered at 0.2 Hz. *Arrows* mark intracellular stimulation of the triggering neurone only in **A** and **B**. **D** Method 9.3.2: logarithm of EPSP amplitude (*ordinate*) is plotted against stimulus number (*abscissa*). **E** Method 9.3.3: EPSP amplitude (*ordinate*, mV) is plotted against the sum of amplitudes of previous responses (*abscissa*). **F** Method 9.4.1: logarithm of "depression" (*ordinate*) is plotted against time (*abscissa*, t, s). **G** Comparison of experimental data (*dots*) with theoretical curves plotted from Eq. (9.4.2–4) with parameters calculated by method 9.4.1 (*interrupted line*) and 9.4.2 (*continuous line*). *Ordinate* amplitude (E_i, mV); *abscissa* sequential stimulus number (i) and time (t, s). Averages from three series of stimulation at 0.2 Hz frequency are plotted in **D**, **E** and **G**. [1025]

vals of 10 to 30 min between series. As shown in Figs. 10.1A, B and 10.2C, the first stimuli from each series evoked high-amplitude EPSPs (E_1 in Table 10.2). Usually, a single stimulus evoked a single spike in the triggering neurone (Figs. 10.1C10 and 10.1C20, interrupted lines; 10.2B, C, lower records). When double triggering spikes appeared (Figs. 10.1C1, 2, 5 and 10.2A, lower records), the EPSP evoked by the second spike was depressed. The PPD was more marked in LPa2 than in RPa3 neurones (compare Fig. 10.1B and A).

LFD is evident in Figs. 10.1A–E and 10.2C during continuous stimulation. In rare cases (3 out of 20 neurones), initial "sensitization" was observed (E_2>E_1) before the LFD appeared. There was complete or partial restoration of the response amplitude after an interval of 10–15 min without stimulation compared with its value in the initial series. However, the time-course of the LFD was not fully restored, especially in LPa2 neurones (Fig. 10.1D, E). Figure 10.1D shows that EPSP amplitude fell faster in series III–VI than during series I and II.

10.5 Calculation of Parameters by Basic Methods of the Depletion Model

Detailed estimates of the parameters of depletion models and their changes in the course of the LFD were made on the basis of 56 stimulation series for 8 neurones. In 9 cases, the results of 2–10 series were averaged (Figs. 10.2, 10.3). Figure 10.2D, E shows that the theoretically equivalent methods 9.3.2 and 9.3.3 give very similar results. Thereupon, only the data obtained by method 9.3.3 will be given below (Table 10.1). Figure 10.2F illustrates a calculation of the model parameters by method 9.4.1. The depression (E_1 – E_2)/E_1 had a time-course which did not correspond to a linear function: the respective correlation was not significant. It can be explained by a violation of some postulates of the depletion model. In fact,

Table 10.1. Mean parameters of synaptic transmission calculated by methods 9.3.3, 9.4.4 and 10.2 for four RPa3 and four LPa2 neurones

Neurone	Parameter	Method of calculation		
		9.3.3	9.4.4	10.2
RPa3	F	0.23±0.03	0.35±0.05	0.19±0.03
		0.17–0.31	0.23–0.44	0.12–0.24
	k(1/s)	–	0.012±0.003	0.010±0.002
			0.007–0.018	0.006–0.015
	qC_1(mV)	22±9	16±6	29±12
		12–49	5–34	10–65
	qM(mV/s)	–	0.14±0.03	0.22±0.05
			0.10–0.24	0.14–0.36
LPa2	F	0.29±0.01	0.35±0.04	0.19±0.03
		0.28–0.31	0.29–0.47	0.16–0.27
	k(1/s)	–	0.004±0.002	0.003±0.001
			0.0002–0.008	0.0001–0.006
	qC_i(mV)	10±5	9±5	17±10
		1–24	1–25	2–46
	qM(mV/s)	–	0.016±0.006	0.020±0.048
			0.005–0.025	0.005–0.034

Notes. The data of this and the following tables were calculated from the first stimulus series for each neurone. Mean ±SEM and ranges are given.

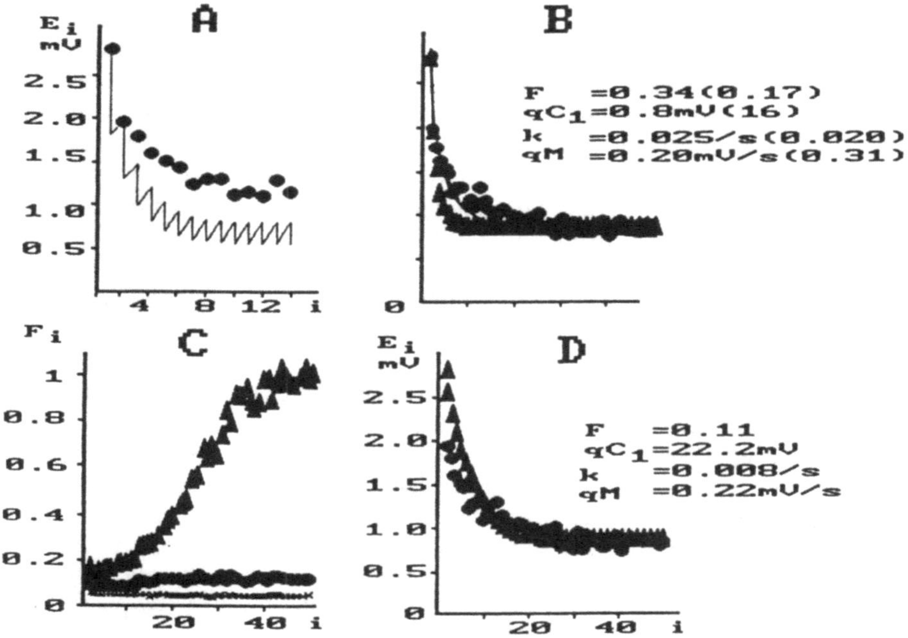

Fig. 10.3A–D. Illustration of methods based on the discrete model of transmitter depletion. Data of the experiment shown in Fig. 10.2 were used. **A** Method 9.4.4: comparison of experimental data from initial part of a stimulus series (*dots*) with theoretical curve. **B** Comparison of experimental data (*dots*) with predictions of methods 9.4.4 (*triangles*) and 10.2 (*continuous line*). *Inset* gives parameters calculated by methods 9.4.4 and 10.2 (*numbers without and with parentheses*, respectively). **C** Changes in F_i calculated by method 10.3 at different F_1: F_1'(*crosses*), F_1'' (*dots*) and F_1''' (*triangles*). **D** Comparison of experimental data (*dots*) with predictions of method 9.4.4 (*triangles*) at $F = F_1''$ estimated by method 10.3. *Abscissae* in **A–C** stimulus number (i). *Ordinates* in **A**, **B**, **D** EPSP amplitudes (E_i, mV). *Ordinate* in **C** values of F_i. [1025]

our conditions were unfavourable for application of method 9.4.1, since continuous stimulation was used, and the restoration was incomplete during intervals between series (see Sect. 10.4). Accordingly, it can be seen from Fig. 10.2G that the theoretical curve (interrupted line) plotted on the basis of the parameters calculated by method 9.4.1 differs from the experimental data (dots) both at the initial and final parts of the stimulus series. Under these circumstances, we did not use method 9.4.1 for systematic calculations. Nevertheless, it should be noted that our calculations (Fig. 10.2F) indicate the usefulness of a verification of the linear approximation corresponding to expression (9.4.1–6). No similar verifications were made by authors who used method 9.4.1 (see references in Sect. 9.4.1 and Table 9.1).

Table 10.1 gives parameters of the depletion model calculated by methods 9.3.3, 9.4.4 and 10.2 from data obtained in the first stimulation series with a frequency of 0.1 Hz for four RPa3 and four LPa2 neurones. At this frequency, methods 9.3.3 and 9.4.4 gave rather similar values of both F and qC_i, especially for LPa2 neurones. At higher f (0.2, 0.5 and 1.5 Hz), larger differences between the results of the two methods were found (Fig. 10.4A). The correlation between values of F calculated by methods 9.3.2 and 9.4.4 was large and highly significant (Fig. 10.4A), and between values of F calculated by methods 9.4.4 and 10.2, it was close to 1 ($r = 0.99$). Method 10.2 gave lower values of F than both methods 9.3.3 and 9.4.4. The mean values for other parameters were not significantly different and some of them, like k (in both RPa3 and LPa2 neurones) and qM (in LPa2 neurone) were almost identical (Table 10.1). Theoretical curves plotted on the basis of method 10.2 (Fig. 10.3B, continuous line) usually corresponded better to the experimental data than those plotted on the basis of methods 9.3.3 and 9.4.4 (Fig. 10.3A, B, triangles).

Table 10.1 shows also that values of F for neurones RPa3 and LPa2 were very similar; they were even more similar than the mean initial amplitudes (E_1 in Table 10.2). However, values of k and qC_i for RPa3 neurones were about two or three times as high as the corresponding parameters of synapses formed by the same presynaptic neurone on LPa2 neurones. The difference for the mean qM was about tenfold. During repeated stimulus series, both F and k increased. The increase in k was especially large in the LPa2 neurone, and this reflected the faster depression of EPSPs. In contrast, qC_i decreased in both identified synapses while qM did not change significantly. When f was increased, both qM and k also increased. In four of six tested cases, F increased with the increase in f and in two cases, it decreased.

10.6 Calculation of Parameters by Method 10.3

Three F parameters were determined (Table 10.2). Similar to methods 9.3.3, 9.4.4 and 10.2, both k and qC_i differed by two to three times for the two identified synaptic connections, qM differed by ten times. Parameters F_1 for the two synapses were practically identical (F_1') or similar (F_1'' and F_1'''). Parameters k and qM increased in repeated series and also with an increase in f. According to the data given in Tables 10.1 and 10.2, the values of F_1' were close to F calculated by method 10.2. However, for all 20 cases analyzed, mean F_1'' (0.19 ± 0.07, ±SEM) and, especially, F_1''' (0.13 ± 0.02) were significantly lower on average than F calculated by methods 9.4.4 (0.41 ± 0.03) and 10.2 (0.23 ± 0.02). Nevertheless, a significant correlation ($r > 0.8$) was observed between the values of F determined by method 10.3 and those obtained by other methods. Parameters qC_i and, especially k and qM estimated by method 10.3 (Table 10.1), were very similar to the respective estimates of the other methods (Table 10.2).

The amplitudes theoretically predicted by method 10.3 coincided with the experimental data by the definition of the method. When values of F_1 obtained by method 10.3 were substituted into the equations of the discrete model 9.4.4, a

Table 10.2. Parameters of synaptic transmission calculated by method 10.3 for four RPa3 and four LPa2 neurones

Parameters	RPa3	LPa2
E_1(mV)	4.6±1.2	2.8±1.5
	2.3–7.9	0.4–7.1
F_1'	0.19±0.04	0.20±0.02
	0.08–0.26	0.17–0.25
F_1''	0.14±0.3	0.18±0.01
	0.06–0.21	0.15–0.21
F_1'''	0.09±0.02	0.13±0.02
	0.04–0.15	0.10–0.18
k(1/s)	005±0.001	0.002±0.001
	0.002–0.006	0.0001–0.004
qC_i(mV)	51±30	16±8
	18–144	2–37
qM(mV/s)	0.16±0.04	0.015±0.005
	0.10–0.26	0.005–0.025

better agreement with the experimental data was observed than for the parameters obtained by the method 9.4.4 itself (compare Fig. 10.3B and D).

10.7 Comparison of Parameters of Depletion and Quantal Models

The binomial parameters were calculated by the two methods described in Chapter 8. The values of p calculated by methods 8.4 and 8.5 will be designated p_1 and p_2, respectively, and the mean value $p = 0.5(p_1 + p_2)$ and the corresponding value of n will be used. A decrease in p and n was found during the low-frequency stimulation (see Chap. 11). Therefore, we used two pairs of each p and n calculated for two different "quasi-stationary" regions (see Sects. 7.4.2 and 11.4) from several series: p_I (and n_I) calculated for initial regions and p_s (and n_s) calculated for late ("steady state") regions. Figure 10.4C shows that on average, p_I was larger than F calculated by method 9.4.4. The differences between p_I and F determined by other methods were even higher. The mean values of p_s also exceeded respective values of F estimated by different methods (Fig. 10.4D). In no case was a significant correlation found between values of p_I (or p_s) and values of F calculated by methods 9.3.3, 9.4.4, 10.2 and 10.3 (Fig. 10.4C, D). Whereas F for the two synapses tested did not differ significantly, p_I and p_s for the LPa2 neurone were always less than for the RPa3 neurone (compare empty and filled symbols in Fig. 106C, D). In contrast, p_1 and p_2 were highly correlated and did not differ significantly from each other, on average (Fig. 10.4B, see also Section 8.6). In addition, values of p and n were compared in different combinations with F and qC_i, but a significant correlation was in no case found.

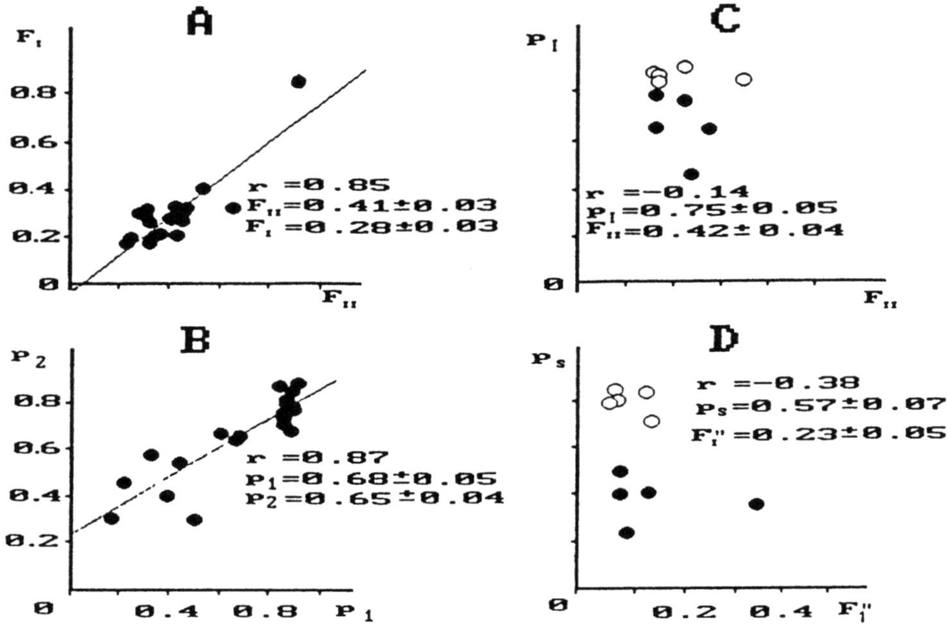

Fig. 10.4A–D. Comparison of the fractional release F with binomial parameter p. **A** Correlation of F estimated by methods 9.3.3 (*ordinate*) and 9.4.4 (*abscissa*). N = 22. **B** Correlation of p estimated by methods 8.5 (*ordinate*) and 8.4 (*abscissa*). N = 20. **C** Lack of correlation between p_I calculated for initial stimulation regions (*ordinate*) and respective values of F_{II} calculated by method 9.4.4 (*abscissa*). Linear regression lines are shown in **A** and **B**. **D** Lack of correlation between p_s calculated for final regions of stimulation (*ordinate*) with F_1" estimated by method 10.3 (*abscissa*). In **C** and **D**, data for RPa3 and LPa2 neurones are plotted by *empty* and *filled circles*, respectively. The values of r were highly significant for **A** and **B** (P<0.0001) and not significant for **C** and **D** (P>0.6 and P>0.25, respectively). [1025]

10.8 Discussion

Parameters of the depletion model, determined by different methods, were found to be of the same order of magnitude as those of neuromuscular and neuro-neuronal synapses studied by other authors (Table 9.1). An exception was qM, which was low (especially for LPa2 neurone), and was similar to the respective value for crayfish neuromuscular junctions (Table 9.1, #5) which is not typical for other junctions. The low mobilization naturally explains the similarity of the values F (and also qC_i) obtained by methods 9.3.3 and 9.4.4.

The two identified synaptic connections made by the same presynaptic neurones were found to be similar in respect to the release mechanisms according to all variants of the depletion model (similar F values). In contrast, the available pool of transmitter (qC_i) and especially its mobilization (qM) from the basic store (or the rate of its synthesis) were different for these synapses. Differences in some properties of branches of the same axon going to different postsynaptic cells have been already noted [344, 345, 703, 746] and explained by the possible retrograde (feedback) influences of the postsynaptic neurones on the presynaptic endings. The synapses investigated in the present study which differ from each other both in parameters of synaptic transmission and in the changes of these parameters during LFD provide a convenient model for the analysis of mechanisms of the suggested feedback influences. Another interesting problem is the role of the abovementioned differences between synapses in the performance of the respective physiological functions by these neurones considered to be "command" ones [63, 656].

Significant disagreements were found here between theoretical predictions based on the known models (9.3.3 and 9.4.4) and the experimental data. The disagreements suggest that the postulates of the transmitter depletion model, especially postulates (9.2–3) and (9.3.1–2), might not be exactly satisfied. A slight improvement resulted from our modification (method 10.2) of the discrete model [205] based on the hypothesis of a "partial" depletion of the transmitter (see Sect. 9.5.1 for references). Method 10.3 gives exact agreement between the theoretical predictions and experimental data by its definition. More important, it enables k to be determined unambiguously and model parameters to be calculated for any assumed type of changes in F during stimulation.

The analysis of the LFD, especially by method 10.3, shows that this phenomenon can be explained on the basis of either of two mechanisms or their combination: the depletion of C at a constant F (or even increasing F) or the decrease in the latter at a constant C. For all these cases, agreement between theoretical and experimental data can be obtained under the assumption that the rate of the transmitter demobilization is constant. We stress again that an agreement between theoretical and experimental points does not prove the uniqueness of a theoretical model. For example, another possibility would be to keep to the postulates (9.2–3) and (9.3.1–2) and explain LFD on the basis of changes in M [226]. We note also that modifications of LFD time-courses during repeated series of stimulation were found to be explained by both a reduction in the amount of the available transmitter and an increase in the rate of exchange between the available and basic pools.

Our data shows a significant difference between parameters of the quantum hypothesis and the transmitter depletion model, and suggest that these two groups of parameters reflect different aspects of the synaptic function. The identification of the binomial parameters with the depletion model parameters made by many authors (see Sect. 9.8) seems to be erroneous. The quantum hypothesis makes use of fewer postulates and gives consistent estimates over a wide range of experimental situations. An important advantage of the quantum hypothesis is that it enables presynaptic as well as postsynaptic and "intermediate" mechanisms to be analyzed (Sect. 6.1). In the framework of the quantum hypothesis, there is no need to postulate the constancy of the coefficient (q) which relates the amount of transmitter to the PSP amplitude [Eq. (9.2–3)]. In the context of the depletion model, the necessity of this postulate can be avoided only when some methods of quantal analysis are used in addition to the methods of the depletion model itself (see Sect. 9.8). Otherwise, the depletion model cannot be used without condition (9.2–3). Note that this condition is stipulated only in rare cases, and there is no strong evidence for its adequacy. Moreover, cases are known (see Sect. 6.1) in which response amplitude (E) might be changed at a constant amount of the transmitter released (R). For example, changes in v were observed in many cases (see Sect. 7.10.3 for references and Sect. 11.4 for our data) and they might reflect, in principle, an involvement of presynaptic mechanisms.

Nevertheless, it should be pointed out that there are cases when the use of the quantum hypothesis is virtually impracticable, for example, while recording high-amplitude EPSPs with small amplitude fluctuations and especially with LFD, which prevents collecting sufficiently large samples under stationary conditions. In these cases, the methods based on the depletion model and its modifications may prove to be useful. It is also worth noting again that the depletion model and the quantum hypothesis might be not mutually exclusive and might describe different mechanisms of synaptic transmission. Additional investigations are necessary to describe these differences more precisely.

The model of transmitter depletion is widely used in studies of mechanisms of synaptic plasticity. However, both the survey of the literary data (Sect. 9.7) and our present data indicate definite difficulties in the interpretation and the ambiguity of the conclusions on underlying mechanisms.

10.9 Summary

Unitary excitatory postsynaptic potentials (EPSPs) were evoked in two identified neurones (RPa3 and LPa2) of the snail parietal ganglia by stimulation of an identified triggering LPa7 neurone. Repeated stimulation at a frequency of 0.1 to 1.5 Hz produced a low-frequency depression (LFD) of the EPSPs. Parameters which characterize transmitter release (F), storage (C), mobilization (M) and breakdown (k) were determined, using known and modified methods of the model of transmitter depletion. The fraction (F) of the available pool which was released in two different synapses of the same triggering neurone did not differ significantly. The available pool of transmitter (C) and the demobilization constant (k) for LPa7–RPa3 synapse were two to three times higher, while the mobilization (M) was ten times higher than the respective values for LPa7–LPa2 synapse. These differences underlie more pronounced LFD in LPa2 neurones. Predictions of the depletion model showed deviations from the experimental data. A method of calculation of k was devised which allowed any changes in parameter F during the stimulation to be adopted. No correlation was found between the parameters of the depletion model (F and C) and the parameters (p and n) calculated on the basis of the binomial model of the quantal release. This suggests that the depletion and quantal models describe different synaptic mechanisms. Comparison of these two quantitative approaches shows that the quantum hypothesis can be more effective and give less ambiguous conclusions about mechanisms of synaptic plasticities than the depletion model.

11 Quantal Analysis of Short-Term Plasticities at Central Synapses

11.1 Introduction

Analysis of the transmitter depletion model allowed us to obtain data related to mechanisms of the LFD at central synapses. On the other hand, our analysis revealed some imperfections in the depletion model. The depletion model (Sect. 10.8), in contrast to the quantum hypothesis (see Sect. 6.1), is unsuitable for resolving the problem of localization of plasticity mechanisms (presynaptic versus postsynaptic). It was concluded that the depletion model could not give additional methods of determination of the quantal parameters. The hippocampal responses appeared to be inconvenient for the application of the depletion model in view of complicated postactivation changes which include PPF and FF in addition to LFD. Therefore, it seemed reasonable to concentrate our further work on the analysis of plasticity mechanisms using the methods of quantal analysis (Chap. 8). Before starting an analysis of LTP mechanisms, we decided to consider more simple phenomena of PPF and LFD [88, 621, 1018, 1045–1047]. We hoped that this analysis would, firstly, be of interest by itself (Sect. 7.10), and secondly, test the sensitivity of these methods for distinguishing pre- versus postsynaptic localization of plasticity mechanisms.

11.2 Paired-Pulse Facilitation of Hippocampal Responses

The method of detection of minimal EPSPs of area CA3 (Sect. 4.4) was used. However, in distinction from the experiments of Chapter 4, not one but two weak testing stimuli with interstimulus interval of 15 to 30 ms were applied after the strong conditioning stimulus. Therefore, the testing stimuli produced two EPSPs (EPSPI and EPSPII) against the background of the IPSP created by the strong conditioning stimulus (Fig. 11.1A, oscillogram). In one case, a microstimulation with paired pulses of 2.5 μA was used. Quantal parameters (m_2 and $v_{2)}$ were calculated by the variance method (Sect. 8.5).

Smoothed experimental amplitude histograms are exemplified by Fig. 11.1A, B. To control the estimated parameters, theoretical distributions (Fig. 11.1 A, B, dotted lines) were constructed according to Eq. (8.7–3). In most cases (14 out of 20), experimental and theoretical histograms did not differ significantly (P>0.1, Chi-square test). Calculated parameters are given in Table 11.1. For all cases ex-

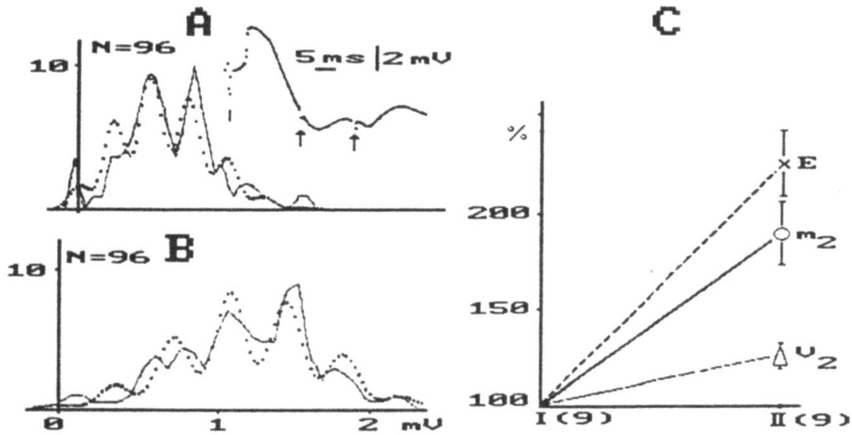

Fig. 11.1A–C. Quantal analysis of paired-pulse facilitation in vivo. **A, B** Experimental (*continuous lines*) and theoretical (*dotted lines*) amplitude distributions for the first (**A**) and second (**B**) testing EPSPs recorded in the CA3 area after the septo-fimbrial stimulation. *Ordinate* number of occurrences; *abscissa* amplitude (mV). *N* Number of measurements. Experimental distributions were plotted according to the "moving bin" method [813]. The oscillogram in **A** represents the averaged response; paired testing pulses (*arrows*) were delivered after strong conditioning stimulus (*bar*). C Changes in the mean EPSP amplitude (*E, cross*), mean quantal content (m_2, *circle*) and quantal size (v_2, *triangle*). Values calculated for EPSPII were normalized to the respective values for EPSPI taken as 100% and plotted as the percentage of the latter. *Vertical bars* represent ±SEM. *Values in parentheses under abscissa* refer to the number of measurements. [1018]

cept one (Table 11.1, #1, 68 μA), a significant PPF was observed, i.e. the amplitude of EPSPII increased for >30% (and up to 350%) in comparison with the amplitude of EPSPI (Fig. 11.1C, cross). It is clear from Table 11.1 and Fig. 11.1C that the PPF was due to an increase in both m and v. With the measurements of EPSPs superimposed on conditioning IPSP, v could be changed, in principle, due to differences in R_{in}. To check this possibility, parameters of facilitated EPSPII were compared with control EPSPI evoked by a single testing stimulus delivered at the same time relative to the conditioning IPSP as EPSPII. For both neurones tested (Table 11.1, asterisks), m and v were increased when the EPSPII was preceded by EPSPI and was facilitated. For one neurone (Table 11.1, #4), calculations for non-facilitated EPSPI in two positions relative to the conditioning IPSP was made and v appeared to be similar (311±60 and 298±65 μV). EPSPs of one neurone (Table 11.1, #6) were evoked by paired-pulse microstimulation through a microelectrode inserted between the CA3 and DG (see Fig. 12.4 and Table 12.2). The changes in m and v were similar to those in the experiments with macrostimulation.

Changes in m during the PPF were more prominent than changes in v. Parameter m increased in all nine cases when the amplitude of EPSPII increased for more than 30% (Table 11.1) and, therefore, they were statistically significant by

Table 11.1. Quantal analysis of paired-pulse facilitation of EPSPs from the in vivo hippocampus

Neurone	Current(μA)	N	v(μV)	m	p	n
1	68	49	253/405	1.8/1.4	0.49/0.41	3.7/3.3
1	75	96	256/325	2.6/3.8	0.42/0.57	6.8/6.8
2	100	72	193/297	0.7/1.5	0.21/0.43	4.2/3.3
3	–	49	295/408	2.9/3.8	0.41/0.60	7.0/6.5
4	180	76	311/371	5.5/6.2	0.60/0.73	9.1/8.6
4*	180	60	298/371	3.9/6.2	0.57/0.73	6.8/8.6
5*	130	58	323/349	1.3/1.7	0.24/0.31	5.5/5.4
6	2.5	66	180/271	0.6/1.4	0.13/0.42	5.1/3.3
6**	2.5	53	235/171	3.1/11.2	0.60/0.75	5.2/14.8
6**	2.5	41	230/275	1.8/3.8	0.28/0.56	6.4/6.3

Notes. For each case, parameters of control and facilitated EPSPs are given to the left and right of the slashes, respectively. One asterisk is explained in text. Two asterisks mark responses recorded after tetanic stimulation (Table 12.2, neurone #2).

the criterion of signs (P<0.03). Figure 11.1C illustrates mean changes in the calculated parameters. One can see that the increase in EPSP amplitude was mainly due to changes in m. In addition to the increase in m, a smaller average increase in v was also evident (Fig. 11.1C, triangle). The changes in v were observed in practically all cases. It is illustrated by the fact that dots and circles in Fig. 11.2 are situated above the equality line. Comparison of the binomial parameters for the first and second EPSPs might suggest that m was increased due to an increase in p (Table 11.1). Changes in n were not statistically significant. However, the estimates of binomial parameters could not be considered reliable under our conditions (Chap. 8).

11.3 Low-Frequency Depression of Hippocampal Responses

Quantal parameters were compared for minimal EPSPs recorded at initial and later periods of low-frequency stimulation, i.e. before and after development of the LFD described in Chapter 4. Parameters calculated by the histogram method (Sect. 8.4) will be indexed by 1; parameters calculated by the variance method (Sect. 8.5) will be indexed by 2. Figure 11.3 illustrates amplitude distributions constructed for the responses already shown in Fig. 4.4. Histograms for the initial 9 to 20 responses (control, "non-depressed" EPSPs) from each of three series of the low-frequency stimulation and for the later "depressed" EPSPs are shown in Fig. 11.3A and B, respectively. Table 11.2 (#3, 48 μA) showed decreases in both m_1 and m_2 with a smaller diminution of v_1 and v_2 during the LFD.

Comparison of all estimates revealed a decrease in m_1 in six and in m_2 in five out of seven pairs (Table 11.2). The decrease in the average m of for the pooled data of the two methods (Fig. 11.3, circle) was statistically significant (P = 0.01,

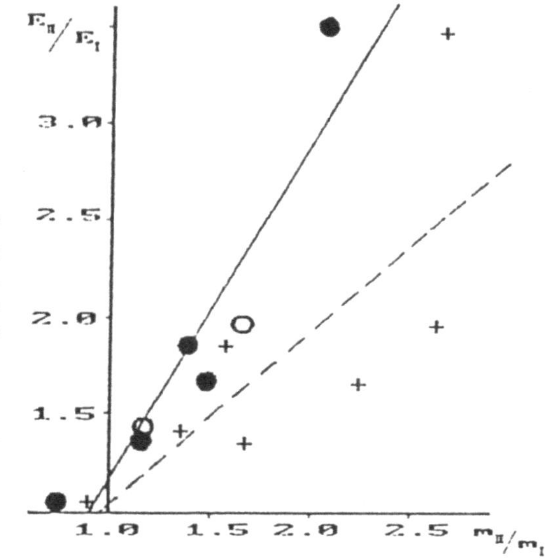

Fig. 11.2. Relative values of the PPF (EII/EI) plotted against changes in the mean quantal content (mII/mI). *Dots* and *circles* were obtained from the estimates of m given in Table 11.1: *dots* represent neurones 1–4, *circles* represent neurones 4* and 5*. *Crosses* represent m values calculated according to the respective Poisson equation. Linear regressions for circles and dots (*continuous line*) and for crosses (*interrupted line*) are shown. The latter is close to the equality line. [88]

N = 14, Wilcoxon test for matched pairs, here and below). Similarly to the PPF, changes in m might be explained by changes in p (decreases in five out of the seven cases for each method, Table 11.2). Parameter n was not changed, on average.

Comparison of Fig. 11.3A and B shows a decrease in the distance between peaks. Decreases in both v_1 and v_2 were observed in most cases (Table 11.2), so that the average v decreased (Fig. 11.3C, triangle). The decrease was at the border of the statistical significance (P = 0.05, N = 14). On the whole, the results ob-

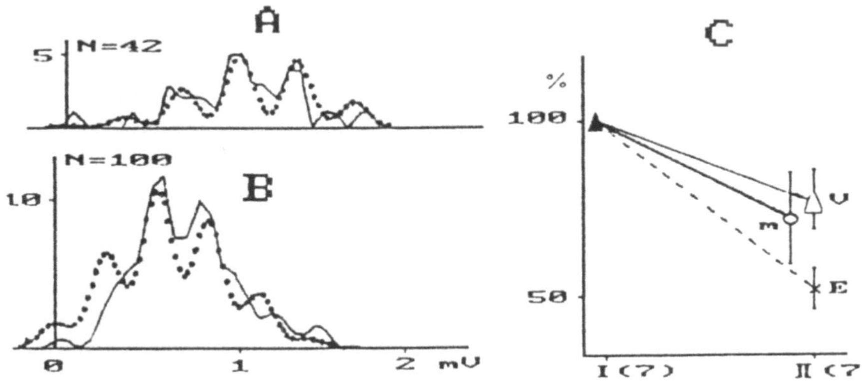

Fig. 11.3A–C. Quantal analysis of low-frequency depression of hippocampal EPSPs. **A, B** Amplitude distributions built for non-depressed (**A**) and depressed (**B**) EPSPs recorded during initial and late parts of low-frequency stimulus series, respectively, in the experiment illustrated in Fig. 4.4. Theoretical distributions (*dotted lines*) are based on parameters calculated by method 1 (Sect. 8.4). **C** Mean changes in the EPSP amplitude (*E*) and quantal parameters (*m*, *v*) as a result of the LFD. Parameters *m* and *v* are the mean values for estimates of the two methods (Sects. 8.4 and 8.5). See Fig. 11.1 for further explanation. [1018]

Table 11.2. Statistical analysis of low-frequency depression of EPSPs recorded from the in vivo hippocampus

Neurone	Current(μA)	N	v(μV)	m	p	n
1	200	29	233/327	2.4/1.4	0.13*/0.42	18.9*/3.42
		27	192/107	1.5/2.4	0.68/0.52	2.2/4.7
2	150	30	300/150	2.1/3.9	0.84/0.60	2.5/6.5
		21	300/152	1.5/2.4	0.68/0.52	2.2/4.7
3	36	23	264/279	2.1/2.1	0.68/0.52	2.2/4.7
		23	251/181	1.4/2.2	0.52/0.53	2.5/4.2
3	48	42	317/201	3.2/5.1	0.67/0.60	4.8/8.6
		100	280/167	2.6/4.6	0.60/0.49	4.4/9.2
4	36	15	344/256	3.5/4.3	0.78/0.66	4.4/6.5
		30	215/220	3.8/3.4	0.52/0.52	7.4/6.2
		62	215/301	2.0/1.8	0.02*/0.39	128*/4.7

Notes. For each neurone parameters of non-depressed (upper two rows) and depressed (lower rows) are given. Values to the left and to the right of the slashes were calculated by method 8.4 and 8.5, respectively. Values of p and n marked by asterisks are not significantly different from zero (0.13±0.24 and 0.02±0.20 for p; 18.9±12.4 and 128±197 for n).

tained by two independent methods showed significant changes in m and a strong tendency to a decrease in v during LFD.

11.4 Low-Frequency Depression at Snail Central Synapses

Quantal analysis suggested a purely presynaptic localization of LFD mechanisms at the CNS of *Aplysia* [213, 475]. However, these data were obtained under conditions of suppressed synaptic transmission (see Sect. 7.10.3). Moreover, the analysis was based exclusively on the Poisson equations. Our aim was to analyze unitary EPSPs of the snail central neurones recorded in normal ionic medium (see Chap. 10) using statistical methods based on the binomial assumption.

The methods of the UPSP recording were similar to that described in Chapter 10. Each series consisted of 5 to 120 stimuli (usually 20–30 stimuli) delivered at 0.1, 0.2 or 1 Hz (the 0.1 Hz frequency was common). Intervals between series were from 10 to 15 min. Each recorded series was divided into "quasi-stationary" plateau regions (Sect. 8.6). To extract such regions, consecutive average amplitude (N=5 or 10) was plotted against stimulus number. The "quasi-stationary" plateau region was defined as a continuous part of this plot without significant change in the averaged amplitudes (P>0.01, t-test). Plateau regions with similar mean amplitudes from several different series were combined into one region. For most neurones, the initial "combined regions" were excluded from a further analysis because of a small number of measurements (N<16). For other "combined regions", quantal parameters were calculated and theoretical distributions were built (see Sect. 8.7). In addition to the two methods from the previous Section, the fail-

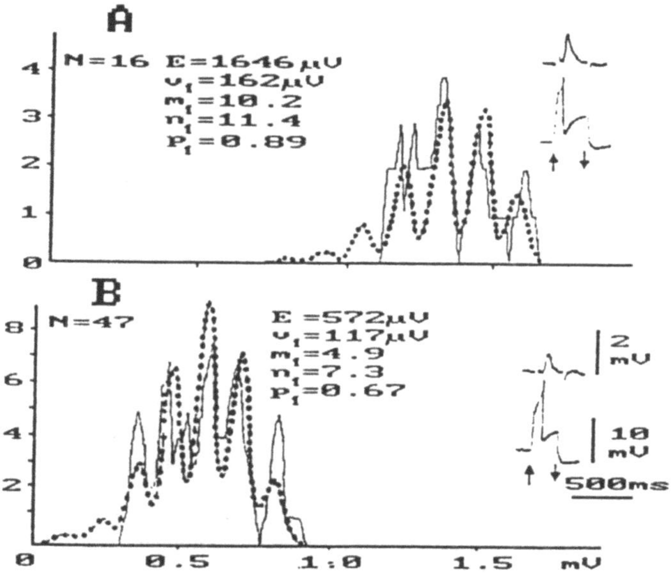

Fig. 11.4A,B. Changes in EPSP amplitude distributions during low-frequency activation of an identified synaptic connection in the snail CNS. **A** Experimental (*continuous line*) and theoretical (*dotted line*) amplitude distributions for initial plateau regions from several series of 0.1 Hz stimulation. **B** The same for the last regions from the same series. Oscillograms show EPSPs of the target RPa3 neurone (*top beam*) and spikes of the triggering LPa7 neurone (*bottom beam*). Only lower parts of the spikes are shown. *Arrows* mark switching the intracellular depolarizing pulse on and off. Parameters calculated by method 1 (Sect. 8.4) are given in *insets*. [621]

ure method [Eq. (7.3.4–2)] was used for neurone LPa2 to calculate quantal parameters.

Insets in Fig. 11.4 illustrate examples of unitary EPSPs of a RPa3 neurone (top beam) induced by the spikes of a triggering LPa7 neurone (bottom beam). The experimental histogram for the first combined region (Fig. 11.4A) shows clear peaks. The sample size was very small because of the fast LFD in initial parts of the testing series (see Figs. 8.4, 10.1). The dotted curves in Fig. 11.4A, B show theoretical distributions based on the binomial model. Corresponding quantal parameters are given in the inset. Comparison of the first (Fig. 11.4A) and the last (Fig. 11.4B) combined plateau regions shows a more than twofold diminution in m_1. Parameter v_1 also tended to decrease. A more detailed analysis showed differentiated changes in the quantal parameters at different stages of the LFD. Figure 11.5A summarizes data of seven experiments on six RPa3 neurones and shows a decrease in averaged v_1 and v_2 during the initial stages of the LFD (regions II and III). Parameters m_1 and m_2 were not significantly changed at the same stages of LFD (Fig. 11.5A, regions I–III). In contrast, the subsequent LFD (Fig. 11.5A, region IV) was accompanied by a diminution in m_1 and m_2 without any further de-

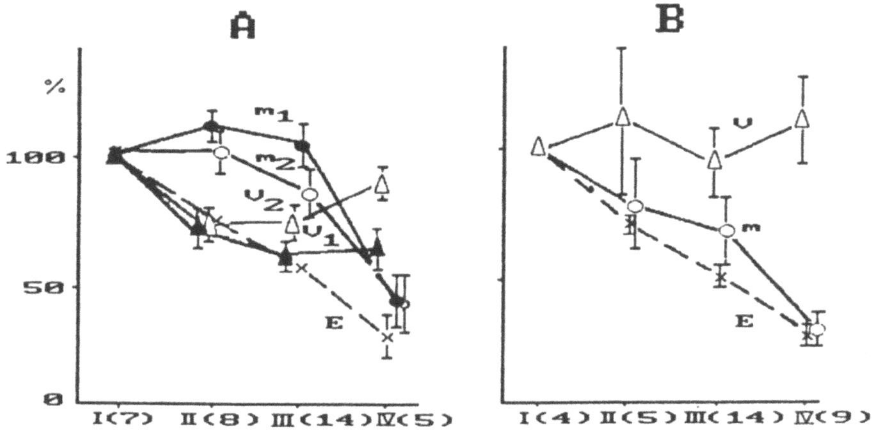

Fig. 11.5A,B. Mean changes in relative EPSP amplitudes (*E*) and quantal parameters (*m* and *v*) during LFD in RPa3 (**A**) and LPa2 (**B**) neurones. Estimates for later regions (Roman numbers *II* to *IV under abscissae*) were normalized to the respective values of the first "quasi-stationary" region (*I*). The number of "combined quasi-stationary" regions are given *in parentheses*. See Fig. 11.1C and text for additional explanations. [621]

crease in v_1 or v_2. Changes in the quantal parameters estimated for three LPa3 neurones were similar to those for the neurones RPa3.

Figure 11.5B shows that, distinct from neurones RPa3 (Fig. 11.5A) and LPa3, v was not significantly changed during LFD in five experiments on LPa2 neurones, while changes in m paralleled the amplitude depression. For all three identified neurones, binomial parameters tended to decrease. However, the only significant decrease was that of p when regions I and IV were compared ($P<0.05$ for RPa3 neurones and for all neurones studied). In some LPa3 and RPa3 neurones, as well as in most LPa2 neurones, p fell below 0.3 or became uncertain for regions IV, so that the experimental histograms could be fitted by Poisson distributions.

11.5 Paired-Pulse Facilitation in Hippocampal Slices

Only one method of quantal analysis was used in our initial studies of the hippocampal PPF (Sect. 11.2). Both the number of neurones and sample sizes were small. Although we tried to control the influence of the conditioning IPSP on the changes in v, the results might be questioned because of the complicated procedure used to detect the minimal EPSPs. The changes in v were of statistical significance only at the border. Moreover, no changes in v were found more recently by Yamamoto [1126] in three CA3 neurones with recordings of presumably unitary EPSPs in vitro. Therefore, we repeated analysis of hippocampal PPF using additional methods and more stable in vitro slice preparations [425].

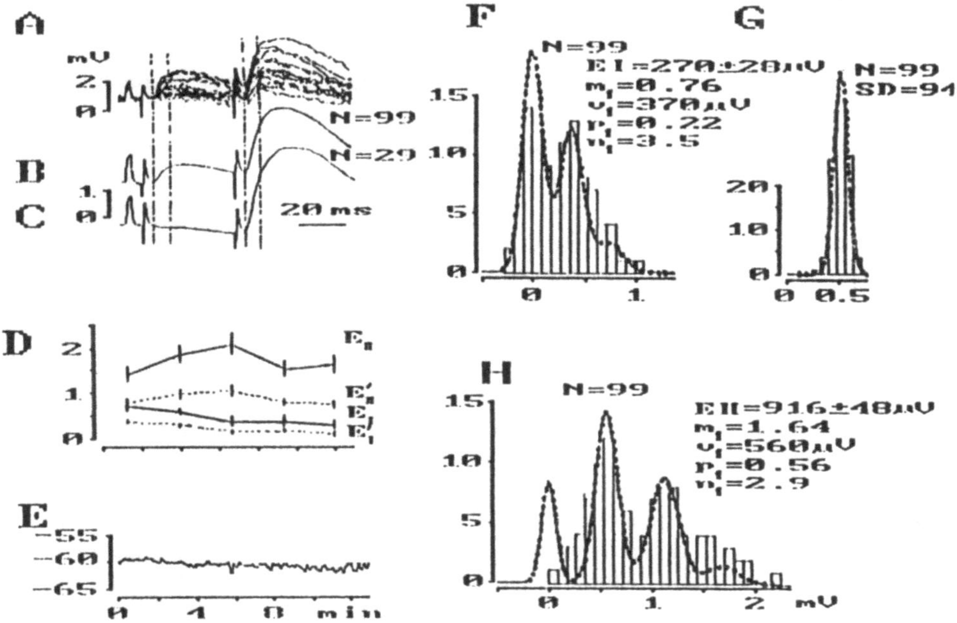

Fig. 11.6A–H. Minimal EPSPs evoked by paired-pulse stimulation in a hippocampal slice. **A** Single responses (N = 15). **B** Average of 99 responses. **C** Average of sweeps taken as EPSPI failures; note calibration pulses in **A–C**; *vertical dotted lines* show the window chosen for the amplitude measurements. **D** Time course of the first (E_I) and second (E_{II}) EPSP amplitudes and mean window amplitudes (E_I' and E_{II}', respectively); *abscissa* time from the beginning of the registration, in min (see **E**); *ordinate* mean value (±SEM, N = 20), in mV. **E** Resting membrane potential (RMP) plotted against time of the experiment. **F–H** Experimental (*columns*) and theoretical (*dotted lines*) distributions of mean window amplitudes for EPSPI (**F**), EPSPII (**H**) and calibration pulse (**G**). Estimates of the histogram method are shown in **F** and **H**. Poisson and binomial models were used to calculate theoretical distributions for **F** and **H**, respectively. [425]

Adult guinea pigs (400–600 g) were used in these experiments. Transverse hippocampal slices (400–500 μm thick) were prepared as previously described [565]. The fluid in the chamber contained 124 mM NaCl, 5 mM KCl, 2 mM $CaCl_2$, 2 mM $MgSO_4$, 1.25 mM $NaHCO_3$ and 10 mM glucose. O_2/CO_2 (95%/5%) was continuously bubbled through the medium kept at 35±0.5 °C. The intracellular glass recording electrode (4 M K-acetate, 20–100 MOhm) was inserted in the *stratum pyramidale* of CA1. Two stimulation electrodes (tungsten in glass, 0.1–1 MOhm) were inserted in the *stratum radiatum* and/or *oriens*. Each of the two inputs was tested every 8–10 s with one input following the other by 4 s. One input (usually from *stratum radiatum*) was stimulated by paired pulses with an interval 40 or 50 ms, the other input serving as control. Extracellular control responses were routinely recorded after withdrawal from the neurone. In average records corresponding in time to the "window" taken for the measurements (see below), no FPs which could significantly influence amplitude measurements were detected. EPSPs were stored on a computer system for off-line analysis, high frequencies being usually cut at 300 Hz. Two kinds of measures were used: the amplitude difference between two fixed points of the window and the "mean window amplitude". The first point of the window was set at the onset of the averaged EPSP (Fig. 11.6A–C, dotted lines) and the second point near the EPSP maximum (or 1–6 ms before it) at a distance (window width) of 3–10 ms from the first point. Both measures provided highly correlated values (Fig. 11.6D) and led to the same main conclusions. Most data presented are based on amplitude measurements with the number of measurements (N) between 72 and 102 for each region. S_n was evaluated from the measurements of the calibration pulse (Fig. 11.6G) or background noise before the stimulus artefact with the same window width. Since amplitudes were <5 mV, no correction was made for the non-linear summation. The histogram (Method 1) and variance (Method 2) methods were used to estimate the quantal parameters. In the frame of the latter method, we used here a simple correction for the noise by converting negative amplitudes into 0. With $N_0>2$, the method of failures was used as an additional variant of Method 2 (Method 3 in Fig. 11.7A). An impaled neurone was accepted for recording if it had no spontaneous discharges and a RMP <–60 mV. Stimulus intensity (0.7–1.6 V, 20–40 μs) was set to evoke minimal EPSPs with average amplitudes of EPSPI between 0.2 and 0.6 mV and with failures (Fig. 11.6A, C). All neurones showed a clear PPF so that the amplitude of EPSPII increased for 26–355% in comparison with EPSPI (Fig. 11.7A, ordinate) and ranged 0.4–1.3 mV.

Most amplitude distributions showed clear peaks and troughs. The comparison of m calculated by different methods for 19 regions with $N_0>2$ showed coefficients of correlation (r) from 0.80 to 0.93. Values of m_2 were highly correlated with both m_1 (r = 0.80) and m calculated by the failures method (r = 0.92), means (±SEM, here and below) being 1.55±0.15 for the method of failures and 1.49±0.14 for the variance method. In view of the good agreement for values estimated by the variance and failures methods, their averages were used for regions with failures (Method 4, in Fig. 11.7A, m_2' in Fig. 11.7B).

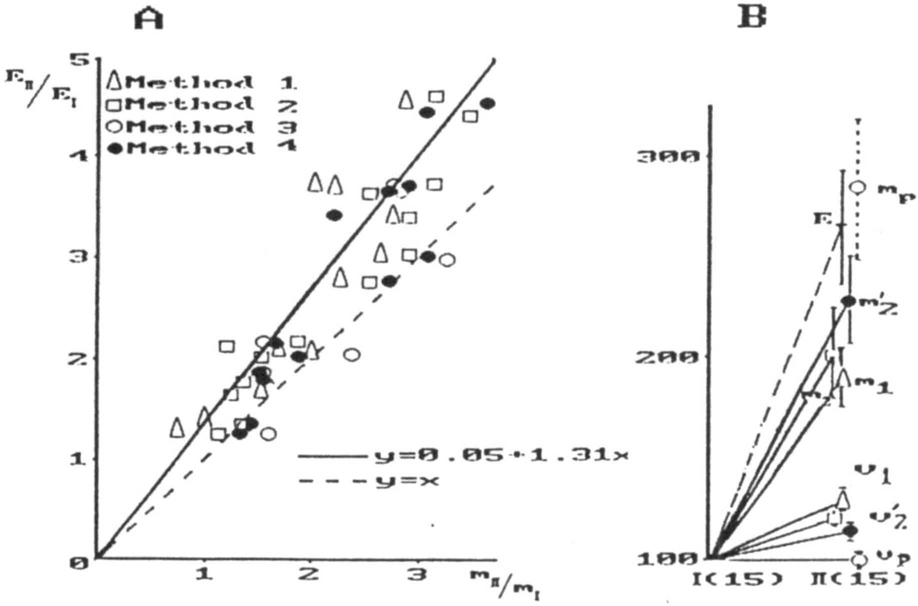

Fig. 11.7A,B. A Relative magnitude of PPF (E_{II}/E_I) in hippocampal slices plotted against changes in m (m_{II}/m_I) measured by different methods (see *inset* and text). *Continuous line* represents the regression for data obtained by the histogram and "mixed" methods; *interrupted line* is the equality line. **B** Mean changes in relative EPSP amplitude (E), mean quantal content (m_1, m_2, m_2') and quantal size (v_1, v_2, v_2'). Values m_p and v_p were calculated according to the "mixed" method from the formulae of the Poisson distribution. Values *in parentheses* show the number of measurements for the histogram and variance method. See Fig. 11.1C for additional explanations. [425]

Ratios of m for 12 neurones (15 plateau regions) are plotted against amplitude ratios in Fig. 11.7A. An increase in m with the PPF is evident from this plot as well as from the data of the illustrated experiment (Fig. 11.6F and H, insets) and from the mean data for all experiments (Table 11.3, Fig. 11.7B). The regression line in Fig. 11.7A has a slope which is more than 1 ($P<0.001$, t-test), indicating an increase in v. Correspondingly, Table 11.3 and Fig. 11.7B show a significant increase in v calculated by different methods. The change in v estimated by the method of failures (Table 11.3) also tended in the same direction, but it was not statistically significant. It was already noted that evaluation of changes in binomial parameters might be less reliable than that for the basic quantal parameters. Most of experimental histograms for EPSPI and EPSPII were fitted on the basis of the Poisson and binomial assumptions, respectively. This might indicate an increase in p due to the PPF. Mean relative parameters for EPSPII as ratios to the corresponding values for EPSPI for the Method 2 were $p_2 = 1.54\pm0.13$ and $n_2 = 1.45\pm0.22$. These data probably underestimate changes in p: the shift to the binomial distribution should indicate a more significant increase in p.

Table 11.3. Quantal analysis of paired-pulse facilitation of hippocampal responses in vitro

Method	N	m	v(mV)	p	n
1(Histogram)	15	1.50±0.57	240±140	0.49±0.17	3.7±2.3(3)
		2.61±0.65**	306±175**	0.40±0.14	7.0±3.2(10)
2(Variance)	15	1.29±0.56	286±147	0.32±0.07	3.9±1.0
		2.36±0.58**	341±187*	0.43±0.06**	5.5±1.3**
3(Failures)	8	1.13±0.37	297±97	0.30±0.08	3.9±0.8
		2.23±0.26**	329±153	0.41±0.04*	5.4±0.7*
4(Mixed)	13	1.41±0.37	279±129	0.31±0.08	3.9±1.0
		2.39±0.51**	348±193*	0.42±0.06**	5.5±1.9**

Note. Each value represents the mean ±SD for EPSPI (upper row) and EPSPII (lower row for each method). Amplitudes of EPSPI and EPSPII were 314±155 μV and 778**±408 μV, respectively. *) and **): significant increase in comparison with EPSPI values, P<0.025 and P<0.005, respectively, the sign test. The number of estimates for n and p is given in parenthesis.

11.6 Discussion

The results demonstrate the applicability of the proposed methods to analysis of plasticity mechanisms. The methods were found to be sufficiently sensitive to reveal changes in both m and v. Two basic methods gave consistent results for all phenomena analyzed. Determination of the binomial parameters should be considered with reservations. Correspondingly, the changes in p and n were not as consistent as those in m and v. Nevertheless, our data may suggest that reported changes in m were due mainly to changes in p. In addition to the testing of the methods, the results are also of interest by themselves for understanding the mechanisms of short-term synaptic plasticities. In fact, for the first time, statistical analysis of plasticity mechanisms was performed at the level of mammalian brain synapses and synapses of the mollusc CNS at normal ionic medium. The data of Korn et al. [528, 529] on the goldfish Mauthner cells as well as data of McNaughton et al. [700] and Yamamoto [1126] on hippocampal neurones in *in vitro* preparations (Sect. 7.10) became known after our basic results had been published. In general, our results demonstrate increases in m and, therefore, indicate a significant participation of presynaptic mechanisms in all plasticity phenomena studied. This conclusion is consistent with known data for peripheral and central synapses (Sect. 7.10.2).

In addition, changes in v were also found. They were especially prominent during LFD in two types of the identified neurones (RPa3 and LPa3) of the snail CNS, so that only changes in v might be responsible for the initial stage of the LFD. A seemingly different result for the third neurone (LPa2) deserves additional comments. Differences in properties of the two identified synaptic junctions were already described (Chap. 10). Quantal analysis revealed some additional differences. In fact, values of m for the neurone LPa2 were within 0.4–3.7 (and usually

< 1) with small and often uncertain p to the end of the LFD as compared to m within 3–20 and p = 0.47–0.89 for the majority of RPa3 and LPa3 neurones. A faster EPSP depression was noted in LPa2 neurones in comparison with RPa3 neurones (Sect. 10.4). The depression in LPa2 neurones was especially prominent immediately after the beginning of the low-frequency stimulation. The result was that initial regions with small N were excluded from the statistical analysis. Therefore, one might suggest that presumed changes in v in these neurones at earlier stages of the LFD were overlooked. At first sight, changes in v found in our studies are in conflict with the known data (see Sect. 7.10). One of the reason for such discrepancies might be the use of Poisson statistics instead of binomial statistics in most publications cited with rare exceptions [528, 529]. To test the possible influence of the application of the Poisson statistics, we recalculated our data on the PPF (Sects. 11.2 and 11.5) using the Poisson equations (7.3.4–1) and (7.3.5–2). In this case, no significant change in v was found (Fig. 11.2, crosses; Fig. 11.7B, m_p and v_p. Therefore, if we assumed the Poisson model, we would conclude that v was not changed during the PPF, similarly to other authors who used this model in studies of the PPF, LFD and HSF [212, 213, 475, 571, 577, 1126]. It should be noted also that some published distributions actually showed decreases in distances between peaks in the course of LFD, especially at initial periods of the LFD [212, 475]. It might be suggested that the presumed changes in v were not as evident in the publications cited as in our experiments because these authors dealt with responses initially depressed by ionic changes. In fact, we observed no significant changes in v at later periods of the LFD when EPSPs had been already depressed (Fig. 11.5AIII, IV). There were no changes in v in the LPa2 neurone, which demonstrated a fast depression after only a few stimuli.

As already noted (Sect. 6.1), changes in v can depend on presynaptic as well as on other factors. An essential participation of the "membrane" postsynaptic factors [such as changes in V_0 and R_{in}, see Eq. (6.1–3)] in snail central neurones seems to be hardly probable, since even a stronger repeated direct and polysynaptic activation produced no measurable changes in these parameters [612]. The available data may provide some additional hints on probable localization of mechanisms of the changes in v during PPF. In fact, postsynaptic changes in v are expected to be dependent on the amount of transmitter released by the first presynaptic impulse [577] and, therefore, one should expect a significant correlation between amplitudes of EPSPI and EPSPII. However, the correlation was found to be significant only in two of nine tested cases for both in vivo (r = 0.25 and 0.43) and in vitro experiments (r = 0.42 and 0.45). Interestingly, the correlation of EPSPI with the EPSP evoked through the other (control) pathways in the in vitro experiments was not significant in either case. The conclusion of the correlation analysis is that either mechanisms of changes in v are localized mainly presynaptically or they are not dependent on the release of a given transmitter. The presumed presynaptic mechanisms might be related to changes in a balance between free and bound transmitter resulting from ionic shifts produced by the first presynaptic spike during the paired-pulse stimulation (see Sect. 9.7.3 with references to the "residual calcium" hypothesis). Another explanation might be a difference in the filling of synaptic vesicles situated at different distances from the release sites. Differences in the diameters of vesicles located at different distances from the synaptic cleft were found in the neocortex [665], and similar measurements could be made in the hippocampus. The presumed presynaptic changes in v are in conflict with the idea about the constancy of the transmitter portion released and with the suggestion that the portion can be changed only by extraordinary interventions such as very long-lasting stimulation after suppression of transmitter synthesis (Sect. 7.10.3). The constancy of v was accepted after the formulation of the quantum hypothesis and maintained for many years (see Sect. 7.10). However, more recent observations of multicomponent quanta and related explanation of these observations seem to provide a natural explanation for presynaptic changes in v (see Sects. 7.4.4 and 7.11). A probable mechanism of the decrease in v during LFD might be a depletion of the available transmitter so that an incomplete portion is released by the second volley. This means a release of either a not completely filled vesicle or a smaller number of vesicles (or other hypothetical "subunits"). The explanation based on transmitter depletion is indirectly supported by the data showing the most considerable decrease in v in the initial period of the LFD in the snail neurones when the transmitter depletion could be especially high due to large EPSPs. The other support is data obtained in the context of the transmitter depletion model (Chap. 10).

The LFD described here possesses some characteristic features of the behavioural habituation (see Sect. 4.7). Moreover, a LFD with similar properties was found to participate in the habituation of behavioural defensive responses in invertebrates and possibly in extinction of related CRs (see Sect. 2.5). In view of these data, the idea of the depletion of the available transmitter might be regarded as a specification at the synaptic level of the Pavlovian concept about "exhaustion of nerve cells" as a basis for the extinction of the CR [797]. Such "passive" exhaustion should not be a mechanism unique to either habituation or extinction, bearing in mind that these phenomena might involve complicated polysynaptic neuronal nets and include elaboration of antagonistic habits [520]. Moreover, the transmitter depletion is unlikely to be the sole mechanism even for the simplest forms of the synaptic LFD: some other, "active" alterations might be indicated by changes in m and p as well as by possible changes in the depletion model parameter F. Presynaptic mechanisms of changes in v during LFD in the snail central neurones are also possible but further experiments are necessary to test other possibilities. For example, changes in subsynaptic receptor sensitivity were found in the same identified neurones under some conditions [46, 676, 918].

11.7 Summary

Statistical methods of the quantal hypothesis based on the binomial model and described in Chapter 8 were used to analyze mechanisms of the following short-term phenomena of synaptic plasticity: paired-pulse facilitation (PPF) and low-frequency depression (LFD) of minimal hippocampal EPSPs recorded from unanaesthetized rabbits, LFD of unitary EPSPs recorded from snail central neurones and PPF of minimal EPSPs recorded from guinea pig hippocampal slices. Two basic methods (histogram and variance methods) gave similar results. Changes in the mean quantal content and binomial parameter p were found for all phenomena tested. This indicates presynaptic mechanisms for both FF and LFD. Changes in quantal size (v) were also found which were especially prominent in the initial period of the LFD in two types of identified neurones of the snail CNS. Literary data, which demonstrated a constant v during PPF and LFD in neuro-neuronal synapses, were discussed. It is suggested that the conclusion about the constancy of v could be explained by recordings of responses under conditions of suppressed release and by inadequate use of the Poisson statistics. Lack of correlation between amplitudes of the first and the second EPSPs suggests presynaptic mechanisms for changes in v during hippocampal PPF. In general, the studies show that the statistical methods used are consistent and are sensitive enough to reveal changes in both basic quantal parameters during plastic modifications at central synapses.

12 Quantal Analysis of Hippocampal Long-Term Potentiation in Vivo

12.1 Introduction

The results of our previous experiments indicate, firstly, that major mechanisms of hippocampal LTP consist in an increase in the efficacy of excitatory synapses (Chaps. 1–6) and, secondly, provide adequate statistical methods to study more precise localization of plasticity mechanisms in central synapses (Chaps. 7–11). Therefore, the quantal analysis of LTP appeared to become a real task. On the other hand, a decrease or even disappearance of IPSPs during LTP was occasionally found both in our experiments (Chap. 5) and by other authors in more recent in vitro experiments [723, 1127, 1128]. These data might suggest a suppression of inhibitory systems to be an essential mechanism of hippocampal LTP. Therefore, it seemed to be necessary to reinvestigate the changes in intracellular PSPs before proceeding to the statistical analysis of LTP.

The fluctuations of hippocampal minimal EPSPs allow in principle a "non-quantal" interpretation (Sect. 8.8). Therefore, it seemed reasonable to compare recordings of minimal and unitary hippocampal PSPs. Only one record of a unitary EPSP at the level of the mammalian cortex [47] was published, without any quantitative description, before our initial experiments [90, 1047]. Therefore, an initial step in our following studies was the attempt to record hippocampal UPSPs and to compare statistical parameters of the UPSPs with parameters of minimal EPSPs recorded previously.

12.2 Methods

Recording electrodes were inserted stereotaxically in the CA3–CA4 region and stimulating electrodes were placed in the DG. However, morphological controls were not carried out in all experiments, and in one case, the stimulating electrode was found to be situated within the CA1 area. Both stimulating and recording electrodes were moved during the experiments so that one could not be sure that all our recordings were made from the DG–CA3 system. On the other hand, electrophysiological controls were used (Sect. 1.2) which showed FPs with a wave form typical for the hippocampus. These controls, together with measurements of the electrode depth, indicated that responses of hippocampal neurones were recorded in all our experiments with activation of the DG–CA3 (or CA4) system in the

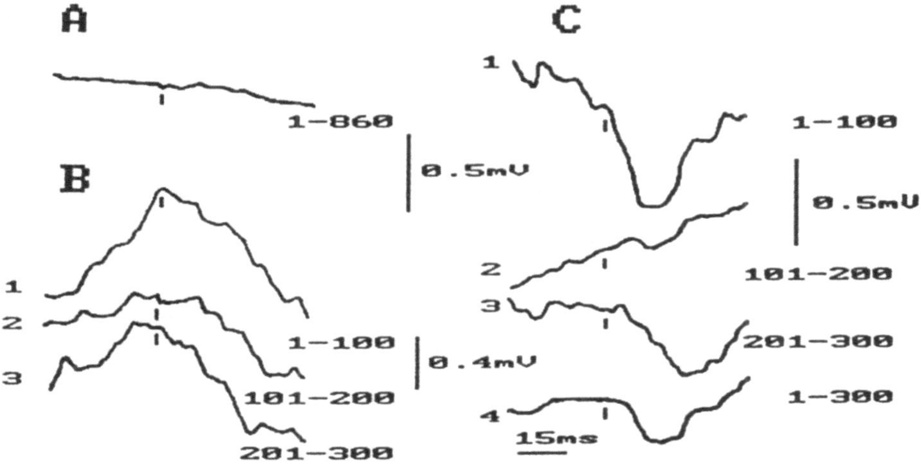

Fig. 12.1A–C. Examples of intracellular activity recorded at in CA3 area with spike-triggered averaging. **A** No correlated activity. **B** Common input: the depolarizing wave begins before spikes of the "triggering" neurone. **C** Presumed unitary IPSP. Here and in Figs. 12.2 and 12.3, *numbers near traces* are sequential numbers of averaged sweeps. *Short vertical bars* indicate the time of occurrence of the triggering spike. [90]

most cases. In the first series, an electrode holder with two electrodes was fixed to the cranial bones at the site of projection of the DG and CA3–CA4 regions with the coordinates P4.0, L5.5 [336]. The first electrode was used for extracellular recording of a triggering neurone and the second for intracellular or "quasi-intracellular" recordings of a target neurone. The electrodes were placed at an angle relative to each other so that at a depth of 7–8 mm, the distance between their tips was expected to be <1 mm. Bridge circuits were used to inject currents. The currents served to increase the frequency of spikes of the triggering neurone and to suppress discharges of the target neurone. The activity was tape recorded and then processed by PDP–8A computer using an algorithm written by Dr. A.G.Gusev. The algorithm (1) detected triggering spikes; (2) averaged activity of a target neurone in an assigned time intervals around spikes of a triggering neurone; (3) rejected intervals of activity of the target neurone with spikes and artefacts; (4) measured areas within an assigned interval and convert it into amplitudes; (5) plotted amplitude histograms. Subsequent analysis consisted of calculations of quantal parameters by methods 8.4 and 8.5 and the construction of theoretical distributions (Sect. 8.7).

12.3 Unitary Postsynaptic Potentials of Hippocampal Neurones

Altogether 96 pairs of triggering and target neurones were recorded. Of these, recordings of 26 pairs were selected for a detailed analysis on the basis of the following criteria: sample size with at least 300 discharges of the triggering neurone and sufficient stability of the target neurone as judged by the steady potential level (20–50 mV) and spike amplitude (10–40 mV). The whole recorded activity of the target neurone (up to 800 responses) and successive regions (N=100) were averaged. Hence, three or more successive regions were distinguished for each pair. An arbitrary criterion of the existence of a correlation between spikes of the triggering neurone and the activity was adopted. The criterion was an appearance of wave forms which were similar in time of onset and shape to each other and to the average of the total activity of the target neurone in all three regions (or in at least 75% of regions if they numbered >3). According to this criterion, no correlation was found in 18 of 26 pairs (Fig. 12.1A). Waves, starting before the triggering spikes, were observed in five pairs (Fig. 12.1B). These waves presumably reflected common inputs to the recorded neurones [1, 368]. Wave forms with appropriate delays (a few ms) after the triggering spikes were found in averaged records of three target neurones (Figs. 12.1C1–4, 12.2A1–5). These waves exceeded both the average background activity before the triggering spike and control activity recorded after the target neurone was lost (Fig. 12.2A6, 7). In two out of the three cases, hyperpolarizing waves were observed (Fig. 12.1C) which presumably represented unitary IPSPs. The IPSPs were recorded from two target neurones following the spikes of the same triggering neurone. Predominantly depolarizing waves, which presumably represented unitary EPSPs, were recorded in one experiment (Fig. 12.2).

Since a detection of averaged polysynaptic UPSPs with amplitudes of higher than several μV seems unlikely at the mammalian CNS (see Sect. 2.2.7), EPSPs and IPSPs with the average amplitudes of 150 and 250 μV, respectively, were regarded as monosynaptic. The relatively long latency (3.5 ms for EPSP and about

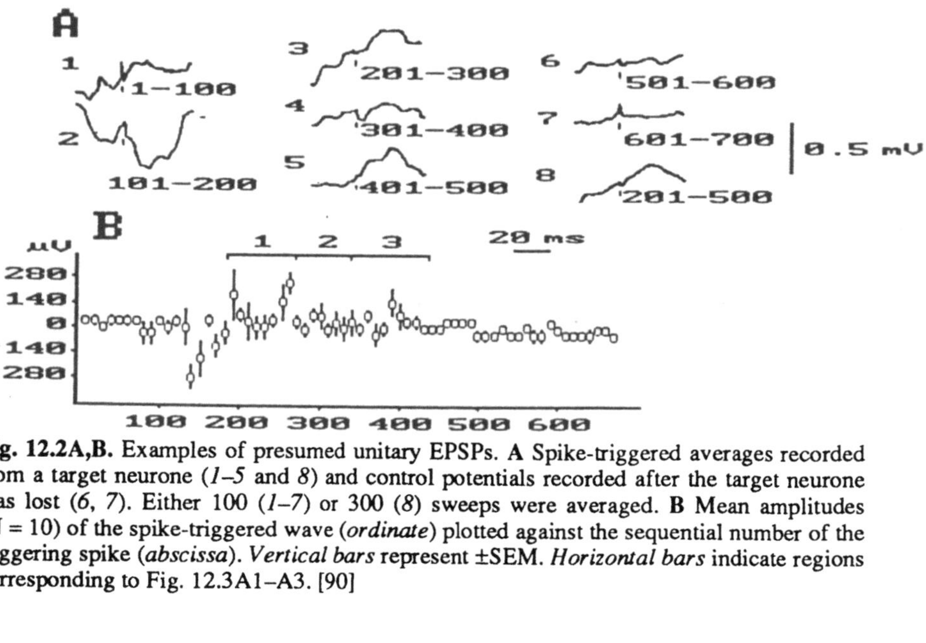

Fig. 12.2A,B. Examples of presumed unitary EPSPs. **A** Spike-triggered averages recorded from a target neurone (*1–5* and *8*) and control potentials recorded after the target neurone was lost (*6, 7*). Either 100 (*1–7*) or 300 (*8*) sweeps were averaged. **B** Mean amplitudes (N = 10) of the spike-triggered wave (*ordinate*) plotted against the sequential number of the triggering spike (*abscissa*). *Vertical bars* represent ±SEM. *Horizontal bars* indicate regions corresponding to Fig. 12.3A1–A3. [90]

4 ms for both IPSPs) matches the slow conduction velocity (Sect. 4.7) for the distance of about 1–2 mm. This latency is in the range of the latencies of MPSPs which were assumed to be monosynaptic (Sect. 4.7). According to the selection criteria, the UPSPs were observed in all or most of the recorded regions. Nevertheless, significant variations in their amplitudes were found in different regions (compare 1–3 in Fig. 12.1C and 1–5 in Fig. 12.2A), so that a reliable averaged wave with suitable time course might even disappear in some regions (Fig. 12.2A2). The variability created difficulties for the statistical analysis, which requires a stationarity (Sect. 7.4.2). We used the same approach as described in Section 11.4 and extracted "quasi-stationary" plateau regions on the basis of mean amplitudes from 10 consecutive measurements, so that every plateau region contained about 100 measurements without a significant drift (Fig. 12.2B). Three such regions were analyzed for the presumed unitary EPSP (Fig. 12.2B, 1–3), and one region was chosen for the presumed unitary IPSP (Fig. 12.1C1). Fig. 12.3 shows respective amplitude distributions for the EPSP (Fig. 12.3A, B1), for the control region (Fig. 12.3B2) obtained after the target neurone was lost and for the IPSPs (Fig. 12.3C). One can see that this theoretical distribution based on the histogram method and Poisson statistics fitted the right hand sides of the experimental histograms but sometimes failed to fit the left hand sides (Fig. 12.3A1, B1). One possible cause for the disagreements might be a domination of spontaneous IPSPs.

Only parameters m and v were included in Table 12.1, because values of binomial parameters were less than estimates of their errors in all cases. The cases with p close to 0 and uncertain n correspond to the Poisson model. Table 12.1 shows that parameters calculated for the consecutive regions were usually of the

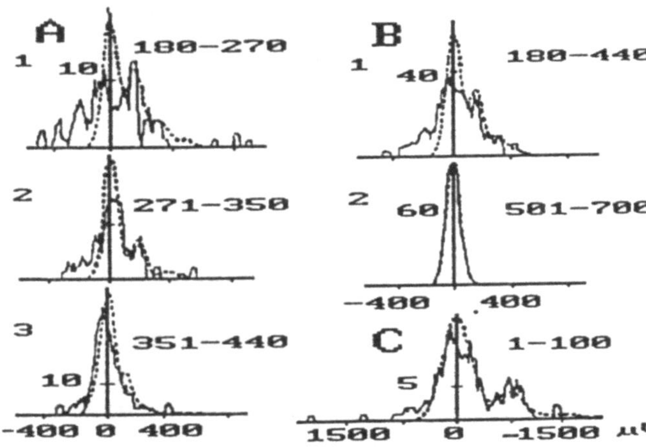

Fig. 12.3A–C. Experimental (*continuous curves*) and theoretical (*dotted curves*) distributions of amplitudes of presumed hippocampal UPSPs. **A** EPSPs from three regions (*1–3*) marked in Fig. 12.2B. **B** EPSPs for pooled regions 1–3 (*1*) and for a control region recorded after the neurone was lost (*2*). The control histogram **B2** was fitted by a Gaussian distribution. **C** IPSPs corresponding to Fig. 12.1C1. Axes in **A–C** are the same as in Fig. 8.1. Numbers to the *right* of the histograms represent sequential numbers of the responses. [90]

Table 12.1. Quantal parameters of presumed unitary postsynaptic potentials recorded from in vivo hippocampus

Response	N	v(μV)	m
EPSP	261	165/79	0.54/0.40
EPSP (1)	91	175/74	0.65/0.49
EPSP (2)	80	200/135	0.36/0.28
EPSP (3)	90	150/34	0.34/0.28
IPSP	100	750/360	0.26/0.28

Notes. Parameters to the left and to the right of the slashes were calculated by method 8.4 and 8.5, respectively. Numbers in parentheses correspond to respective regions in Fig. 12.2.

same order and m<1 for all cases. Quantal sizes of the presumed unitary EPSP (mostly about 100–200 μV) were within the range obtained for other central synapses (Sect. 7.8.3) and for minimal EPSPs of hippocampal neurones according to our (Chaps. 8, 11) and more recent literature data (Table 7.2). For the IPSPs, v was higher which corresponds to the estimate of other authors for one CA3 neurone in vitro (Table 7.2, #6).

Summarizing, our data are in general agreement with the quantal interpretation of fluctuations of hippocampal minimal EPSPs. However, small values of m and p, as well as the Poisson fit do not agree with the majority of previous results obtained with minimal hippocampal EPSPs. This discrepancy forced us to improve the methods of stimulation and to use weak currents (2 to 10 μA at stimulus duration 0.15–0.30 ms) delivered through a microelectrode instead of the macrostimulation usually used in our previous in vivo experiments.

12.4 Post-Tetanic Changes in Postsynaptic Potentials Evoked by Microstimulation

Similar to our previous experiments (Chap. 5), macroelectrodes were inserted into the DG for the conditioning tetanization (a train at 20 Hz, 0.2–2 mA, of 5–10-s duration). However, the testing stimuli were delivered through a tungsten microelectrode inserted between the recording microelectrode and the stimulating macroelectrode at 1 to 2 mm distance from the recording site. The position of the testing stimulating microelectrode was adjusted during the insertion of the recording microelectrode. The criterion of the proper position of the stimulating microelectrode was an appearance of prominent FPs evoked by paired (intervals of 10–30 ms) current pulses of several dozen μA. The testing stimuli were delivered with intervals of 3 to 5 s. After a stable intracellular or quasi-intracellular activity was found, the stimulating current was varied until an averaged EPSP was evoked by the second pulse with amplitude significantly larger than that of the FP obser-

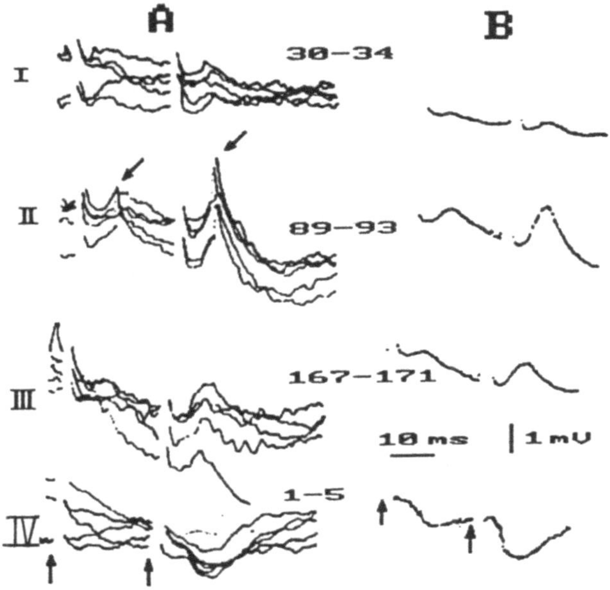

Fig. 12.4A,B. Postsynaptic potentials evoked by microstimulation before and after tetanization. **A** Single responses. **B** Corresponding averages of 33–66 responses. *I* Before tetanization; *II*, *III* after tetanization (from 1.5 to 5 min and from 7.5 to 10 min, respectively); *IV* extracellular records obtained after the neurone was lost. The stimuli are marked by *vertical arrows*. *Oblique arrows* in **AII** show partial spikes which were eliminated from the averaged record **BII**

ved before the cell penetration at comparable stimulus strengths. The other condition was that failures were frequent in response to the first stimulus but very rare in response to the second stimulus within the testing pair. This condition suggests that as few presynaptic fibres as possible are excited but at least one fibre is excited almost every time.

Examples of minimal EPSPs revealed by this method are given in Fig. 12.4AI, BI. No PSPs were distinguishable in most of the single sweeps following the first stimulus in the paired-pulse testing paradigm (Fig. 12.4AI). However, the averaging revealed an EPSP with the amplitude of about 0.2 mV (Fig. 12.4BI). Following the second stimulus, small EPSPs were distinguishable not only in the averaged record (Fig. 12.4BI) but also in almost every single sweep (Fig. 12.4AI). From 50 to 85 single or paired testing stimuli were delivered as a pretetanic control, upon which the tetanization was applied. We avoided delivering the tetanic stimuli through the testing microelectrode to be sure in the constancy of its properties before and after the tetanus. In addition, the current strength was recorded continuously. The strength of the testing current changed after tetanization in one experiment, and the corresponding records were discarded. One min after tetanization, the testing stimulation was resumed by series of 75 to 225 stimuli with about 1 min intervals between the series until the neurone was lost. In some cases, the conditioning tetanization was repeated after two or three testing series. FPs were recorded once more after the neurone was lost, and only those cases were analyzed when the averaged PSP amplitude after the tetanus was considerably larger than the amplitude of the related FP. The exception was the experiment with the polarity of the extracellular FPs (Fig. 12.4AIV and BIV) opposite to that of EPSPs (Fig. 12.4AI and BI). Since the FP was negligible before cell penetration, we believe that the exceptionally high negative amplitude of the FP in this particular experiment was explained either by a small shift of the microelectrode at the moment when the neurone was lost or/and by effects of several subsequent tetanizations delivered during prolonged recording of this neurone lasting for about 2 h. Only responses recorded after the first tetanization were taken for statistical analysis.

On the whole, 43 neurones were recorded. Of these, 20 neurones were excluded from the analysis mostly because the neurone was lost within <3 min after tetanization. Activity of 23 neurones was recorded for a period from 3 min to about 2 h after the first tetanization. In 15 of 23 neurones, amplitudes of EPSPs, or IPSP, or both (Fig 12.4) were increased in comparison with pretetanic responses ($P<0.01$, t-test). Such an increase will be called LTP. No statistically significant decrease in IPSP amplitude was found. The mean latency of the averaged minimal EPSPs in tested neurones was 3.3±0.4 ms, and it was not changed after tetanization.

12.5 Statistical Analysis of Minimal EPSPs Evoked by Microstimulation

For the quantal analysis, we used the most precise methods from those elaborated and tested during our previous experiments with recordings of both minimal and unitary PSPs. Specifically, we used computer programmes both to measure EPSP areas, which increased signal-to-noise ratios, and to construct "moving bin" histograms, which facilitated the subdivision of the amplitude histograms according to method 8.4. Another computer programme was developed to average responses within an assigned interval of amplitudes, which facilitated determination of N_0 by the trial and error method.

In addition, a programme enabled us to analyze responses corresponding to different peaks of amplitude histograms. In the example (Fig. 12.5), the averaging of responses of <140 μV (Fig. 12.5A, region "0"), revealed an IPSP with a latency of about 8 ms and without EPSPs (Fig. 12.5B, upper record). Therefore, the number of averaged responses (i.e. 7) could be taken as N_0. The averaging of responses from regions I–III around the corresponding peaks of the histogram (Fig. 12.5A) showed that they contained EPSP-IPSP sequences (Fig. 12.5B). The EPSP latency (5 ms) and rise time (3 ms) were equal to the corresponding time characteristics of the averaged responses from all three regions (Fig. 12.4BI, second response). This

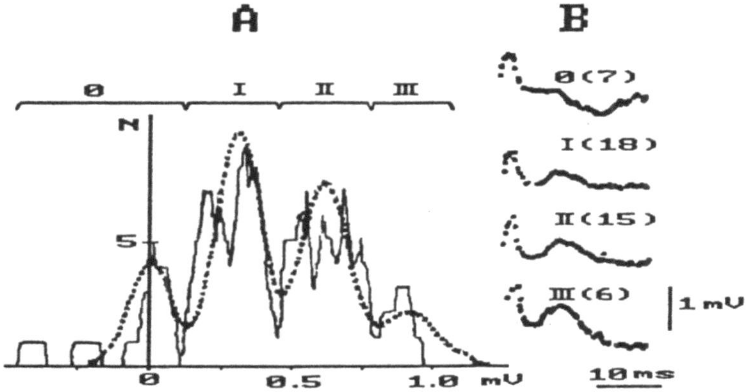

Fig. 12.5A, B. An example of analysis of EPSP amplitude histograms. **A** Experimental (*continuous curve*) and theoretical (*dotted curve*) distributions. The meaning of the axes is the same as in Fig. 8.1. **B** Averaged responses corresponding to different histogram regions marked in **A** by respective *Roman numbers*. Numbers of responses averaged are given *in parentheses*

observation is in accordance with the interpretation of the EPSP fluctuations based on the quantum hypothesis (see Sect. 8.8). On the basis of several trials of such a preliminary analysis, it was possible to select the interval for EPSP measurements which included only the initial part of the EPSP not contaminated by the IPSP.

The "quasi-stationary" plateau regions were defined as described in Sects. 8.2 and 8.6. Figure 12.6A and B exemplifies the extraction of the plateau regions. No significant changes in mean amplitudes occurred within each of the regions marked by the horizontal bars (t-test, $P>0.01$). The regions were chosen for the pretetanic period (Fig. 12.6A, B, region I), for the "early" (from 1.5 to 5 min after tetanization; Fig. 12.6A, B, region II) and for the "late" (5–12 min; Fig. 12.6A, B, region III) post-tetanic periods.

12.6 Changes in Quantal Parameters After Tetanization

Stable recordings of six EPSPs suitable for the statistical analysis were obtained from five neurones with statistically significant LTP. Each "quasi-stationary" region in these neurones contained from 23 to 79 responses (Table 12.2). Responses to single testing stimuli were recorded in one neurone (#1 in Table 12.2) and responses to paired testing stimuli were recorded in four neurones. However, only in one of these neurones (Table 12.2, #2), both stimuli in the testing stimulus pair produced notable averaged EPSP before tetanic stimulation (Fig. 12.4BI) with amplitude histograms distinct from noise histograms. In the other neurones, distinct averaged EPSPs were seen only in response to the second stimulus from the testing pair.

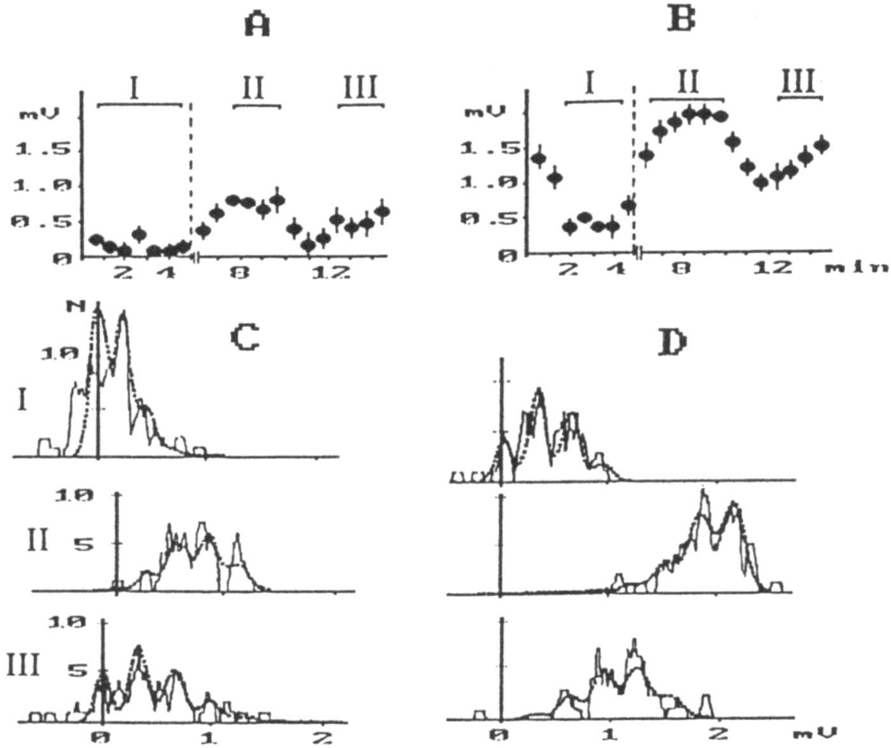

Fig. 12.6A–D. Quantal analysis of post-tetanic changes in hippocampal EPSP. **A, B** Time courses of the amplitude of the first (**A**) and second (**B**) responses evoked by paired-pulse microstimulation (see Fig. 12.4). *Ordinate* mean EPSP amplitude calculated from ten consecutive responses ±SEM (mV). *Abscissa* time (min). About 1 min is omitted after tetanization (*vertical broken lines*). *Horizontal bars* (*I, II, III*) mark "quasi-stationary" regions. **C, D** Experimental (*continuous curves*) and theoretical (*dotted curves*) amplitude distributions for the first (**C**) and second (**D**) responses before (*I*) and after (*II, III*) tetanization. Theoretical distributions were built according to parameters calculated by the histogram method. Sample sizes and quantal parameters are given in Table 12.2 (neurone 2). See Fig. 8.1 for further explanation. [1014]

The quantal parameters were calculated by methods described in Sections 8.4 and 8.5, and theoretical distributions were built in accordance with the parameters calculated by the histogram method 8.4 (dotted curves in Fig. 12.6C, D). In the frame of the method 8.5, m_2 was calculated by Eq. (7.3.4–2) when $N_0>4$, otherwise the variance method was used (Sect. 7.3.5). Despite some differences between the single values calculated by the two methods (Table 12.2), the parameters were very similar or even identical on average: $v_1=297\pm58$ μV (±SD), $v_2=281\pm85$ μV; $m_1=2.3\pm1.4$, $m_2=2.4\pm2.5$; $p_1=0.53\pm18$, $p_2=0.53\pm12$; $n_1=6.2\pm3.0$, $n_2=6.3\pm3.5$ (N=17 for m and v, N=9 for p and n). High and statistically significant correlations were found for m and p estimated by the different methods (r>0.9,

Table 12.2. Quantal analysis of post-tetanic changes of EPSPs evoked by microstimulation in the in vivo hippocampus

Neurone	Stimulus	Region	N	v(μV)	m	n	p
1	1	I	38	220/449	2.7/1.3	8.6/3.6	0.31/0.37
		II	68	220/441	3.5/3.2	12.7/7.6	0.27/0.37
		III	60	251/281	1.9/1.7	–	–
2	1	I	66	223/180	0.8/0.6	–/5.1	–/0.13
		II	36	276/235	2.6/3.1	4.3/5.2	0.61/0.60
		III	41	325/230	1.4/1.8	–/6.4	–/0.28
2	2	I	44	305/271	1.4/1.4	3.0/3.3	0.45/0.42
		II	53	305/171	6.3/11.2	7.5/14.8	0.84/0.75
		III	37	310/275	3.5/3.8	5.4/6.3	0.64/0.56
3	2	I	79	375/160	0.4/0.4	–	–
		II	67	303/370	1.9/1.4	–/4.2	–/0.34
		III	63	261/232	0.7/0.5	–	–
4	2	I	25	355/375	2.2/2.0	4.0/3.8	0.55/0.54
		II	23	385/293	3.1/4.1	4.8/6.6	0.65/0.61
		III	27	390/394	2.7/2.7	5.7/4.8	0.48/0.57
5	2	I	57	320/241	0.8/0.7	–	–
		II	35	225/389	2.9/1.7	–/4.5	–/0.38

Notes. I, II and III are pretetanic, "early" and "late" post-tetanic regions, respectively. Estimates of n and p which are not significantly different from 0 are not given. See Table 11.2 for additional explanations.

P<0.001). In general, the agreement between the methods was better here than for experiments with macrostimulation (see Fig. 8.3).

The time course of the mean amplitudes of two EPSPs evoked by paired pulses and the selection of the plateau regions are illustrated in Fig. 12.6A, B; the corresponding amplitude distributions are given in C and D. The number of failures (see amplitudes around 0) was considerably diminished after tetanus (Fig. 12.6CII,III) or no failures occurred (Fig. 12.6DII). The histograms shifted to the right along abscissa without systematic changes in the distances between the peaks. This indicated post-tetanic changes in m with a constant v.

These visual evaluations were supported by the quantitative estimates presented in Table 12.2 and summarized in Fig. 12.7. Table 12.2 shows increases in both m_1 and m_2 after tetanic stimulation in all cases except one. This means a statistically significant increase ($P = 0.012$ for m_1 and $P<0.004$ for m_2, $N = 11$, Wilcoxon test for matched pairs here and below). The post-tetanic increase in the mean m_1 and m_2 is illustrated by Fig. 12.7. In contrast, both v_1 and v_2 varied without any common trend. There was no significant changes either in v_1 ($P>0.6$) or in v_2 ($P>0.8$). The mean values remained unaffected by the tetanic stimulation (Fig. 12.7, v_1 and v_2). It is interesting to note that the comparison of EPSPs evoked by the first and second testing stimuli in the same neurone (Table 12.2, #2) revealed a clear tendency to increases in both v_1 and v_2 (see Sect. 11.2).

Parameter p also showed a tendency to an increase after tetanization. The value of p_1 did not differ significantly from 0 in half of the control regions and were ex-

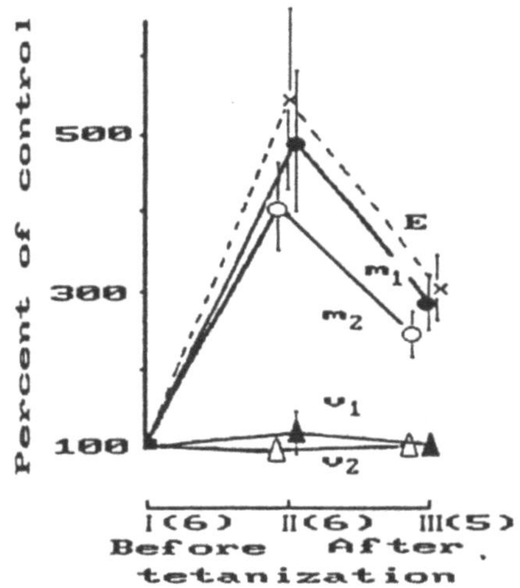

Fig. 12.7. Post-tetanic changes of mean EPSP amplitude and mean quantal parameters. Post-tetanic values (regions *II* and *III*) were normalized to the pretetanic control (region *I*) and plotted as percent of the control. Mean amplitudes (*E*), *m* and *v* are plotted as *crosses*, *circles* and *triangles*, respectively. Estimates of two different methods (8.4 and 8.5) are indexed by 1 and 2 and plotted as full and empty symbols, respectively. *Vertical bars* represent ±SEM. *Values in parentheses* refer to the number of measurements. Only potentiated EPSPs evoked by microstimulation were measured. [1014]

cluded from Table 12.2. The Poisson statistics gave an adequate description of the corresponding experimental distributions (Fig. 12.6C1). As noted in Sect. 8.5, the estimate of p_2 from Eq. (8.5–1) is inaccurate when amplitude distributions are close to the Poisson one, so that corresponding values of p_2 and n_2 might also be ignored. In all cases given in Table 12.2, parameter n was also increased after tetanic stimulation in comparison with the control regions.

Summarizing, the statistical analysis revealed a definite increase in m while v was not changed after tetanic stimulation. The increase in m was probably due to changes in both binomial parameters but the estimates of the latter were not reliable.

12.7 Discussion

Post-tetanic changes in minimal PSPs evoked by microstimulation confirm the general conclusions made on the basis of our experiments with macrostimulation (Chap. 5). A significant increase in EPSP amplitude was found after tetanization. This result is also in agreement with literary data obtained on hippocampal slices [31, 32, 404, 405, 723, 910, 1127, 1128] (see reviews cited in the Introduction for more recent references). The potentiated EPSP were found to have a tendency to diminution which might represent, firstly, a transition from a short-term PTP to

LTP and secondly, the LFD (Chaps. 1 and 4). Measurements corresponding to the early (1.5–5 min) and later (5–12 min) post-tetanic periods showed qualitatively analogous changes in quantal parameters in comparison with the pretetanic control level. Short-term EPSP potentiation lasted usually <1–3 min under our conditions (Chaps. 1, 5) which is consistent with most literary data [25, 145]. Therefore, EPSP potentiation described in this chapter should be related mostly to LTP rather than PTP.

In this series of experiments, we observed no prolonged (>1 min) post-tetanic IPSP depression (see Chap. 5). In accordance with observations of other authors [4, 404, 405], IPSPs increased after tetanization. This indicates that a suppression of inhibitory systems is not necessary for LTP expression [388]. It may be suggested that the long-lasting suppression of inhibitory systems, which was observed systematically by some authors [723, 1127, 1128] and sometimes in our in vivo experiments (Chap. 5) represents an additional mechanism which can be especially pronounced under some conditions, e.g. after too strong or repeated tetanizations, and it might be related to pathological phenomena such as "spreading depression", epileptiform activity or "kindling" (see Sects. 1.10, 5.8 and General Conclusions for more discussion). The post-tetanic increase in IPSP amplitudes can be explained either by a potentiation of excitatory synapses on inhibitory interneurones or/and by a potentiation of inhibitory connections themselves. The first explanation seems to be natural in view of a longer latency of the IPSPs in comparison with that of EPSPs. It is supported by some indirect evidence [4] as well as by LTP in presumed inhibitory interneurones [193, 784].

The validity of the basic results of the present chapter depends on the argumentation in favour of the adequacy of the methods of the quantum hypothesis under our conditions. Evidence in support of such application has been discussed in previous chapters. Comparison of time characteristics and quantal parameters of UPSPs (Sect. 12.3) with those of minimal EPSPs provides additional support both for the monosynaptic nature of the recorded EPSPs and the "quantal" interpretation of their statistical parameters. Additional support for the monosynaptic nature of the minimal EPSPs is the lack of significant changes of their latencies even with up to 20 times increase in their amplitudes (compare the response to the first stimulus in Fig. 12.4BI with that to the second stimulus in Fig. 12.4BII).

The post-tetanic changes in m at a constant v can be easily interpreted in the context of the quantum hypothesis, similarly to analogous changes for other types of plasticities (see Sect. 7.10). In the context of the alternative, "non-quantal" interpretation (Sect. 8.8), an increase in the calculated parameter m of a monosynaptic response at a constant v would signify changes in the number of presynaptic fibres activated by the same testing stimulus. It would mean a considerable and prolonged post-tetanic increase in excitability of stimulated presynaptic elements or a long-lasting relief from a presynaptic block. All data obtained in our studies and by other authors indicate a constancy of the presynaptic volley and rather a decrease in postsynaptic excitability of hippocampal neurones during LTP (Sects. 1.8, 1.10, 5.7, 6.3).

On the other hand, the comparison of statistical parameters of presumed UPSPs with the related parameters of minimal EPSPs evoked by macrostimulation revealed definite differences in values of m and types of amplitude distributions. It is these differences that forced us to use here a more labour-consuming method of microstimulation instead of the macrostimulation used previously. In the microstimulation experiments, amplitude histograms for pretetanic EPSPs were often found to be close to distributions predicted on the basis of the Poisson assumption and, therefore, they were similar to the amplitude histograms of the UPSPs. Pretetanic values of $m<1$ were often found (Table 12.2). It can thus be concluded that the method of microstimulation is more adequate for quantal analysis than the method of macrostimulation previously used. The procedure of the estimation of the "near threshold" current in experiments with macrostimulation might lead more often to activation of rare synaptic connections of high efficiency, whereas such connections can be less readily revealed under conditions of either UPSP recording or microstimulation. This explains the differences in the estimated m (and also p) values. The situation is analogous to that of an activation of a synaptic connection with a large number of release sites, most of which have a low probability of release. Under these conditions, the amplitude distribution should behave as a binomial distribution because most of the responses contributing to the amplitude histogram are produced by a relatively small number of high probability release sites [806] (see Sect. 7.7 for more discussion).

Our results suggest a low release efficacy for the most of intrahippocampal synapses in "normal", non-facilitated and non-potentiated state. This is in agreement with data obtained by the spike triggered averaging at the rabbit neocortex [392, 1027]. Our recordings of the UPSPs from unanaesthetized animals indicate that the efficacy of both intrahippocampal and intraneocortical connections is typically not only low but also very unstable. The variability of the UPSPs might be related to LFD, PTP and LTP, but exact reasons are not understood in full (see also more recent in vitro observations [660]).

As a result of LTP induction, parameters m, p and n for EPSP evoked by microstimulation became comparable to or even larger than values obtained in macrostimulation experiments (compare full triangles and empty dots in Fig. 8.3). Judged by the severalfold increase in m due to LTP and PPF, it can be suggested that stimulating current activated several fibres (or several release sites), most of which were practically ineffective in non-facilitated and non-potentiated states but became effective after LTP induction (see discussions of the problem of "silent" synapses in the hippocampus [719, 1053] and other structures [322, 420, 531, 707, 759, 823]).

The basic result of our statistical analysis is the demonstration of essential increase in m during LTP without any significant changes in v. In the context of the quantum hypothesis, these data indicate a presynaptic localization of basic LTP mechanisms, i.e. a post-tetanic increase in the mean number of transmitter quanta liberated per presynaptic impulse. The constancy of v indicates that neither the amount of transmitter in one quantum nor the effectiveness of its influence on the postsynaptic membrane changed substantially following LTP induction. These data suggest that other postsynaptic changes such as shifts in RMP or R_{in} might also be insignificant. This is in accordance with our (Chaps. 5, 6) and literary data (see reviews cited in the Introduction and General Conclusions).

In agreement with the increase in m, a long-lasting increase in release of synaptic transmitters was found after tetanization of various hippocampal pathways both in the in vitro [913] and in vivo experiments [146, 149, 278, 369, 645]. The increased release persisted up to several hours or even days [150] and was inhibited when the induction of LTP was blocked. It should be noted, however, that at least one group reported a lack of increase in endogeneous excitatory amino acids during potentiations induced either by tetanic stimulation [37] or phorbol ethers [36] (see, however, [658]). An involvement of presynaptic mechanisms is also indicated by decreased presynaptic fibre excitability [858], increased Ca^{2+} uptake [7] and synaptosomal Ca^{2+} concentration [644], presynaptic localization of phosphorylated proteins [66, 373, 628, 629]. The increase in m with only small or absent changes in v is a common result of studies of short-term plasticities at various peripheral and central junctions (Sect. 7.10) including the hippocampus (Chap. 11) and was found to accompany long-term changes at peripheral synapses (Sect. 7.10.5). On the other hand, an increase in v and in ACh sensitivity was reported for the bullfrog sympathetic ganglion, especially after antidromic stimulation (Sect. 7.10.5). It has already been mentioned that antidromic stimulation produced no LTP. More recent publications related to the quantal analysis at hippocampal slices will be considered in Chapter 14.

Naturally, some small changes in v could remain not revealed if they exceeded the limits of accuracy of the methods used in our study. In particular, the lack of a significant correlation between values of v_1 and v_2 in the present experiments might indicate that the estimates of v were less reliable than estimates of m. It might also be suggested that an increase in v during LTP was compensated by IPSP potentiation and increased membrane conductivity. However, such explanation seems to be not suitable for two cases (Table 12.2, #1 and #2, first stimulus) with EPSPs without conditioning IPSPs.

Some reservations should be made in relation to the straightforward interpretation of post-tetanic changes in quantal parameters of hippocampal EPSPs. One difficulty might be related to possible nonuniformity of parameter p (see Sect. 7.7) and inadequacy of the simple binomial model. However, the histogram method in its present variant could be applied to either simple or compound binomial distributions. We already noted also that the agreement between the quantal parameters determined by two different methods supported the simple binomial model as a first approximation under present conditions. It should be also stressed that a significant decrease in the CV and N_0 strongly suggest a significant post-tetanic increase in m independently on the assumption of either the simple binomial model or less constrained quantal models (see Sect. 7.3.3).

Another reservation is related to the fact that, according to our data, initial efficacy of the hippocampal synapses might be very low. Provided the number of activated fibres is not changed, the comparison of amplitude distributions of control responses with potentiated ones suggested that some ineffective synapses become effective. The analysis of presumed UPSPs also supported the idea of low initial synaptic efficacy of hippocampal and neocortical synapses (see above). The conclusion about purely presynaptic mechanisms based on quantal analysis is strictly applicable only to initially effective synapses. Ineffective synapses, making no essential contribution to the control minimal EPSP, could increase their efficacy not only due to presynaptic changes but, in principle, also (or exclusively) due to postsynaptic or subsynaptic reorganization, e.g. a formation of new subsynaptic receptors [306, 342, 1022]. In this case, our data indicate that the mean efficacy of the transmitter quanta (mean v) for presumable "new" synapses is about the same as that for initially effective ("initially existing") synapses. Therefore, the quantal analysis shows that hippocampal LTP is produced either by presynaptic changes in existing synapses or by formation of "functionally new" synapses with postsynaptic properties similar to those of existing synapses. The first, more traditional explanation based on an increase in efficacy of excitatory synaptic connections due to an increase in the number of transmitter quanta released per presynaptic spike is, nevertheless, consistent with our observations and seems to be attractive.

12.8 Summary

An attempt was made to record hippocampal unitary postsynaptic potentials (UPSPs) to verify the amplitude fluctuations of minimal excitatory postsynaptic potentials (EPSP) on the basis of the quantum hypothesis. Intracellular activity of CA3–CA4 target neurones was averaged synchronously with extracellular spikes of adjacent triggering neurones. In 5 out of 96 recorded pairs of neurones, an activity was found which reflected common inputs to these neurones. Hyperpolarizing or depolarizing waves which presumably represented UPSPs were revealed in three target neurones. Quantal size (v) estimated by two methods (histogram and failures) was usually about 100–200 μV and was close to previous estimates for minimal EPSPs evoked by macrostimulation. However, the mean quantal content (m) was <1, and the amplitude distributions were described by the Poisson statistics, which appeared to be different from the previous data. Based on these observations, microstimulation was used in experiments aimed at the quantal analysis of long-term potentiation (LTP). Currents of 2–10 μA were applied at 1–2 mm distance from recorded CA3–CA4 neurones. Under these conditions, both v and m, as well as the type of amplitude distributions, were usually close to these of UPSPs. On the basis of analysis of UPSPs and minimal EPSPs evoked by microstimulation, the conclusion was drawn that most hippocampal synapses are of a low and unstable efficacy. Prolonged (up to 15 min) increases in amplitudes of EPSPs and inhibitory PSPs were found after tetanization in 15 of 23 neurones recorded in experiments with microstimulation. It is concluded that suppression of inhibitory systems need not be a necessary condition for LTP expression. Quantal analysis of LTP mechanisms was performed by the histogram method and a combination of the failures and variance methods. Statistically significant post-tetanic increase in m which paralleled EPSP amplitude changes was shown by the two methods. No change in v was found. These data suggest that the major mechanism of LTP consists in an increase in efficacy of excitatory synapses. Although the quantal analysis under conditions of low initial release does not exclude a participation of postsynaptic changes, the available data are consistent with presynaptic location of LTP mechanisms which ensure an increase in the number of transmitter quanta released per presynaptic impulse.

13 Quantal Analysis of Minimal Postsynaptic Potentials in Hippocampal Slices: Binomial Model

13.1 Introduction

It has been already noted in Chapter 12 that recording of UPSPs is preferable for a more precise quantal analysis. However, the recording of UPSPs is not practical for studies of long-lasting events in neurones of vertebrate CNS. In fact, simultaneous recordings from two hippocampal neurones in our in vivo experiments [90] revealed UPSPs in only few cases (Sect. 12.3). Even under more stable conditions of recordings from hippocampal slices, the success rate of UPSPs from CA3–CA1 pairs was found to be very low [351, 864, 865]. In addition, several other problems are not resolved for CNS neurones (Chaps. 7, 11). I mention here the high S_n and also small sample sizes (N) because of time limitations of intracellular recordings from healthy cells and necessity of "stationarity" for reliable statistical analysis. The limitation of N might be illustrated by our in vivo studies (Chaps. 11 and 12), as well as by studies of other authors. An example is the publication on unitary IPSPs in the Mauthner cell [528] with the average number of counts equal to 161 and with <70 counts in some cases. All these problems, common to most neurones of the CNS, are heavily aggravated for recordings from higher levels of the mammalian brain (e.g. cortical structures) and especially for in vivo recordings. Therefore, we decided to expand and reinforce the previous quantal analysis by additional experiments on the CA1 area of hippocampal slices. Initial attempts to apply the quantal hypothesis to recordings from hippocampal slices were reviewed in Sections 7.8.5 and 7.9.4 (see also our initial data in Sect. 11.5). Before we started the present experiments, only preliminary data on statistical analysis of EPSPs recorded in region CA1 [33] were published. The related full publications appeared recently [351, 864, 865].

The final aim was a more precise description of the neuronal mechanisms of hippocampal LTP using more stable and more numerous recordings and at longer periods after tetanic stimulation than it was possible to achieve in our in vivo experiments (Chap. 12). As a first step, we evaluated the performance of statistical procedures based on the binomial model in both computer and physiological experiments with various N. Monte Carlo simulations were performed with variations of model parameters in a broad range of physiologically realistic situations. Both methods used in the previous chapters and modified variants of the methods were applied to minimal EPSPs of CA1 area [1054, 1056, 1061].

13.2 Methods

A programme was written to simulate, firstly normally distributed noise values with various S_n, and secondly, amplitude values binomially distributed and convolved with this normally distributed, additive, and independent noise according to Eq. (8.7–3). In most computer and physiological experiments, S_v was assumed to be equal to 0.05v (Sect. 8.8). We believed that a finite S_v value is more realistic than $S_v = 0$ as was assumed by other authors for spinal [464, 568, 1068] and hippocampal [864, 865] neurones. In our simulation experiments, v was generally fixed (v = 100), while the other parameters from Eq. (8.7–3) were varied to study their influence on the estimate of "basic" (m, v) and binomial (p, n) parameters by several methods listed below.

Electrophysiological experiments were done on hippocampal slices of guinea pigs by the methods analogous to those described in Sect. 11.5. Intracellular minimal EPSPs evoked by testing single or paired-pulse stimulation of the *stratum radiatum* and *stratum oriens* (40 or 50 ms interval within the pair) were stored on a computer system for off-line analysis, high frequencies were usually cut at 300 Hz. The filtering did not influence EPSP amplitudes but diminished noise (Fig. 13.3A, B). Data from two series of physiological experiments will be analyzed. In the first series, an attempt was made to record at least 500 responses under stable recording conditions (Fig. 13.3D). The aim was to compare parameters estimated by different methods at varying N. The recording was stopped if the neurone was deteriorating as judged by a deviation of RMP for >5 mV or by spontaneous spiking. Otherwise, either tetanic stimulation was applied after about 500 testing stimuli (about 1.5 h of recording period) or the recording was continued as long as possible. In the second series, after recording about 100–200 responses at stable conditions, tetanic stimulation was delivered and as many as possible post-tetanic responses were collected. In one cell, two conditioning tetani were given during the recording of the same neurone. The conditioning tetanic pattern typically consisted of ten trains at 100 Hz for 200 ms with a stimulus duration equal to three to five times of the testing stimulus duration. Intertrain intervals were equal to that between testing stimuli (8–10 s). The off-line analysis comprised response averaging to determine fixed time points (a time "window", Sect. 11.5), subsequent measurements of amplitudes from single responses, and calculation of quantal parameters. Background noise was measured before the stimulus artifact with the same window width. Data presented here are based on the measurements of the "mean window amplitude" (Sect. 11.5), which will be referred to as "amplitudes" for simplicity. "Quasi-stationary" plateau regions (Sects. 8.2, 8.6) were determined from plots of means of 16 to 64 amplitude measurements (Fig. 13.3D). Two procedures to estimate N_0 were compared. According to the first ("subjective") procedure, N_0 was determined for every plateau region by visually comparing single sweeps with the average over the region. Figure 13.3C exemplifies the averaged failures with no significant response corresponding in time to the first EPSP, i.e. EPSPI (Fig. 13.3A, B). Only a small late hyperpolarization is visible in Fig. 13.3C, well after the time point corresponding to the

EPSPI peak. A similar late hyperpolarization was observed in one more neurone recorded in this study. All doubtful cases (if any) were averaged and sweeps containing failures were separated by successive exclusion of responses during averaging. The determination of N_0 was usually repeated by two observers, and as a rule, it provided parameters (v and m) similar within <10%. In cases of a disagreement, the mean value of two determinations was taken as final N_0 estimate. According to the second ("objective") procedure (Sect. 7.3.4), N_0 was estimated as the double number of negative amplitude values. Four methods based on the binomial assumption were used to calculate quantal parameters. The first (histogram) method was made objective by using a simplified variant of the deconvolution technique (Sect. 7.3.7). EPSP amplitude and noise amplitude histograms were constructed typically with the number of bins set to 30 and the computer algorithm searched for a "best-fitted" binomial distribution. To this aim, the value of v_1 was systematically changed in a given range, which started typically from the value equal to the maximal amplitude down to 20 μV by steps of 1 or 2 μV. At every given v_1, the histogram was divided corresponding to different quantal contents (x) and m was calculated as $m_1 = E/v_1$. Parameters p_1 and n_1 were calculated from Eqs. (7.6.1–1) and (7.2–2), respectively. The expected proportion of responses for each bin was compared with the experimental distribution using the Chi-square test. From the range of the tested v_1 (and the corresponding m_1, n_1, and p_1), those values, which gave the maximal probability from the Chi-square table were taken as final estimates. The value of S_n was either directly calculated from measured noise amplitudes (Fig. 13.3F, $S_n = 85$ μV) or it was extracted from the noise distribution using a similar procedure based on the Chi-square test (Fig. 13.3F, $S_n = 87$ μV). The values of S_n that were found by these two procedures never differed for >5%, and estimates based on these two procedures were practically identical. The Chi-square test showed that in no case was the noise histogram significantly different from a Gaussian distribution (P>0.1). The algorithm enabled an optimization not only of v but also of both v and S_n. In this variant, S_n was not directly estimated from the noise measurements, but it was a free parameter together with v. The variant with S_n optimization was tested in a separate series (Sect. 13.4.6).

The method described in Section 8.5 was used in two variants: the variance methods (parameters will be indexed by 2) and the method of failures (parameters will be indexed by 0). Equations (7.3.5–3) and (7.3.4–2) were used to calculate m_2 and m_0, respectively. The combined variance-failures method described in Section 7.6.6 was tested for individual cases. The combined equation was solved by iteration. Parameters estimated by the combined method will be indexed by 3. We calculated p_3 and m_3 satisfying Eqs. (7.3.5–3) and (7.3.4–2). Parameter v_3 and n_3 were calculated by Eqs. (7.2–3) and (7.2–2), respectively. Because there are two methods to determine N_0, two variants ("subjective" and "objective") of the method of failures and of the combined method will be considered.

13.3 Results of Computer Simulations

13.3.1 Testing Different Methods

Figure 13.1 shows simulated binomial distributions with large N (N = 1000) and two different S_n. As expected, with a low noise (S_n = 0.25v), v can be easily estimated visually as the average distance between the peaks (Fig. 13.1A). The quantal size estimated by the computer programme on the basis of the histogram method (Fig. 13.1A, v) deviated less than 5% from both the visual estimate and the simulated value. More unexpected was a good agreement (within ±10%) of the estimated quantal parameters with the simulated values at larger noise when the histogram did not contain distinct regular peaks (Fig. 13.1B). To simulate more

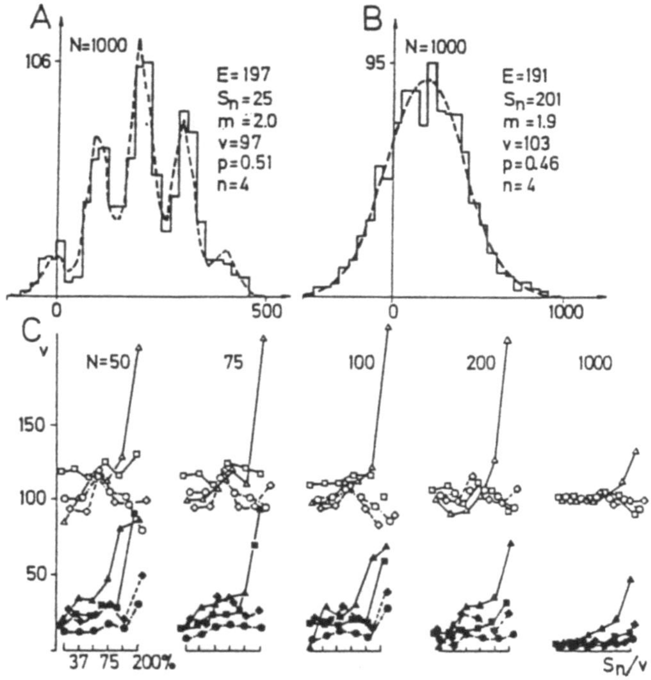

Fig. 13.1A–C. Comparison of various methods of quantal analysis in computer experiments. **A** Example of simulated (*columns*) and algorithm fitted (*interrupted line*) amplitude distributions. *Abscissa* amplitude (in arbitrary units); *ordinate* number of counts per bin. Parameters of the simulated distribution: v = 100, p = 0.5, n = 4, S_n = 25. Parameters found by the programme as the best fit for the stimulated distribution are shown. **B** Same as **A** but with simulated S_n = 200. **C** Mean quantal sizes (*ordinate, empty symbols*) and SD of these means (*ordinate, full symbols*) plotted for different S_n/v ratios (25, 37, 75, 100 and 200% indicated *under abscissae*) and various N (indicated *above each graph*). Parameters of the simulated distribution were v = 100, p = 0.5, n = 4. The experiment was repeated ten times for every combination of S_n/v ratio and N. *Triangles, squares, circles* and *diamonds* represent data calculated by the histogram, variance, failure and combined methods, respectively. [1056]

typical physiological situations, histograms were constructed with smaller N (N = 50–200) and various S_n/v ratios. Every simulation was repeated ten times, and parameters were calculated by four methods in every computer experiment. The results of such experiments are shown in Fig. 13.1C for a simulated distribution (with v = 100, n = 4, p = 0.5, and various S_n and N). Empty triangles in Fig. 13.1C represent mean v_1 values calculated from ten computer experiments. The deviation of the mean v_1 values from the simulated v = 100 is usually less than 10–20% even for the smallest samples (N = 50–100) provided S_núϖ. The estimated v_1 values were significantly different from 0 ($P<0.01$, Student's t-test, see full triangles which represent SD). It is interesting to note that in the range of N = 75–200 (and $S_n \leq v$), the deviation of v_1 from the simulated v value was not strongly dependent on N and S_n. A larger noise level ($S_n = 2v$) prevented a reasonable determination of v_1 at smaller sample sizes (N = 50–200). At $S_n = 2v$, the deviation increased up to 40% of v even with N = 1000. With $S_n \geq v$, multiple integers of the simulated values of v were often found by the algorithm as the best fit. This led to a significant increase of SD of v_1 (Fig. 13.1C, compare full triangles at different S_n).

Results similar to those demonstrated in Fig. 13.1 were obtained in other computer experiments with n equal to 2, 4, 8 or 15 and with p variations from 0.05 to 0.95. Sometimes at $S_n \geq v$, averages of v_1 determinations were smaller than the simulated v values. The estimates of the binomial parameters by the histogram method were less reliable than the estimates of m and v. Both systematic deviations of mean n_1 values from the simulated n and large SD were found at small N. The results were slightly better for p_1. It did not deviate systematically from the simulated p, but at $N \leq 100$, the SD of p_1 was usually large, so that estimated mean p was often not significantly different from 0. Similar for the case with m and v, the agreement between calculated and simulated binomial parameters did not dependent strongly on the S_n, provided $S_n<v$.

13.3.2 Testing Methods of p Determination

One of the major objectives of the computer experiments was to test Eq. (8.5–1). The equation is based on the assumption that E_{max} is known, which is true only for $S_n = 0$ and for sufficiently large p and N. Estimates of the basic quantal parameters both by the variance method and (to a lesser degree) by the methods of failures, are influenced by estimates of p_2. It is clear that with a small p (at a limited N and small S_n), E_{max} will be less than the theoretically predicted value, so that Eq. (8.5–1) will overestimate p. The overestimate was confirmed in simulations with various N (100 and 1000) and S_n (from 0.25v to 2v). An example is given in Fig. 13.2B (dotted bars). However, the influence of this overestimate on the calculated m_2 was small when p was a small fraction of 1 (Fig. 13.2C, D). Moreover, at small N, the difference between the theoretically predicted and the observed E_{max} was partially compensated by the addition of noise, so that estimated p_2 were not significantly different from simulated values (Fig. 13.2A, dotted bars). If p was close to 1, the addition of noise led to an underestimate of p. The un-

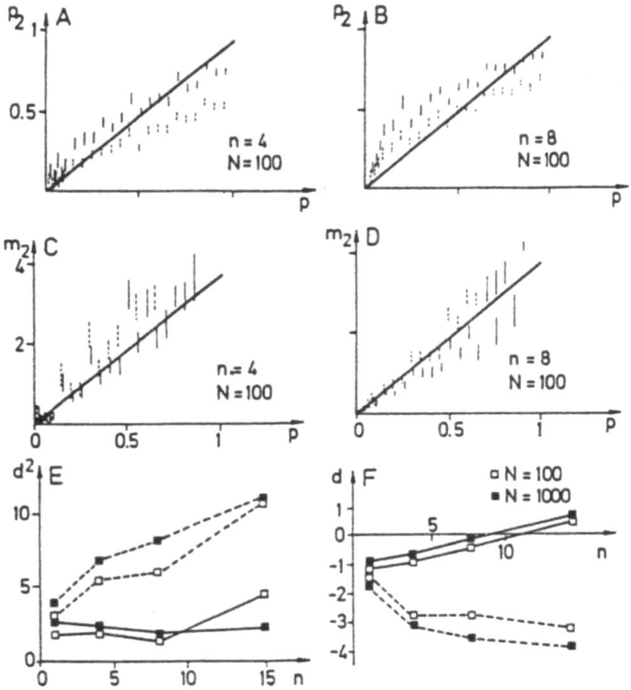

Fig. 13.2A–F. Comparison of parameters estimated by Eqs. (8.5–1) and (13.3–1). **A** Values of estimated p_2 (*ordinate*) are plotted against simulated p (*abscissa*). Values calculated from Eq. (8.5–1) and Eq. (13.3–1) are represented by *interrupted* and *continuous bars*, respectively. The *length* of each bar is equal to two SEM calculated from ten experiments. Simulated parameters: $v = S_n = 100$, $n = 4$. **B** As **A**, but for the simulated $n = 8$. **C, D** As **A** and **B**, but with m_2 plotted as *ordinate*. Some m_2 values are missing because of large deviations from the simulated values. **E** Variances of mean differences between m_2 and m (*ordinate*) for various n (*abscissa*). Variances were calculated from experiments similar to those illustrated in **C** and **D**. **F** Mean differences between m_2 and m (*ordinate*) for various n (*abscissa*). Note a reduction of both variance (**E**) and mean differences (**F**) for estimates based on Eq. (13.3–1) (*continuous lines*) as compared to those based on Eq. (8.5–1) (*interrupted lines*). [1056]

derestimate was especially prominent with small n (compare Fig. 13.2A and B, dotted bars), so that m_2 differed obviously from the simulated m (Fig. 13.2C, D, dotted bars). Some of the determinations exceeded the borders of the graph (Fig. 13.2C, D). Figure 13.2E, F demonstrates that Eq. (8.5–1) led to both random (Fig. 13.2E, interrupted lines) and systematic (Fig. 13.2F, interrupted lines) deviations of m_2 determinations in comparison with simulated m. Computer experiments with variations of p (from 0.05 to 0.95) and n (from 1 to 15) are summarized in Fig. 13.2E, F.

To compensate for these deviations, a new half-empirical formula for p evaluation was chosen using a trial-and-error method in simulation experiments with variable parameters. The best results for n = 4 and v = 100 were provided by the equation

$$p_2 = E/\{M - 0.3S_n * \ln[2N * E/(M - S_n)]\}, \qquad (13.3\text{–}1)$$

where M is the mean of the three largest amplitudes. Equation (13.3–1) was tested in a broad range of n (from 1 to 15), p (0.05–0.95), v (50–500) and S_n/v (0.25–2). It significantly improved the estimate of p and especially of m (Fig. 13.2A–D, continuous bars). The improvement is evident from the comparison of mean variances (Fig. 13.2E) and mean differences (Fig. 13.2F) between simulated m and m_2 calculated on the basis of Eqs. (8.5–1) and (13.3–1) (interrupted and continuous lines, respectively). The mean quantal parameters estimated by both the variance and failures methods based on Eq. (13.3–1) were close to the simulated values even at small N and large S_n (Fig. 13.1C, squares and circles). The method of failures (Fig. 13.1C, open circles) gave better average estimates than the variance method (Fig. 13.1C, squares) with smaller SD of the mean estimates (Fig. 13.1C, full circles). Values of v_3 estimated by the combined method were also in reasonable agreement with the simulated v (Fig. 13.1C, rhombus). However, the combined method gave only coarse estimates of p, and parameter n was estimated even less precisely. It should be noted that in the simulations shown, theoretically predicted N_0 were used.

13.3.3 Testing "Objective" Method of Failures Determination

It is evident that the "objective" N_0 determination is influenced by the noise level. We tested the estimates derived by this method with simulations of distributions with parameters n equal to 2, 4, or 8, p ranging from 0.1 to 0.9, and N = 1000 and various S_n/v (0.25v–1.5v). Eight distributions with the mean m = 1.58 ± 0.27 (±SD), ranging from 0.6 to 3.2, were simulated with $S_n/v = 0.25$. The determined m_0 ranged from 0.70 to 2.91 with the mean value and SD (1.58 ± 0.31) which were practically identical to the simulated values. With $S_n/v = 0.5$ underestimates were less than 15%. The underestimate increased with larger S_n/v so that at S_n = 1.5v, the calculated m_0 was only about half of the simulated m. A good correlation between the estimated m_0 and simulated m was evident for all S_n/v tested (r ranged between 0.97 and 0.99, P<0.0001 for all S_n). Therefore, the method gave correct N_0 only at small S_n when separated peaks were visible in the simulated histogram (Fig. 13.1A). With S_n>0.5v, the method essentially overestimated N_0, evidently due to the "tails" from peaks with x>0. The overestimate of N_0 led to an underestimate of m, and consequently, to an overestimate of v.

13.4 Results of Physiological Experiments

13.4.1 General Description

The results of two series of experiments will be presented. The aim of the first series was to compare quantal parameters estimated by different methods at various N. In this series, responses to paired-pulse stimulation (Fig. 13.3A, B) were either recorded as long as possible (Fig. 13.3D) or the tetanic stimulation was applied after about 500 stimuli. In the second series, the tetanic stimulation was delivered after 100–200 testing stimuli under stable conditions. All responses were divided into groups depending on whether they were evoked by the first or the second stimulus (EPSPI and EPSPII, respectively) and whether they were recorded before or after tetanic stimulation (pre- and post-tetanic EPSP, respectively). In the present chapter, we compare estimates obtained by different methods at different N and consider the parameters of the non-facilitated EPSP (i.e. pretetanic EPSPI). For some comparisons, data from various groups irrespective of the recording conditions will be combined (i.e. pre- and post-tetanic EPSPI and EPSPII).

13.4.2 Testing EPSP Amplitude Stability

During a period of from 50 min to more than 2 h, 300 to about 1800 responses were recorded. Altogether, 21 sets of EPSPs from 8 neurones (13 inputs) were analyzed. Figure 13.3A–C illustrates minimal EPSPs recorded in the experiment with the largest sample (N = 1756). EPSP fluctuated (Fig. 13.3A, B) with occasional failures (Fig. 13.3C) so that it was possible to apply all described methods of the quantal analysis. Suggesting that the amplitude fluctuations (Fig. 13.3A–C) are mainly due to a variable number of transmitter quanta released (see Sect. 13.5), I will use the terminology of the quantal analysis (Chap. 7). All responses were divided conventionally into three groups according to their stationarity. Group 1 consisted of EPSPs without significant trends of averaged amplitudes during the major part of the recording period (6 out of 21 EPSPs). More than 900 responses were collected for 3 out of the 6 EPSPs, the rest of them were recorded from the same neurone (2 different inputs) with N=340. In the second group (9 out of 21 EPSPs), amplitudes declined to the end of the experiment. The decline typically occurred after 100 to 300 stimuli and in two cases after about 600 stimuli. The last six EPSPs composed a third group with more complicated amplitude variations during the experiments. Typically, the amplitude increased after 200 to 600 stimuli. Figure 13.3D demonstrates a case with an initial increase in the average EPSP amplitude followed by a stable plateau region with about 500 responses (Fig. 13.3D, horizontal bar) and subsequent fluctuations of averaged amplitude with a final tendency to the amplitude diminution (Fig. 13.3D). For the statistical treatment (Fig. 13.3E–H), only responses from the plateau region were taken. After recording of about 500 responses, the tetanization was delivered in four neu-

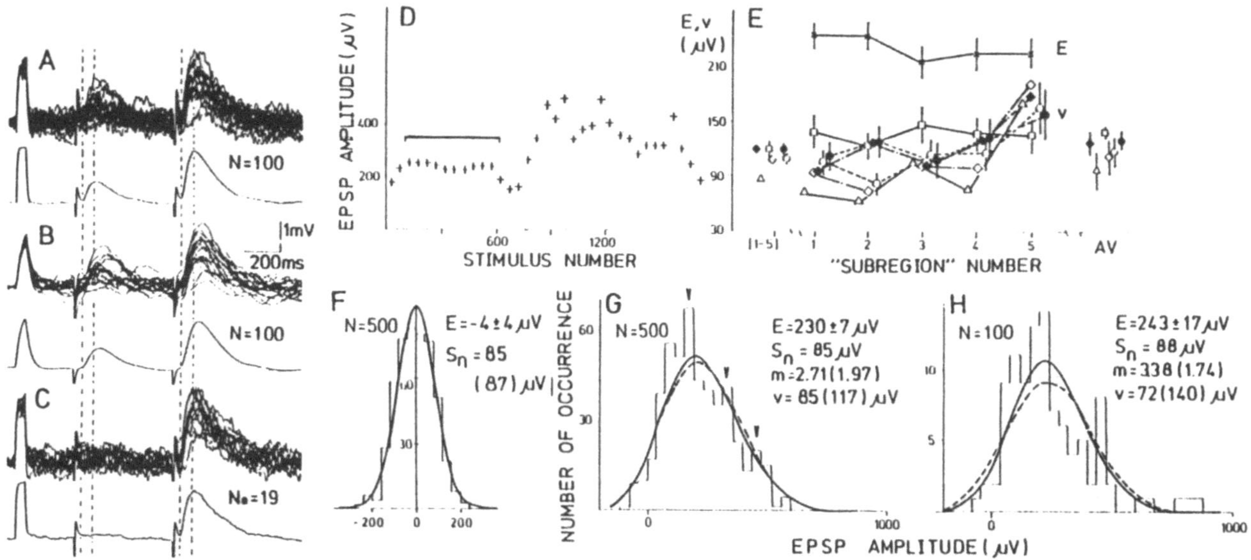

Fig. 13.3A–H. Statistical analysis of minimal EPSPs recorded from a hippocampal slice. **A, B** Single (*upper rows*) and averaged (*lower rows*) non-filtered (**A**) and high-frequency (300 Hz) filtered (**B**) responses to double-pulse stimulation. *N* Number of averaged responses. **C** Examples of sweeps taken as failures in response to the first stimulus. Note calibration pulses in the beginning of the sweeps **A–C**. **D** Changes of mean EPSP amplitudes in the course of the experiment. *Ordinate* mean amplitude ±SEM from 50 consecutive sweeps. *Abscissa* stimulus number. *Horizontal bar* indicates plateau region used for calculations (**E–H**). **E** Mean amplitudes (*E*) and quantal sizes (*v*) calculated by different methods (*vertical axes*) plotted for the whole plateau region with 500 measurements (*1–5 in brackets under horizontal axes*) and for consecutive smaller "subregions" with 100 measurements (*1 to 5 under horizontal axes*). Averages of estimates from small samples are given as right-hand symbols (*AV at horizontal axes*). *Triangles, squares, full circles, empty circles, full diamonds*, and *empty diamonds* represent values estimated by the histogram (binomial deconvolution), variance, failures subjective, failures objective, combined subjective, and combined objective methods, respectively. *Bars* denote ±SEM for amplitudes and averages (*AV under horizontal axes*) or estimates of standard errors of v in other cases. **F–H** Experimental (*columns*) and theoretical (*curves*) distributions of noise (**F**) and EPSP (**G, H**) amplitudes for regions with N = 500 (**F, G**) and with N = 100 (**H**). *Continuous* and *interrupted curves* (**G, H**) were constructed on the basis of the parameters found by the histogram and variance methods, respectively. Parameters for the two methods are given in the *insets* (values in parentheses for the variance method). [1061]

rones. In two neurones, a significant PTD was evident. In the other two neurones, no significant long-lasting changes were induced.

13.4.3 Comparison of Parameters Calculated from Different Samples

Figure 13.3E exemplifies a good correspondence between determinations of v from the whole plateau region (Fig. 13.3E, left) and averaged determinations from the corresponding five "subregions" (Fig. 13.3E, AV). Similar comparisons of the mean calculations from small samples (N = 50–100) and the corresponding calculations from large samples (N = 300–900) were performed for other cases (25 EPSPs for the histogram and variance methods, 20 EPSPs for the failures and combined methods). The similarity of estimated values from small and large samples as well as good correlations between the estimates were evident. The coefficient of correlation was 0.90 for the histogram, 0.98 for the combined method and >0.99 for the other two methods (P<0.0001 for all methods). The largest difference (about 20%) between the estimates from small (v_1 = 163±18 μV, ±SD, here and below) and large (133±102 μV) samples was found for the histogram method. The estimates for the variance method were larger (189±132 and 175±127, respectively). For all methods except the histogram method, the difference between determinations from small and large samples was <8% and it was the smallest (<2%) and not significant for the combined method (P>0.12, Wilcoxon test for matched pairs here and below).

13.4.4 Objective vs. Subjective Method of Failures Determination

Simulation experiments (Fig. 13.3) showed that the objective method of failures determination led to an essential underestimate of m (and therefore an overestimate of v) at S_n>0.5v. The subjective method is advantageous because it is independent of the method of amplitude measurements. Therefore, within the context of the binomial model, this method is completely independent from other methods. Our impression was that the procedure of subjective determinations of N_0 could be applied to all our recordings independent of the noise level. For the comparison of subjective and objective N_0 estimates (and for the comparison of different methods in the following section), 61 plateau regions with recordings of EPSPI and EPSPII of 9 neurones (11 tested inputs) before and after tetanic stimulation were pooled. A single region comprised 59 to 119 measurements (90 ± 17). The probability of failures varied from 0 to 0.41 or to 0.54 as estimated by the subjective or objective methods, respectively. The mean value for the regions with at least 1 failure was 0.11 ± 0.10 (N = 45) for the subjective method and 0.13 ± 0.13 (N = 36) for the objective method. For 32 out of 61 regions, both methods determined N_0>1. There was an agreement between the two methods of N_0 determination within ±1 for 10 regions, and the average N_0 per one region was exactly the same (13.5 ± 9.6 for subjective and 13.5 ± 11.6 for objective method).

The mean estimates of m for the subjective ($m_0 = 1.65 \pm 0.56$, $m_3 = 1.74 \pm 0.69$) and objective ($m_0 = 1.69 \pm 0.64$, $m_3 = 1.84 \pm 0.78$) methods were not significantly different and the estimates were highly correlated (r = 0.75 and 0.61 for the failure and combined methods, respectively. The correlation was even larger for v estimates (r = 0.89 and 0.80, P<0.0001), and the respective mean values were practically identical: $v_0 = 130 \pm 59$, $v_3 = 129 \pm 62$ μV for the subjective and $v_0 = 131 \pm 64$, $v_3 = 128 \pm 68$ μV for the objective method.

13.4.5 Comparison of Different Methods

Figure 13.3E illustrates estimates of v by different methods for the same neurone. Amplitude histograms and the best fit theoretical distribution for the plateau region with N = 500 are presented in Fig. 13.3G. The inset shows parameters calculated by the histogram and variance methods (numbers without parenthesis and in parenthesis, respectively). Theoretical distributions constructed on the basis of these methods (continuous and interrupted lines, respectively) showed only a slight difference for this neurone, while m_1 was about 27% smaller than m_2. Figure 13.3E demonstrates that the most consistent determinations of v from small samples (N = 100) were obtained by the variance method: v_2 (Fig. 13.3E, squares) estimated for different subregions were not significantly different (P>0.1), while v_1 were more variable (Fig. 13.3E, triangles). In this experiment, the S_n/v ratio was comparatively large (about 75% as judged by v_2). Smaller v_1 (and larger m_1 values), if compared with other methods, were common for most "subregions" of this neurone as well as for other neurones (Table 13.1) as mentioned above (Sect. 13.4.3).

A systematic comparison of different methods (Fig. 13.4) was made on the basis of estimates for 61 plateau regions from the second series of physiological experiments. Despite some differences, the mean estimates were of the same order (Fig. 13.4). The largest difference was found between determinations by the histogram and other methods (Fig. 13.4A, D). Estimates from different methods were highly correlated, the lowest r were found for combinations with m_1 (Fig. 13.4A, D), the highest for the comparisons of failures and combined methods (Fig. 13.4E, F). The correlation of the binomial parameters calculated by different methods was generally lower and for some combinations not significant: e.g. for p_1 and p_2, r = 0.70, p_2 and p_3, r = 0.24.

13.4.6 Optimization of Noise Standard Deviation

In about half of the observed histograms, regular peaks and troughs can be distinguished by eye (Fig. 13.5A, B) including histograms (Fig. 13.5C and D) constructed by the "moving bin" method [813]. It was a common observation that even for these favourable cases, the best fit theoretical distributions were usually smooth and contained no regular peaks (Fig. 13.5A; see Sect. 8.7 for similar in

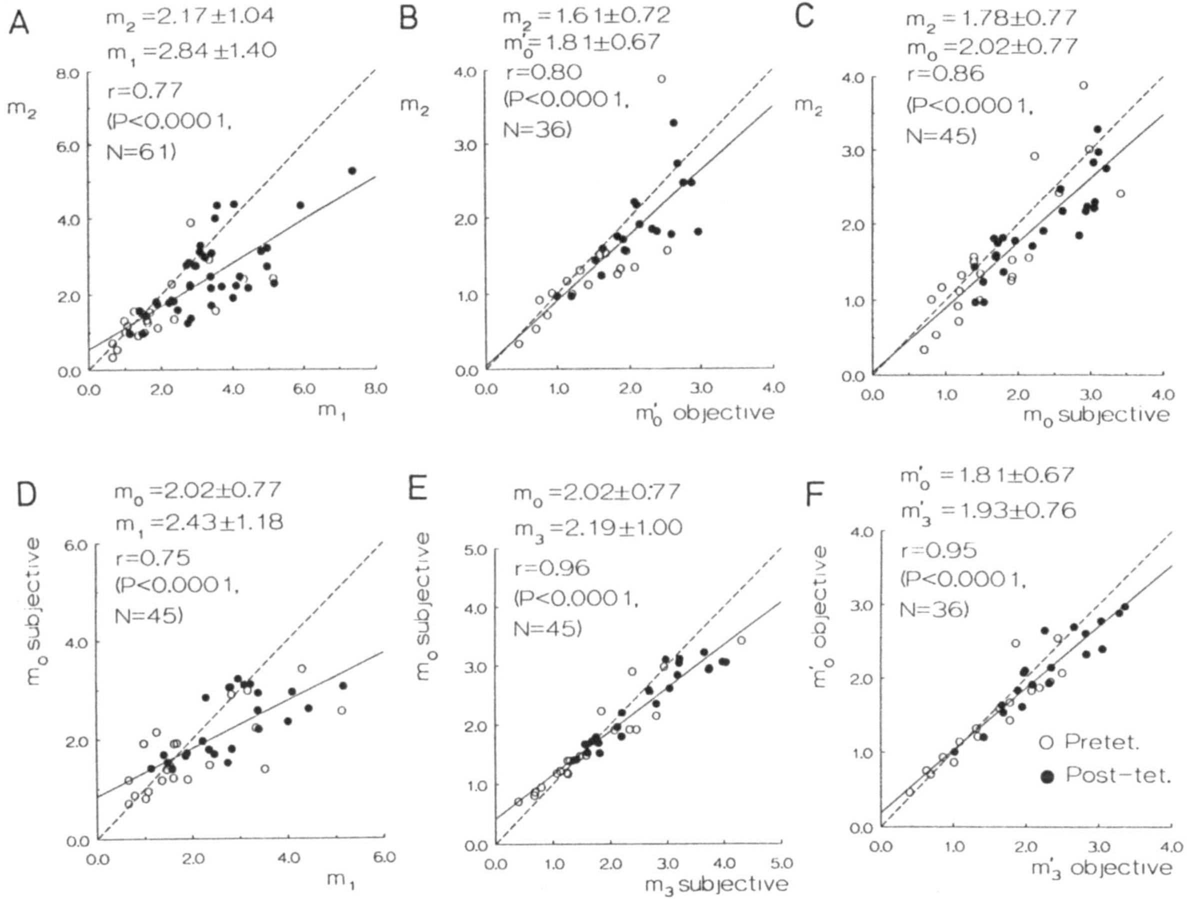

Fig. 13.4A–F. Correlation of mean quantal contents estimated by different methods. **A–C** Correlations for m_2 (*ordinates*) with m_1 (**A**), m_0 calculated on the basis of objective (**B**) and subjective (**C**) N_0 determinations (abscissae). **D–F** Correlation of m_0 based on subjective (**D**) and objective N_0 determinations (*ordinates*) with m_1 (**D**), m_3 based on subjective (**E**) and on objective (**F**) N_0 determinations. [1061]

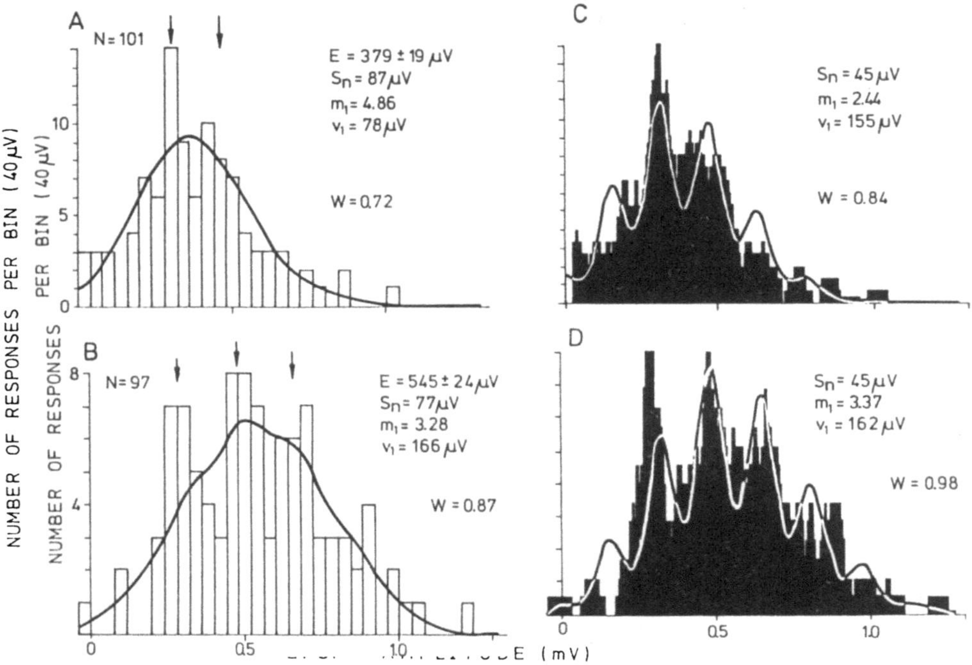

Fig. 13.5A–D. Optimization of noise standard deviation. **A–D** Examples of experimental (*columns*) and theoretical (*curves*) distributions recorded from the same neurone before (**A**, **C**) and after (**B**, **D**) tetanic stimulation. Theoretical distributions in **A** and **B** were constructed by using S_n measured directly from experimental data while those in **C**, **D** were constructed with S_n found by the optimization procedure. Experimental distributions in **C**, **D** were built by the "moving bin" method [813] with a larger bin equal to 40 μV and a smaller bin equal to 4 μV. Probability values for the Chi-square tests (*W*) are given in insets in addition to the mean amplitude (*E*), noise SD (S_n), and quantal parameter values (m_1, v_1). [1061]

vivo observations). This discrepancy between observed and predicted histograms might be due to small samples of either noise or response measurements or both. Alternatively, they might indicate a significant reduction of S_n during EPSP. We collected larger samples to estimate S_n. Not infrequently, S_n determined from large samples (N = 500–1000) were larger than those from small samples (N = 100). S_n was never reduced to a sufficiently low value to account for the appearance of distinct visible peaks in the histograms.

To test the second possibility, we applied a variant of the histogram method which used S_n as a free parameter (Sect. 13.2) instead of S_n from real noise measurements. Figure 13.5 illustrates applications of the two variants of the histogram method to two distributions recorded from the same neurone before and after tetanic stimulation. Theoretical distributions constructed with the measured S_n (Fig. 13.5A, B) and with best fit ("optimized") S_n (Fig. 13.5C, D) are shown. Figure

Table 13.1. Quantal parameters of non-facilitated minimal EPSPs of CA1 area of hippocampal slices

Method	m	v(μV)	p	n
Histogram	1.54±0.84	114±91	0.43±0.15	3.2±1.7(6)
	0.45–3.52	31–390	0.27–0.70	2–6
Variance	1.13±0.48	147±100	0.38±0.08	3.2±1.0(14)
	0.34–2.11	52–462	0.28–0.62	2–6
Failures	1.21±0.46	130±86	0.38±0.08	3.6±1.0(14)
	0.48–2.08	50–371	0.28–0.62	2–6
Combined	1.19±0.55	138±91	0.47±0.12	2.8±1.8(9)
	0.40–2.09	60–396	0.30–0.63	1–6

Notes. Means ±SD and ranges for pretetanic EPSPs evoked by the first stimulus in paired pulse paradigm are given for different methods. The number of neurones was equal to 16. The number of estimates for p and n is given in parenthesis. Mean EPSP amplitude was 136±53 (41–235) μV.

13.5 shows that for cases with distinct peaks, the optimized S_n values (Fig. 13.5C, D, inset, S_n = 45 μV) might be almost two times less than the actually measured value (Fig. 13.5A, B, S_n = 78 and 77 μV). The average optimized S_n (60 μV) was about 13% less than actually measured S_n (69 μV). The difference was significant at P = 0.003. There was a correlation between measured and optimized S_n (r = 0.73, P<0.0001). Comparison of Fig. 13.5A and C reveals an essential difference in estimates of basic quantal parameters for the two variants of the search algorithm: v_1 estimated by the first variant (Fig. 13.5A) was about half of that estimated by the second variant (Fig. 13.5C). The latter value seems to be more realistic as judged from visual estimates of the mean distances between major peaks in Fig. 13.5A and results of other methods (see also Fig. 13.5B, D where v_1 was about the same as for Fig. 13.5C). However, for Fig. 13.5C and D, the difference between estimates from the two variants was negligible. Only a very small difference was found for the pooled data from 63 regions: $v_1 = 129 \pm 66$ and $v_1 = 140 \pm 74$ μV for the measured and optimized S_n, respectively (P>0.05, Wilcoxon test for matched pairs). A significant correlation was found between parameters estimated by the two variants (e.g. r = 0.84 for v_1).

13.4.7 Mean Quantal Parameters of Hippocampal Synapses

Table 13.1 summarizes quantal parameters for the non-facilitated EPSPs of 16 neurones. "Non-facilitated" means that only EPSPs evoked by the first testing stimulus in the paired-pulse paradigm during control (pretetanic) periods were taken from the both series of physiological experiments. Only EPSPs evoked by the stimulation of the *stratum radiatum* were included but we did not find any significant difference with parameters determined for several EPSPs evoked by stimulation of the *stratum oriens*. Table 13.1 shows again the similarity of values calculated by different methods. The exception was the histogram method, which gave significantly larger m and lower v than the other three methods. Based on our simulation experiments, we might expect that estimates of quantal parameters by the histogram method with $S_n \geq v_1$ would be unreliable. We selected a group of eight neurones with $S_n < 0.9v_1$. For this group, the histogram method gave parameters more similar to those determined by the other methods. For example, the mean value of v_1 was equal to 152 μV (ranged from 58 to 390 μV) with v_2 = 174 μV (52–462 μV), v_0 = 142 μV (50–371 μV), and v_3 = 155 μV (60–396 μV) for the same group. This is consistent with the suggestion that the histogram method underestimated v for our sample due to a larger sensitivity to S_n.

In many cases, especially for non-facilitated EPSPs, $p < 0.25$. In these cases, $n > 10$ and often $n > 25$. Predicted distributions calculated on the basis of the binomial statistics [Eq. (8.7–3)] were not distinguishable from the distributions calculated on the basis of the Poisson statistics. Similarly to our in vivo observations, the estimates of SE for the binomial parameters (Sect. 8.2) showed that for cases with $p < 0.25$, the binomial parameters were not significantly different from 0. Therefore, the related distributions were considered to be based on Poisson statistics with uncertain p and n. Similarly to the previous sections, corresponding binomial parameters were excluded from the results (numbers in brackets in Table 13.1). In general, both simulation and physiological experiments showed that the estimates of binomial parameters were less reliable than the estimates of v and m.

13.5 Discussion

13.5.1 Simulation Experiments

Monte Carlo simulations allowed us to refine and test methods used in the previous chapters and add two variants of an additional (combined) method. Assuming the binomial model (see below), the described methods can be applied under realistic conditions of physiological experiments even with comparatively small samples (N=100). It was found that low S_n were more important than large N for reliable estimates of quantal parameters. With comparatively large samples (N = 500–1000), satisfactory estimates of basic quantal parameters (within <10–20% deviations from the simulated m and v values) were obtained up to S_n comparable to v (all four methods) and even exceeding the latter (variance, failures and combined methods). Our results shows that when large N is not available, it is possible to obtain similarly satisfactory estimates by repeating experiments with N = 50 or 100 for about ten times. According to both simulation and physiological experiments, the histogram method appeared to be less reliable and more sensitive to S_n and N than the other methods. The variance method is more universal than the failures and combined methods, because it is independent from the presence of failures and does not include subjective procedures like N_0 determinations. The "objective" method of N_0 estimate [763] was shown to be valid only for comparatively small S_n/v ratios. Both simulation and physiological experiments demonstrated that the histogram method, besides giving more variable estimates, tended to give systemic deviations from the simulated v at S_n>v, when estimates of other methods were still reliable. Nevertheless, the histogram method was often found to give satisfactory estimates even for histograms without distinct regular peaks (S_n>0.5v). Similarly to other methods, the variability of the mean estimates of v_1 and m_1 did not strongly depend on N. The general conclusion of our simulation experiments is that relatively small N = 50–100 can be used for the estimates of m and v in the context of the binomial model. The variance method is preferable. A comparison of results obtained by several methods is useful to evaluate the variability of the estimates of the quantal parameters and S_n/v ratios and, therefore, to evaluate the reliability of the estimates.

13.5.2 EPSP Measurements and Noise Reduction

Usually, peak amplitudes measured either at a single point [528] or as the average over a brief period around the peak [464, 823] were used for statistical analysis. On the basis of our previous studies ([425], see Sect. 11.5), we preferred the "mean window amplitude" rather than amplitude measurements, and chose the right margin of the window before the EPSP maximum to exclude a possible influence of polysynaptic PSPs and early IPSPs. In fact, averages from failures never showed any significant early hyperpolarization and only occasionally revealed a weak late hyperpolarization which was much smaller and had longer latencies than that found in our in vivo experiments (Sect. 12.5). Signs of IPSPs were sometimes evident in facilitated EPSPs (due to PPF or/and LTP). The late parts of EPSPs are sometimes clearly shortened in these cases , so that averaging over a period corresponding to the peak of the control (non-facilitated) EPSP would produce a distortion of the magnitude of the facilitated EPSP unlike the measurements of the "mean window amplitude" corresponding to the initial EPSP slope. Our measurement procedures and electrical filtering were beneficial for an essential noise reduction. They give a partial explanation for the more common appearance of histograms with regular peaks in our recordings in comparison with publications of other authors [864, 865]. Another reason for the difference might be drifts in quantal parameters during prolonged testing stimulations used by the cited authors to collect large samples. They also used an "intermediate" frequency of stimulation (1 or 2 Hz) which can produce either FF or LFD according to our observations. Therefore, except for stochastic variations, changes in v related to FF and LFD might be suspected during the testing (Sects. 11.2, 11.4, 11.5). A decline in the EPSP amplitude was often observed in the present experiments after prolonged testing stimulation and was demonstrated with FP recordings even with interstimulus intervals in the order of 10 s in both in vivo and in vitro preparations (Chap. 1). Preliminary observations made here indicated not only a significant diminution in m during the decline in EPSP amplitude but also a decrease in v. The decrease in v will strongly influence the regularity of peaks in amplitude histograms.

13.5.3 Testing Procedures and Sample Sizes

We only infrequently obtained large samples (N>500) under steady-state conditions. We used comparatively long interstimulus intervals of about 10 s rather than 0.5 or 1 s used by some other authors [351, 864, 865] having in mind the prominent plasticity of hippocampal responses. Longer interstimulus intervals naturally led to a prolongation of time necessary to collect large samples. Even with these rather long interstimulus intervals, a decline of EPSP amplitudes was observed several times. In some neurones, spontaneous drifts might take place in different directions so that large samples in steady-state conditions were difficult to collect. Moreover, we had the impression that prolonged low-frequency stimulation or prolonged intracellular recording interfered with LTP induction: we were not able to induce LTP in 4 neurones tested with >500 stimuli (recording time of >1 h). The observation is in agreement with recent data for the DG [560] and with observations of Friedlander et al. [351], who found that tetanization did not lead to LTP in the majority of unitary and minimal EPSPs in CA1. These authors used prolonged stimulation prior to tetanization. According to our experience, tetanic stimulation after comparatively short testing (100–200 stimuli with about 10-s interstimulus intervals) led to LTP of minimal EPSPs in about 75% of neurones recorded in vitro ([1058], Chaps. 14, 15).

Both simulation and physiological experiments indicated the possibility of substituting estimates of basic quantal parameters from large samples by averages of several estimates from smaller samples (N = 50–100). Therefore, this approach might be used for the evaluation of the quantal parameters and their changes across experiments with successive recordings from several neurones.

13.5.4 Applicability of the Quantum Hypothesis and Simple Binomial Model

I already discussed the applicability of the basic postulates of the quantum hypothesis to the analysis of minimal EPSPs (Sect. 12.7). In accordance with our in vivo data and with in vitro data of other authors [343, 864, 865, 983], we found that amplitudes of hippocampal minimal EPSPs fluctuated more than could be predicted from the noise variations. Moreover, regular peaks and troughs were often distinguishable in the amplitude histograms. This confirms that hippocampal EPSPs may obey the basic postulates of the quantal hypothesis (Chap. 7). The interpretation based on the quantum hypothesis is supported by reasonable values of parameters estimated in the present chapter which are comparable with parameters known for other CNS synapses (Sect. 7.8). In none of the neurones analyzed in the present chapter could a theoretical fit based on the binomial law be rejected by the Chi-square test. Therefore, the present variant of the histogram method (deconvolution) might be considered as a test for the applicability of the simple binomial model. Estimates based on other methods usually also cannot be rejected (Fig. 13.3G, H, interrupted curves). However, the good fit between the observed data and the predicted binomial (or Poisson) distributions (especially with small

N) does not prove that the given solution is unique or the best one. In fact, more complicated models (Sect. 7.7) could be more adequate. Nevertheless, the agreement with the binomial model, as well as the internal consistency of different methods, justify the description of the experimental data on the basis of the binomial statistics as a first approximation. It has to be mentioned that, in principle, different methods could be similarly biased and therefore, produce a significant correlation; but known estimates of quantal parameters of the hippocampal synapses based of the unconstrained discrete model [864, 865] were not substantially different from predictions of the binomial model. This is considered as another argument for the application of the latter.

The major aim of our experiments was not the estimate of the binomial parameters but rather m and v in an attempt to provide a basis for distinguishing between pre- and postsynaptic mechanisms of synaptic plasticity. It should be stressed again that in cases with distinct regular peaks, estimates of v (and consequently m) by the histogram method are not dependent on whether the distribution is a simple binomial or not. In all cases with distinct peaks, different methods gave very similar v. This is considered as a further argument in favour of the application of the simple binomial model as a first approximation for the determination of v and m. Methods based on the binomial model can estimate seemingly good-fitting (but meaningless) n and p from simulated data which could not be described by the binomial statistics (Sect. 7.7). Moreover, our computer experiments showed that estimates of binomial parameters even from simulated binomial distributions are less reliable than estimates of m and v. The correlations for binomial parameters estimated by different methods were usually lower than for the basic parameters and might even be not significant. Therefore, the given values of binomial parameters (Table 13.1) have to be considered as rude evaluations.

13.5.5 Noise Reduction During EPSP

All common methods postulate an independence of noise from PSP generation (Chap. 7). However, in many cases, the peaks in the observed histograms were more clearly visible than one can expect for the given S_n/v (see also Sect. 8.7). This might be interpreted as a diminution of the noise level during the generation of EPSPs. The procedure of S_n optimization confirmed this suggestion and indicated that the average diminution of S_n could be of the order of 15%. Similar diminution of S_n have been previously reported for unitary IPSPs recorded from the sensorimotor cortex in vivo [1056] and recently found for hippocampal EPSPs [588] (see also [920]). The diminution of S_n during evoked EPSP was explained, suggesting that spontaneous release from the testing input generated an essential part of the general intracellular synaptic noise [588]. In fact, increased spontaneous release was described immediately after evoked EPSPs [91] especially in the presence of Ba^{2+} [222]. If the explanation is valid for our case, it signifies that a large and sustained increase in spontaneous release has to take place even during low-frequency (0.1 Hz) presynaptic activation. The activation of the spontaneous release can be measured with a computer programme searching for spontaneous EPSPs of adequate wave forms.

13.5.6 Quantal Parameters of Hippocampal Synapses

The present experiments provided some data related to the quantitative description of synapses formed by *stratum radiatum* fibres on CA1 pyramidal cells. The parameters presented in Table 13.1 are very similar to the parameters of non-facilitated EPSPs shown in Table 11.3 calculated by slightly different methods. Considering the present data, it is necessary to have in mind that we measured the "mean window amplitudes" with the right window border before the EPSP peak. This means that v calculated from EPSP peak amplitudes would be two to three times larger than v in Figs. 13.3, 13.5 and Table 13.1. In Section 11.5, the amplitudes were measured 1–3 ms before the EPSP maximum, and comparisons were given for measurements of amplitude and "mean window amplitude". On the basis of these comparisons, the mean v for non-facilitated *stratum radiatum* synapses might be estimated as about 350 μV with a broad range from about 100 μV to about 750 μV for different neurones. These values are consistent with the data from our measurements of peak amplitudes [566, 1057] and with our estimates from in vivo recordings at CA3 (Table 12.3). These values are of the same order, but larger and with larger ranges, than v obtained by other authors for other excitatory CNS synapses in vertebrates (Sect. 7.8) including hippocampal synapses in vitro (Table 7.2; see also [379a]) but are on average smaller than v reported for presumed unitary EPSPs in DG–CA3 synapses [430, 1126]. Under our experimental conditions, m should depend on the number of stimulated fibres (Sect. 12.7). For the cases with the lowest amplitudes in the non-facilitated state, which correspond presumably to the activation of a single efficient fibre [1053], m was estimated to be in the range of 0.5 to 1. These values are also consistent with the data of our in vivo experiments (Chap. 12). These values are also close to the estimates for the DG [700] but they are less than those found for both recurrent collateral EPSPs [717] and mossy fibre EPSPs [430, 1126] of CA3 neurones. However, the assumed Poisson statistics might overestimate m in these publications. On the other hand, the overestimates should not be very large for the non-facilitated EPSPs because, according to our data, p is not significantly different from 0 for about half of the cases judged by the histogram and combined methods (see also Chap. 12). It was noted that p_2 from Eq. (13.3–1) gave overestimates at $p<0.5$ which did not strongly influence estimates of m (Fig. 13.2C, D). About half of the nonfacilitated EPSPs typically had p in the range of 0.4–0.6 and n equal to 1 or 2 for the cases with the smallest m and presumed activation of one effective fibre. It is tempting to conclude that a typical synapse in the *stratum radiatum* has one or two release sites with variable probability of release from very low to about 0.6. Although this conclusion seems to be reasonable in view of known electron microscopy data [34], it should be considered as preliminary having in mind the above-mentioned reservations concerning the determination of the binomial parameters.

13.6 Summary

Binomial distributions of amplitudes of excitatory postsynaptic potentials (EPSPs) mixed with Gaussian noise were simulated. The objective of Monte Carlo simulations was, firstly, to study influences of sampling size (N) and noise standard deviation (S_n) on estimates of the mean quantal content (m), quantal size (v) and binomial parameters (n and p) by four methods of quantal analysis (histogram, variance, failures and combined methods) based on the binomial model and, secondly, to modify these methods on the basis of comparisons of estimated and simulated parameters. Reliable estimates (within ±10% of the simulated values) were obtained for large N = 500–1000 with $S_n \leq v$ by the histogram (deconvolution) method and with $S_n \leq 2v$ by the other three methods. Similar results were obtained as averages from about ten computer experiments if smaller N = 50–200 were used. In electrophysiological experiments on slices, "minimal" EPSPs were recorded from CA1 pyramidal cells after low-intensity stimulation of the *stratum radiatum* or *stratum oriens*. Amplitudes of minimal EPSPs were shown to fluctuate in a manner predicted by the quantum hypothesis. Amplitude distributions of EPSPs in the nonfacilitated state could be adequately described either on the basis of the binomial statistics with an average p equal to about 0.4 (a range of 0.3–0.7) and an average n of about 3 (2–6) or by Poisson statistics with m of about 1. The quantal analysis suggests that typical values of m and v for a single Schaffer collateral might be about 0.5–1 and 300–400 μV, respectively, with low p (0.1–0.3) and n (2–4). However, the estimates of binomial parameters should be considered as coarse approximations in view of the simulation results and a possible non-uniformity of p. The comparison of results of various methods based on the binomial model in both simulations and physiological experiments indicates the reliability of estimates of m and v under realistic conditions of physiological experiments. The methods are considered to be sufficiently sensitive to make use of them for studies of mechanisms of long-term synaptic plasticity in hippocampal slices.

14 Binomial Analysis of Long-Term Potentiation of Minimal EPSPs in Hippocampal Slices

14.1 Introduction

Our studies of hippocampal LTP (Chap. 12) revealed an increase in m without changes in v, the result traditionally considered to be a strong indication of presynaptic location of underlying mechanisms (Sect. 6.1). However, the number of neurones recorded was small and only a relatively short post-tetanic period (up to 12 min) was studied because of difficulties related to recordings from the in vivo preparation. More recent quantal analysis of long-term postactivation phenomena at peripheral junctions (see Sects. 7.10 and 12.7) also showed mainly an increase in m, although an increase in v was noted in some cases. The in vitro slice preparation provides favourable conditions for long-lasting intracellular recordings from neurones of the mammalian brain. Of the several authors who attempted a quantal analysis in hippocampal slices (see Sect. 7.8.5), Yamamoto et al. [1129] were the first to analyze long-lasting changes. They studied a potentiating action of phorbol esters with recording from CA3 pyramidal cells. However, they based their analysis on Poisson statistics which might lead to erroneous conclusions about changes in quantal parameters (Sects. 11.2, 11.5). More recently, when our in vitro studies were partly published [1022, 1023, 1052, 1057, 1058], several groups [100, 500, 662] reported results based on the analysis of the variability of PSCs recorded after tetanic stimulation [100] and during LTP-like enhancement produced by pairing afferent stimulation with intracellular depolarization of CA1 neurones [500, 662]. The results favour a presynaptic location of the mechanisms of LTP maintenance (see also reports on the quantal analysis of LTP in CA3 [715] and DG [91, 431]). However, one research group [342, 343, 697] demonstrated an increase in v without changes in m after LTP induction in CA1 neurones.

The aim of our following study was to specify changes of quantal parameters during LTP with recording of minimal EPSPs in area CA1 of hippocampal slices and to compare changes in quantal parameters during earlier (5–15 min) and later (16–55 min) post-tetanic periods [1022, 1023, 1052, 1058, 1060, 1062).

14.2 Methods

Methods of recording from guinea pig hippocampal slices were described in Chapter 13. Testing stimulation of *stratum radiatum* or *stratum oriens* consisted usually of double pulses separated by 40 or 50 ms. Stimulus intensity (0.7–1.5 V, 20–40 μs) was set to evoke minimal EPSPs with average amplitudes <1 mV in response to the first pulse (Fig. 14.1A–C). Only recordings from neurones with failures in response to the first pulse before the tetanic stimulation (Fig. 14.1A, C) were analyzed. The analysis included response averaging to determine fixed time points for measurements (a time "window", Fig. 14.1A–D, dotted lines), subsequent measurements of single responses and calculation of the mean window amplitude (Sect. 11.5) which will be referred to as "amplitude" for simplicity. Stable plateau regions were determined from plots of means of 20 or 30 measurements (Fig. 14.1E) for statistical treatment. In theory, the reliability of statistical analysis increases with increased N. However, significant amplitude drifts were observed in the majority of recordings with N>200–300 (Chap. 13). Moreover, an impression was created on the basis of our (Chap. 13) and other studies [351] that LTP is difficult to induce after long-lasting low-frequency testing. On the other hand, computer simulations (Chap. 13) showed that estimates from large samples (N = 1000) can be substituted by several (about 10) estimates from smaller samples (N = 50–100). Therefore, the protocol of the present experiments was the following. About 20 min after the RMP stabilized at the level more negative than 60 mV (Fig. 14.1F), 99 to 198 testing stimuli were given as pretetanic control. The conditioning procedure consisted of ten trains at 100 Hz for 200 ms with a single pulse duration being three to five times larger than that of the testing stimulus. Intertrain intervals were equal to those between testing stimuli (8–10 s). After the tetanic stimulation, test stimulation was continued and as many as possible post-tetanic responses were collected. One neurone (Fig. 14.1E), was tetanized twice with a sufficiently long testing period in between. For the statistical treatment, two or three plateau regions were determined (Fig. 14.1E): a pretetanic (region I) and one or two post-tetanic regions (5 to 15 min, region II and more 15–55 min, region III). Each plateau region consisted of 59 to 119 measurements (average N = 88). Four basic methods (see Chap. 13) were used to calculate parameters (m, v, n, p) for each plateau region. Parameters calculated by the histogram (deconvolution), variance, combined, and failures methods will be indexed by 1, 2, 3 and 0, respectively, as in Chapter 13. Since failures were estimated "subjectively" (Fig. 14.1D) and "objectively", the failures and combined methods will each offer two solutions (Sect. 14.3.3).

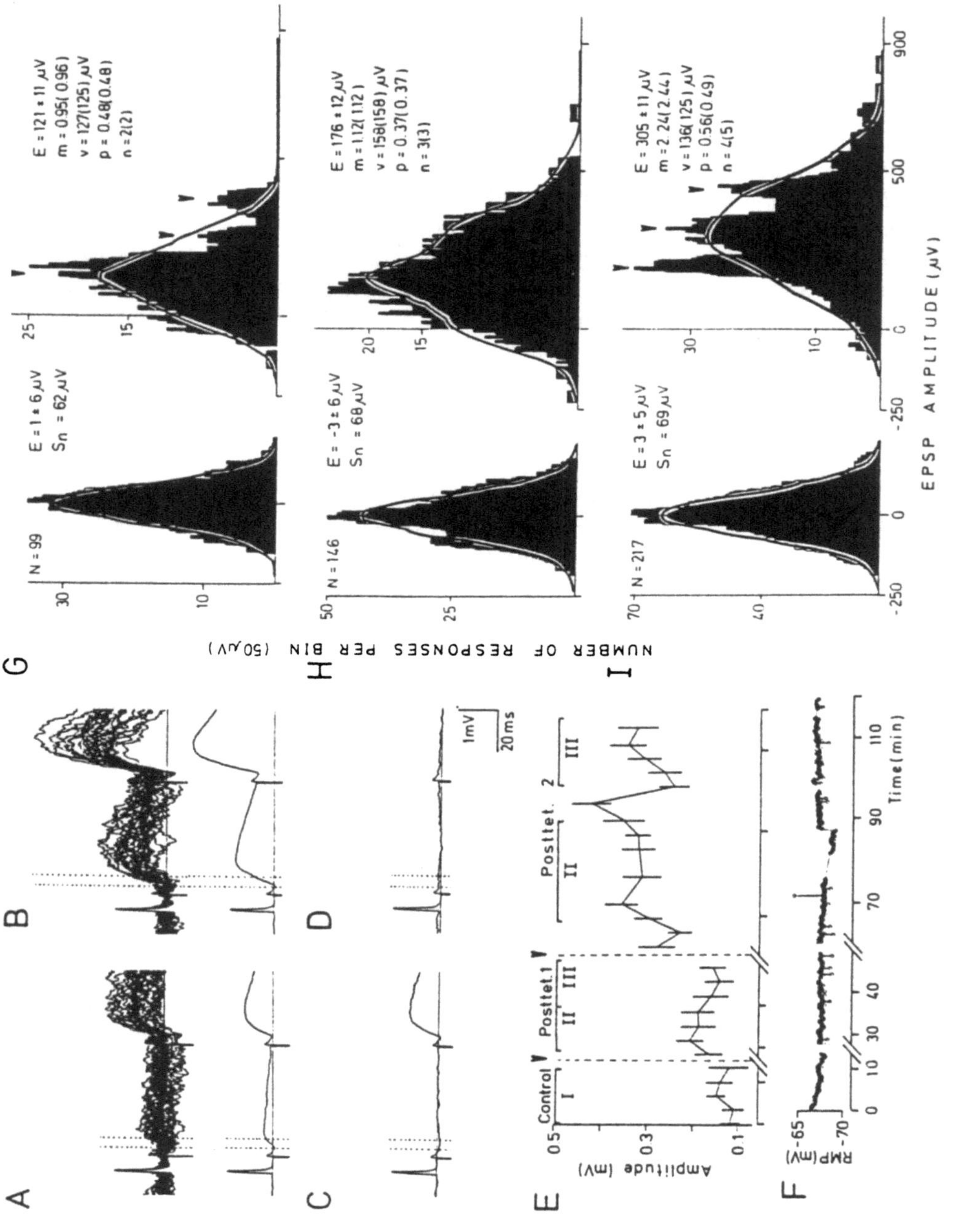
A
B
C
D
1mV
20ms
E
Amplitude (mV)
Control
I
Posttet.1
II
III
Posttet. 2
II
III
F
RMP(mV)
-65
-70
Time(min)
G
H
I
NUMBER OF RESPONSES PER BIN (50μV)
EPSP AMPLITUDE (μV)
N = 99
E = 1 ± 6 μV
Sn = 62 μV
E = 121 ± 11 μV
m = 0.95(0.96)
v = 127(125) μV
p = 0.48(0.48)
n = 2(2)
N = 146
E = -3 ± 6 μV
Sn = 68 μV
E = 176 ± 12 μV
m = 1.12(1.12)
v = 158(158) μV
p = 0.37(0.37)
n = 3(3)
N = 217
E = 3 ± 5 μV
Sn = 69 μV
E = 305 ± 11 μV
m = 2.24(2.44)
v = 136(125) μV
p = 0.56(0.49)
n = 4(5)

14.3 Results

14.3.1 Long-Term Potentiation of Minimal EPSPs

EPSPs of 13 neurones were recorded for at least 15 min after tetanic stimulation. Amplitudes of 26 EPSPs (from 8 neurones activated by 11 pathways), showed a statistically significant increase ($P<0.01$, Student's t-test) for region II if compared with the region I (Fig. 14.1E). In all 18 EPSPs recorded for more than 15 min, the increase persisted during the whole stable recording period (Fig. 14.1B, E). Figure 14.1 illustrates the case where the tetanization was repeated about 30 min after the first one. Each tetanization produced approximately a doubling of the EPSP amplitude (Fig. 14.1E). "LTP magnitude" was defined as the ratio of the mean post-tetanic amplitude to the mean pretetanic amplitude (Chap. 12). The LTP magnitude ranged from 118% to 441% for region II and from 144% to 418% for region III. Mean and single values of LTP magnitudes are represented in Figs. 14.2 and 14.3 (ordinates). There was a negative correlation between LTP magnitude and initial EPSP amplitude ($r = -0.55$, $N = 17$, $P = 0.02$) for 17 EPSPI.

14.3.2 Estimates of Quantal Parameters

We calculated "basic" (m and v) and binomial (p and n) parameters for every region. However, as was shown (Sect. 13.3), estimates of p and n were less reliable than those of m and v. In some cases, p_1 and n_1 were uncertain (Chap. 13) and corresponding histograms were adequately described by Poisson statistics. Such values of the binomial parameters were excluded from Table 14.1. Two variants of the histogram (deconvolution) method were tested as described in Chapter 14: S_n was either measured (Fig. 14.1G–I, left) or was searched as a free parameter together with v. Because the results of both variants were similar (Chap. 14), only the values based on direct S_n measurements will be considered below.

Single values of v and m estimated by different methods were frequently similar: compare parameters in Fig. 14.1G–I estimated by the histogram and variance

Fig. 14.1A–I. Long-term potentiation of minimal EPSPs after two conditioning tetanizations. **A, B** Single (*upper row*) and averaged (*lower row*) EPSPs before (**A**) and after two tetanic series (**B**). **C** Average of sweeps taken as failures in response to the first pulse before tetanization. **D** Average of extracellular potential after the neurone was lost. **E** Amplitude time course; tetanizations are indicated by vertical interrupted lines; *abscissa* time from the beginning of the registration, in min (see **F**, note the time breaks during the conditioning procedures); *ordinate* mean amplitudes ±SEM calculated from 20 consecutive responses, in mV; control (*I*) and post-tetanic (*II* and *III*) plateau regions are marked by *horizontal bars*. **F** Time course (in min) of resting membrane potential (*RMP*, in mV). **G–I** Experimental (*columns*) and theoretical (*curves*) distributions of noise (*left*) and EPSP amplitudes (*right*) for pretetanic (**G**) and for pooled post-tetanic data (regions *II+III*) after the first (**H**) and second (**I**) tetanizations. Parameters calculated by the histogram method are given in **G-I** *without parentheses*; parameters calculated by the variance method are given *in parentheses*. [1062]

Table 14.1. Quantal parameters of minimal EPSPs recorded in the hippocampal slice preparation before (region I) and after (regions II and III) tetanic stimulation

Method	Region	N	m	v(μV)	p	n
Histo-	I	26	2.02±1.18	142±67	0.47±0.09	4.1±1.9(19)
gram	II	26	3.36±1.41***	149±70	0.50±0.10	6.5±2.5**(19)
	III	18	3.79±1.76***	129±56	0.45±0.14	10.3±7.7*(10)
Histo-	I	23	1.85±1.00	154±61	0.47±0.09	4.1±1.9(19)
gram	II	23	3.34±1.50***	158±68	0.49±0.10	6.6±2.6**(17)
S_n<0.9v	III	14	3.93±1.93**	141±57	0.45±0.14	10.3±7.7*(10)
Variance	I	26	1.70±0.93	153±48	0.46±0.09	3.6±1.3(25)
	II	26	2.61±1.05***	184±75*	0.53±0.08*	4.7±1.5***
	III	18	2.59±0.98***	184±72*	0.52±0.08**	4.9±1.1*
Failures	I	22	1.58±0.64	135±56	0.44±0.08	3.7±1.1(21)
subj.	II	15	2.20±0.64**	145±64*	0.48±0.06*	4.3±1.2**
	III	12	2.29±0.66**	155±73	0.48±0.06	4.8±1.0**
Failures	I	20	1.55±0.71	135±49	0.43±0.08	3.4±1.4(19)
obj.	II	12	2.13±0.60**	158±56*	0.47±0.05*	4.4±1.2**(12)
	III	11	2.21±0.45**	170±70	0.47±0.05	4.8±1.1*(10)
Combined	I	22	1.64±0.80	138±68	0.46±0.12	3.5±1.6(14)
subj.	II	15	2.40±0.88**	139±70	0.48±0.10	5.0±3.4*(10)
	III	12	2.43±0.78**	145±66	0.43±0.10	5.6±1.9*(10)
Combined	I	20	1.67±0.85	132±58	0.47±0.12	3.1±1.3(9)
obj.	II	12	2.30±0.72**	149±53	0.44±0.11	5.4±2.6*(9)
	III	11	2.33±0.48*	165±69	0.50±0.10	4.7±1.2(7)

Notes. The numbers are means ±SD. Two variants of the failures and combined methods are given, based on subjective (subj.) and objective (obj.) determinations of N_0. Regions II and III: 5 to 15 and 16 to 55 min after tetanization, respectively. Amplitudes: 270±189 μV (region I, N = 26), 491±330 μV (region II, N = 26), 471±233 μV (region III, N = 18). Number of measurements (N) for n and p is given in parentheses when it differs from N for m and v. *), **) and ***) Significance levels for increases in comparison with pretetanic (region I) values, P<0.05, P<0.01 and P<0.001, respectively, Wilcoxon test for matched pairs. Number of matched pairs in several cases was less than N.

methods (without parentheses and in parentheses, respectively). The mean values estimated by different method are given in Table 14.1.

14.3.3 Post-Tetanic Changes in Mean Quantal Content

The most reproducible post-tetanic change in quantal parameters was an increase in m estimated by different methods (Fig. 14.2A–F). An increase in m_1 was observed for 42 out of 44 post-tetanic regions (Fig. 14.5A, symbols to the right of the ordinate). For solutions with S_n<0.9v which might be considered to be the most reliable (Chap. 13), m_1 increased in all 37 post-tetanic regions. The mean relative post-tetanic value of m_1 did not significantly differ from LTP magnitude (N = 44, P>0.8, Wilcoxon test for matched pairs, here and below, see also Fig. 14.2A).

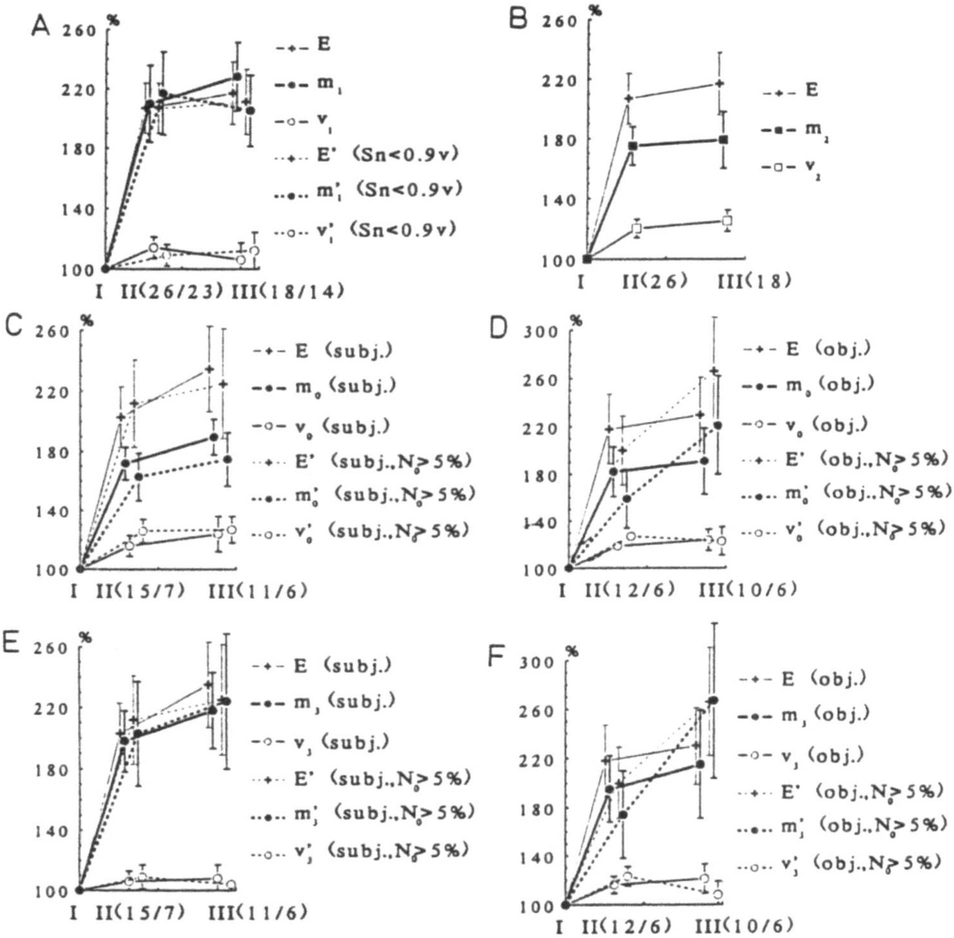

Fig. 14.2A–F. Changes in mean EPSP amplitude (*E*) and quantal parameters (*m* and *v*) estimated by the histogram (**A**), variance (**B**), failures (**C**, **D**) and combined (**E**, **F**) methods. **C–F** Variants based on subjective (**C**, **E**) and objective (**D**, **F**) estimates of N_0. All values calculated for regions II and III were normalized to the respective pretetanic values (point I at horizontal axes) and plotted as percentage of the latter ±SEM (points II and III at horizontal axes). Subset of cases with $S_n<0.9v$ (**A**) and $N_0>5\%$ (**C–F**) are indicated by *dotted lines*. Values in parentheses show the number of measurements for *continuous* and *dotted lines*, respectively. [1062]

Quantal contents estimated by other methods also showed a consistent increase after tetanization (Fig. 14.2B–F). It is clear from Fig. 14.2 and Table 14.1 that m increased significantly for all methods if earlier (region II) and later (region III) post-tetanic values were compared with the pretetanic values (region I). According to all methods used, the changes in m for both region II and III were not significantly different from changes in EPSP amplitudes. Both variants of the failures

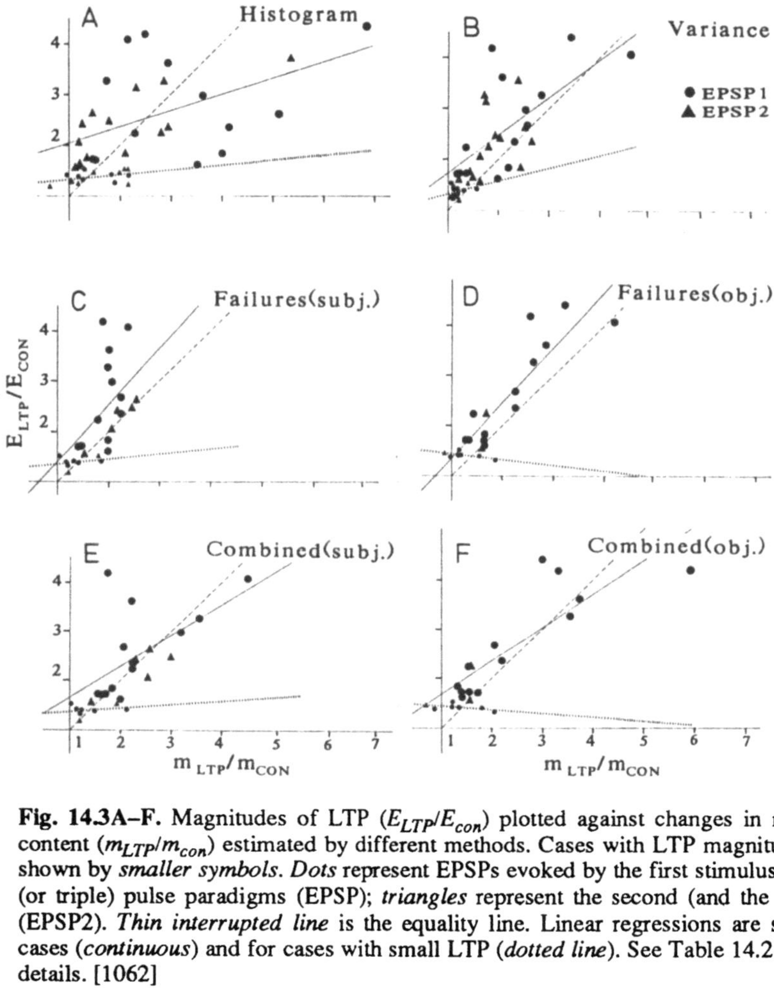

Fig. 14.3A–F. Magnitudes of LTP (E_{LTP}/E_{con}) plotted against changes in mean quantal content (m_{LTP}/m_{con}) estimated by different methods. Cases with LTP magnitude <1.55 are shown by *smaller symbols*. *Dots* represent EPSPs evoked by the first stimulus in the paired (or triple) pulse paradigms (EPSP); *triangles* represent the second (and the third) EPSPs (EPSP2). *Thin interrupted line* is the equality line. Linear regressions are shown for all cases (*continuous*) and for cases with small LTP (*dotted line*). See Table 14.2 for statistical details. [1062]

and combined method (Fig. 14.2C–F) gave very similar changes in m. To obtain more reliable estimates of quantal parameters, cases with N_0>5% of the total number of measurements were separated (Fig. 14.2C–F, interrupted lines). In comparison with the whole sample, the difference in post-tetanic changes of either m or v for this subset was not significant.

LTP magnitudes and changes of m correlated, both for estimates from pooled data of all methods and for estimates obtained by every method separately (Fig. 14.3, Table 14.2, r(m/E), upper rows for different methods).

14.3.4 Post-Tetanic Changes in Quantal Size

The increase in v was smaller than the increase in m and statistically not significant for the histogram (Fig. 14.2A, P>0.2, N = 44) and "combined subjective" method (Fig. 14.2E, P>0.4, N = 26). Changes in v_1 did not correlate significantly with LTP magnitude (r = 0.21, Table 14.2), whereas the increase in v_2 (Fig. 14.2B, variance method) correlated positively with LTP magnitude (r = 0.45, Table 14.2). The highest correlations were found for the two methods with subjective estimates of N_0: for v_0 (Fig. 14.2C, r = 0.79, Table 14.2) and for small changes in v_3 (Fig. 14.2E, r = 0.41, Table 14.2). Comparable (Fig. 14.2D) or even larger (Fig. 14.2F) changes in v_0 and v_3 calculated with objectively estimated N_0 did not correlate with LTP magnitude (Table 14.2). Pooled data from the four methods (see below) showed only a weak correlation at the border of statistical significance (Fig. 14.4B, r = 0.22, N = 88, P<0.05); pooled estimates of v by three methods, excluding the histogram method, did not correlate significantly with LTP magnitude (r = 0.17, P>0.1, N = 66).

In general, the post-tetanic increase in EPSP amplitude was accompanied by an almost equal increase in m estimated by all methods used. The increase in v was small, was not significant for some method and did not correlate with LTP magnitude.

14.3.5 Post-Tetanic Changes in Binomial Parameters

Estimates of parameters p and n are summarized in Table 14.1. Because of the exclusion of uncertain estimates of p and n (see Chap. 13), the sample sizes were

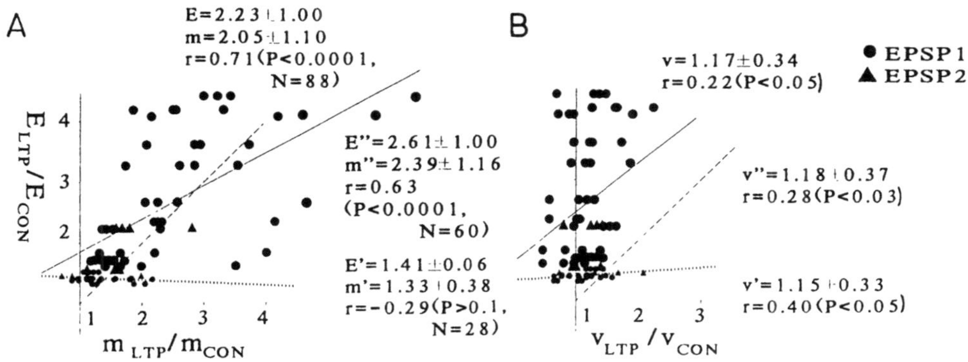

Fig. 14.4A,B. Magnitudes of LTP plotted against changes in mean quantal content (**A**) and quantal size (**B**) for the pooled data from four (histogram, variance, failures, and combined) methods. Variants of the failures and combined methods based on objective estimates of N_0 were used. Means (±SD), coefficients of correlation (*r*), their significance levels (*P*) and sample sizes (*N*) are given for the whole set of measurements (*E, m, v*), for the EPSP subset with large LTP (*E″, m″, v″*), and for the EPSP subset with small LTP (*E′, m′, v′*). See Fig. 14.3 for other notations. [1062]

smaller than for m and v (Table 14.1). Excluded cases constituted a considerable fraction of the whole sample obtained by the histogram method. This fraction tended to decrease after tetanization which corresponded to a post-tetanic increase in p. If estimated values of p and n for all methods were considered, both values tended to increase after tetanization. I stress again the preliminary character of this statement, taking into account a low reliability of the estimates of the binomial parameters under the present conditions.

14.3.6 Differential Changes in m and v for Different LTP Magnitudes

We divided all post-tetanic EPSPs into subsets with "large" and "small" LTP. The border was set at an LTP magnitude equal to 1.55. Changes in m and v and correlations of relative m and v with LTP magnitude were separately calculated for the two subsets. The results are shown as correlation fields for m (Fig. 14.3). The regression line for the whole sample and for the subset with large LTP magnitude were similar (Fig. 14.3, continuous lines). However, the subset with small LTP magnitude (Fig. 14.3, dotted lines) differed strongly. Even though the results somewhat varied for the different methods, the general trend was a lower correlation of changes in m with LTP magnitude in the EPSP subset with small LTP magnitude [Table 14.2, r(m/E)]. Relative increases in v for the subset with small LTP magnitude were similar or even larger than for the whole set of EPSPs (Table 14.2). Thus in the subset with small LTP magnitude, v contributed relatively more strongly to LTP. In the subset with small LTP magnitude, the correlation between changes in v and LTP magnitude tended to be positive for all methods [Table 14.2, r(v/E)], except the histogram method.

In order to obtain a sample with equal representation of all four methods, estimates from different methods were pooled for those cases where the histogram, variance, failures and combined methods could be used for the same EPSPs. For the failures and combined methods, the variants with objective N_0 estimates were used. The analysis showed that post-tetanic changes in m (Fig. 14.4A) and v (Fig. 14.4B) were not significantly different for the subset with large LTP and for the whole set. However, for the subset with small LTP magnitude, changes in m were not correlated with LTP magnitude (Fig. 14.4A, dotted line, $r = -0.29$), whereas changes in v were correlated (Fig. 14.4B, $r = 0.40$). This correlation was higher than for the whole set and for the subset with large LTP (Fig. 14.4B). Without the estimates of the histogram method, the differences between the two EPSP subset was even more pronounced: the correlation of changes in v with LTP magnitude was not significant either for the whole set ($r = 0.17$, $P>0.1$, $N = 66$) or for the subset with large LTP ($r = 0.23$, $P>0.1$, $N = 45$), but it was statistically significant for the EPSP subset with small LTP ($r = 0.53$, $P<0.02$, $N = 21$). Because the absolute changes in v were similar for the EPSP subsets with different LTP magnitudes, the relative contribution of the changes in v to LTP (i.e. the ratio of the net increase in v to the net increase in EPSP amplitude) was about three times larger

Table 14.2. Changes in quantal parameters estimated by different methods for the whole EPSP set (W) and for two subsets with large (L) and small (S) LTP magnitude

Method	Set	N	E	m	v(μV)	r(m/E)	r(v/E)
Histogram	W	44	211±89	217±122***	111±42	0.61***	0.21
	L	28	253±86	254±50***	116±45	0.50**	0.17
	S	16	137±11	152±51**	100±35	0.42	–0.33
Variance	W	44	211±89	177±73***	122±30***	0.77***	0.45**
	L	28	253±86	208±75***	126±36***	0.65***	0.44*
	S	16	137±11	122±18***	114±15**	0.41	0.27
Failures	W	26	216±86	180±43***	119±34**	0.65***	0.79***
subj.	L	18	251±83	199±34***	126±36**	0.47*	0.87**
	S	8	139±11	138±30*	105±22	0.28	0.21
Failures	W	22	223±101	186±82***	122±24**	0.89***	0.23
obj.	L	15	261±102	215±82***	122±21**	0.88***	0.35
	S	7	141±6	125±35	121±32	–0.62	0.65
Combined	W	26	216±86	206±81***	107±28	0.71***	0.41*
subj.	L	18	251±83	235±79***	108±29	0.60**	0.53*
	S	8	139±11	142±40*	104±26	0.27	0.15
Combined	W	22	223±101	204±118***	118±32*	0.84***	–0.09
obj.	L	15	261±102	238±127***	117±22*	0.84***	–0.12
	S	7	141±6	131±48	122±49	–0.55	0.49

Notes. Means ±SD are given for E, m and v calculated from post-tetanic values which were taken as percentages of the respective pretetanic values. r(m/E) and r(v/E): coefficients of correlation of m and v with E. *), **) and ***) Significance levels for r and for post-tetanic increases of m and v, $P<0.05$, $P<0.001$, and $P<0.0001$, respectively. See Table 14.1 for other explanations.

for the subset with small LTP (37%) in comparison with the whole EPSP set (14% for N = 88) and with the subset with large LTP (14%).

Changes in quantal parameters for early and late post-tetanic regions did not differ significantly (Fig. 14.2). For some methods (Fig. 14.2B–F), however, there was a trend to a larger increase in v for region III compared with region II. This trend was more evident for the EPSP subset with small LTP for the pooled data of all methods.

14.4 Discussion

Different from Friedlander et al. [351] and in agreement with our previous observations in the in vivo preparation (Chap. 12), reliable LTP of minimal EPSPs was induced in the majority of neurones recorded in area CA1 [100, 343]. The LTP magnitude was similar to that observed for minimal EPSPs recorded from CA3 neurones in the in vivo preparation (Chap. 12). The failure to induce LTP [351] might be explained by the long-lasting testing stimulation (2 Hz) which could produce a LFD (Chap. 1) and interfere with LTP induction (see Sect. 13.5). According to our observations, LTP of minimal EPSPs was more prominent than

LTP of large compound EPSPs [566, 1057]. The difference might be understood in view of the negative correlation between EPSP amplitude and LTP magnitude. Another possibility is that slice preparations are kept in a "more potentiated" state in experiments with recordings of large EPSP.

In the present experiments, post-tetanic changes in quantal parameters for two comparatively long periods (up to 55 min) were compared and additional methods were applied. Nevertheless, the basic result was similar to that obtained previously with in vivo recordings (Chap. 12). The main result was the significant post-tetanic increase in m which correlated with LTP magnitude. The increase in m suggests a presynaptic location of a major mechanism responsible for LTP maintenance. In addition, distinct from the previous in vivo studies (Chap. 12) and from majority of other studies of LTP and LTP-like changes [91, 98, 100, 249, 431, 662, 715, 1129], a small but definite post-tetanic increase in v was revealed here and in some other publications [343, 697, 562]. Several mechanisms can account in principle for the increase in v including both pre- and postsynaptic ones (see Sect. 6.1). No changes or even some post-tetanic decreases in the sensitivity to iontophoretically applied transmitters were reported [236, 638, 955, 1014, 1048] at least during initial 30–40 min after tetanic stimulation [251]. No appropriate changes in postsynaptic characteristics (R_{in}, RMP) were found in either in vivo (Chap. 5) or in vitro [32, 80, 955] preparations including the present experiments (Fig. 14.1F). A delayed post-tetanic increase in the quisqualate receptor sensitivity was reported [251]. Here, we found only a tendency for a larger increase in v during the later post-tetanic period in comparison with the earlier post-tetanic period. However, the tendency was more clear for the subset with a small LTP. The time course of changes in v should be studied in special experiments with more prolonged recordings.

One of the possible mechanisms of the increase in v is an increase in the amount of transmitter in one quantum. Since the post-tetanic changes in v were quantitatively similar to those found during PPF (Chap. 11), mechanisms of the changes might be related. The lack of correlation between amplitudes of two EPSPs evoked by the paired pulses suggested that the mechanisms of changes in v might be localized predominantly presynaptically (Sect. 11.6). There are also at least two postsynaptic mechanisms which do not exclude each other: changes in spine shapes, which could affect the spine to dendrite coupling and an increase in receptor sensitivity, which could directly influence v (see Sects. 6.1 and General Conclusions).

According to our data, changes in v cover only a small fraction (<20%) of the average LTP magnitude. Large LTP could be induced in some cases without a detectable increase in v. However, the changes in v constitute a more significant fraction of LTP when the LTP magnitude was small. The differential changes of quantal parameters for cases with small and large LTP was demonstrated for the pooled estimates derived by several methods. An exception was the histogram method which revealed only an insignificant post-tetanic increase in v for the whole EPSP set and no increase in v for the EPSP subset with small LTP. However, as shown previously (Chap. 13), the histogram method is less reliable than other methods used here. The net increase in v was similar for different EPSP subsets, therefore, the relative contribution of the mechanism related to the increase in v was larger for cases with small LTP than for cases with large LTP. This makes a saturation of the underlying mechanisms conceivable.

The applicability of the quantum hypothesis to analyze minimal hippocampal EPSPs and the problem of the detection of significant average changes across experiments from individual experiments was discussed previously (Sect. 13.5; see also [343]). We also discussed the reliability of the simple binomial model, which postulates a uniform p. We noted that the histogram method by itself is a test for the validity of this model. We stressed that the lack of significant differences between observed and predicted histograms did not mean that this model had a preference over other (more sophisticated) models, especially if one takes into account the comparatively small N for individual cases used in the present study. The approach based on the binomial assumption is justified as a first approximation if the major aim is the evaluation of changes in m and v rather than in p and n. The simple binomial model was used by many authors for different preparations (Chap. 7). Generally, the application of less restrained and more sophisticated models (Sect. 7.3.7) demands larger N and v/S_n ratios than those available in the present study. The application of less restrained models might not be necessary in every case. Thus, a statistical analysis of effects of anaesthetics on synaptic transmission in the spinal cord using an unconstrained deconvolution procedure [568] gave the same basic conclusion as more traditional approach [1077]. It was already stressed (Sect. 12.7) that visual estimates from the experimental histograms with regularly spaced peaks and especially significant decreases in the CV and N_0 strongly suggest increases in m for a broad range of assumed models (scc [323, 531] for reservations).

Some reservations in relation to the interpretation of m changes in a situation with initially ineffective synapses were made in Sect. 12.7. Ineffective synapses, without contributing essentially to the pretetanic EPSPs, might increase their efficacy not only due to presynaptic changes but in principle also due to subsynaptic reorganizations (formation of synaptic receptors). In this case, our data would indicate that the mean efficacy of one transmitter quantum for presumed "new" synapses is approximately the same as that for initially effective synapses, and furthermore that the sensitivity of existing receptors as well as other postsynaptic characteristics are not changed significantly after LTP induction. However, the traditional explanation based on a presynaptic mechanism which increases the number of quanta released per presynaptic discharge seems to be most attractive in view of other evidences available in the literature (Sect. 12.7 and General Conclusions). Among the supportive evidences are recent results of statistical analysis cited in Sect. 14.1 [100, 660, 662, 715]. The disagreement of these data with the results of Foster and McNaughton [342, 343] can be explained by our finding of essential contribution of the increased v in LTP of small magnitude.

14.5 Summary

Minimal excitatory postsynaptic potentials (EPSPs) were recorded in CA1 area of guinea pig hippocampal slices from 13 neurones after double-pulse stimulation of *stratum radiatum* (Schaffer collaterals) and *stratum oriens*. Amplitudes of EPSPs significantly increased in eight neurones 5 to 55 min after nine tetanizations of the Schaffer collaterals. The increase was considered to represent long-term potentiation (LTP). Altogether 26 EPSPs (11 pathways, 44 post-tetanic regions) were statistically analyzed by four methods of the quantum hypothesis assuming the binomial model of the transmitter release: the deconvolution (histogram), variance, failures and combine (variance-failures) methods. The mean quantal content (m) significantly increased after LTP induction according to all methods used. Quantal size (v) also tended to increase but, according to some methods, the increase was not statistically significant and did not correlate with LTP magnitude. However, for an EPSP subset with a relatively small post-tetanic increase in EPSP amplitude (< 55% above the pretetanic level), an increase in v was correlated with the LTP magnitude, whereas an increase in m was not. A relative contribution of the increase in v to LTP magnitude was essentially larger for cases with small LTP than for the whole EPSP set. In general, the increase in m corresponds to our previous data and favours the presynaptic location of major mechanisms of LTP maintenance, i.e. an increase in the average number of transmitter quanta released by each presynaptic volley. The post-tetanic increase in v might reflect some additional mechanisms which might include an increase in the amount of transmitter in one quantum.

15 Analysis of Fluctuations of Minimal EPSPs in Vitro: Quantal Model

15.1 Introduction

Traditional methods of the quantum hypothesis (Chap. 7) were used in previous chapters which assumed the simple binomial model. However, deviations from the binomial predictions were described in some cases and they were explained by a non-uniform release probability for different release sites (Sect. 7.7). In the case of a spatially non-uniform p, the simple binomial model might lead to an erroneous description of the release statistics [85, 175, 635, 806]. The technique of "noise deconvolution" was mentioned in Section 7.3.7 as an alternative to the methods based on the binomial or Poisson models. The deconvolution procedure, although having its own difficulties and limitations [81, 525, 823] (see especially [230] and Sect. 7.3.7), allows to obtain "noise-free" discrete distributions of EPSP amplitudes from experimental noisy histograms without assumptions regarding the distribution of underlying release probabilities and it allows both non-uniform p and different v at different release sites. With application of this technique to the spinal cord synapses (Sect. 7.3.7), the quantal nature of synaptic release was confirmed, but it was found that the amplitudes of spinal EPSPs were not distributed according to the Poisson or (in general) binomial law. A "compound" binomial model was proposed with a uniform v but with a variety of probabilities (p_i) associated with different release sites [823, 1068]. The aim of our last experimental series [1024, 1059, 1060, 1063] was to test the conclusions related to the mechanisms of LTP maintenance, using a variant of the deconvolution technique which does not assume a simple binomial model of transmitter release.

15.2 Methods

The methods of recording of "minimal" EPSPs from CA1 neurones of guinea pig slices were described in Chapter 14. The same amplitude measurements were used here. The definition of the plateau regions without a significant trend was also similar to that used previously, but we selected only regions with $N \geq 90$.

As a first step, we tested our deconvolution algorithm with Monte Carlo simulations. In distinction from Chapter 13, dealing with the simple binomial model, "compound" binomial distributions convolved with Gaussian noise were simulated here. The compound binomial model [823] assumes that the release is quantized and at any given release site, a quantal PSP is generated with amplitude v and with probability p_i, independently of whether release occurs at any other site. Both in the simulation and physiological experiments, we assumed S_v=0.05 (Sects. 8.8, 13.2). To estimate parameters of amplitude fluctuations, an optimization procedure based on the noise "deconvolution" was used (Sects. 7.3.7, 7.6.5). The task of the statistical analysis was to obtain discrete amplitude distributions (Figs. 15.1A, C, vertical bars) from EPSP amplitudes and noise measurements. The discrete amplitudes were obtained from the noise-contaminated EPSP and from noise measurements (Fig. 15.3C, D, continuous lines) using a deconvolution procedure similar to those described in recent publications [567, 568, 823]. Accordingly, the maximum likelihood estimator (MLE, Fig. 15.1B, D) was used for the optimization procedure. In addition to the usual postulates that EPSPs add linearly and that noise and EPSPs are statistically independent, we assumed that discrete "noise-free" EPSP amplitudes were equally spaced with the quantal interval v as in the "unconstrained quantal model" [823, 1068]. Unlike some authors [568, 864, 865], we used one (and not two) Gaussian curve for the approximation of the noise histogram. For illustrative proposes, amplitude distributions were smoothed and given as probability density functions (Fig. 15.1A, C). After the estimate of v and the minimum number of underlying components (n), we calculated m = E/v and the "average probability" p = m/n. Solutions with probabilities P_i<1/N were considered to be unreliable and neglected when n was estimated. It was possible to find n best fitted compound binomial probabilities p_i from probabilities P_i of the discrete components of the MLE solution [1068]. The procedure, developed by M.L.Ezrokhi and E.E.Martsinkevich, solved the following system of non-linear equations:

$$P_0 = (1-p_1)*(1-p_2)*...*(1-p_n)$$

$$P_1 = p_1*(1-p_2)*...*(1-p_n) + (1-p_1)*p_2*(1-p_3)*...*(1-p_n) +...$$

....

$$P_n = p_1*p_2*...*p_n.$$

A control group without tetanic stimulation from data of Chapter 13 was analyzed to test reproducibility of the deconvolution procedure for large and small N. The analysis of post-tetanic changes was based on the recordings of minimal EPSPs from Chapter 14 with one additional neurone.

15.3 Results of Simulation Experiments

The aim of the Monte Carlo simulations was to test the reliability of our deconvolution algorithm for large (N = 1000) and small (N = 100) samples under various S_n.

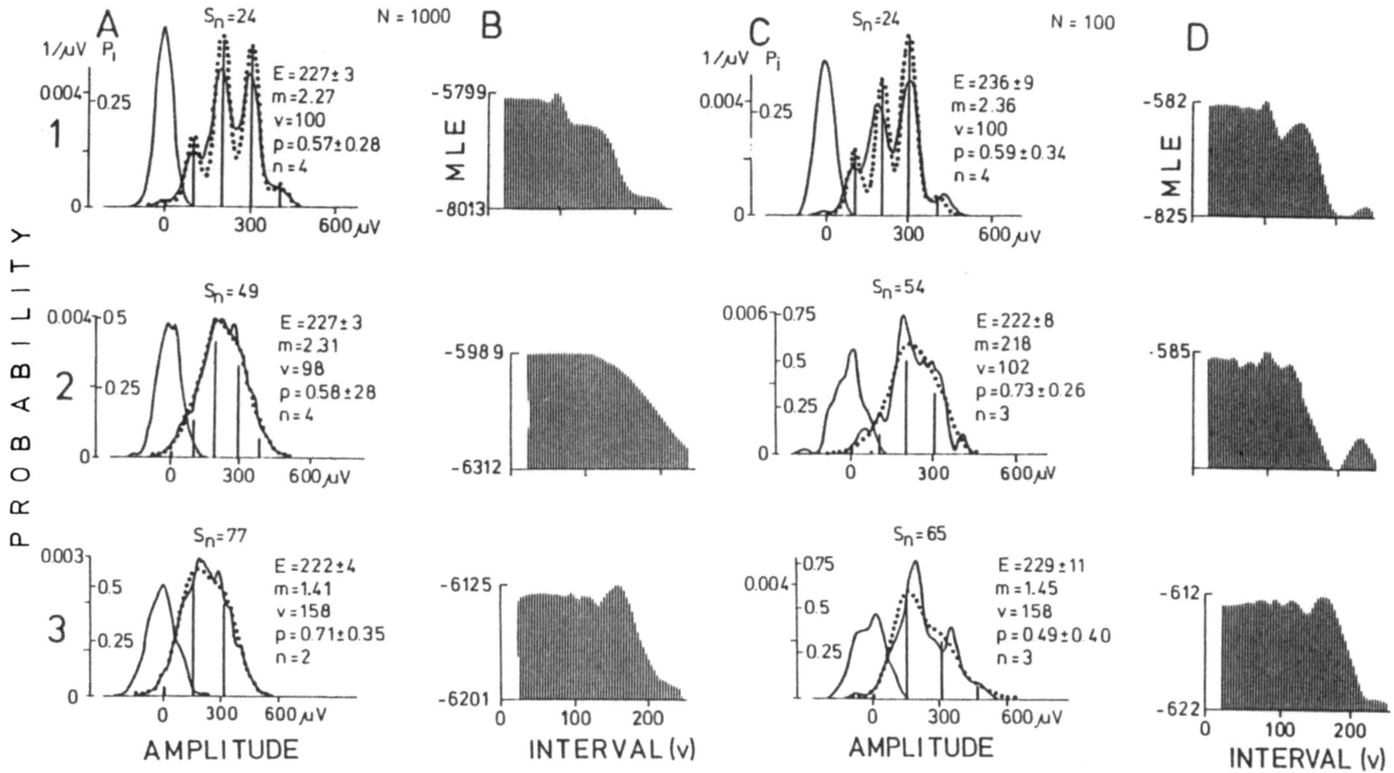

Fig. 15.1A–D. Testing the estimate of quantal parameters by Monte Carlo simulations with compound binomial distributions at different noise levels and sample sizes. **A, C** Smoothed simulated noise and EPSP amplitude histograms (*continuous curves*) and predicted distributions (*dotted curves*). Deconvolution results (discrete amplitude values) are shown by *vertical bars* as probabilities (*ordinate*, P_i) versus amplitude (*abscissa*, µV). *Left* scale of ordinate (1/µV) is for curves (probability density functions); noise curves were reduced vertically by a factor of 2. Compound binomial distributions were simulated with v = 100, $p_1 = 0.2$, $p_2 = 0.5$, $p_3 = 0.7$, $p_4 = 0.9$, with two N (1000 for **A, B**; 100 for **C, D**) and three S_n (25 for *1*, 50 for *2*, and 75 for *3*). Mean amplitudes (E ± SEM) and estimates of m, v, mean p (±SD), and n are given as *insets*. **B, D** Plots of the logarithm of the maximum likelihood function (*ordinate*, MLE) for a range of values of quantal interval (*abscissa*, µV). [1063]

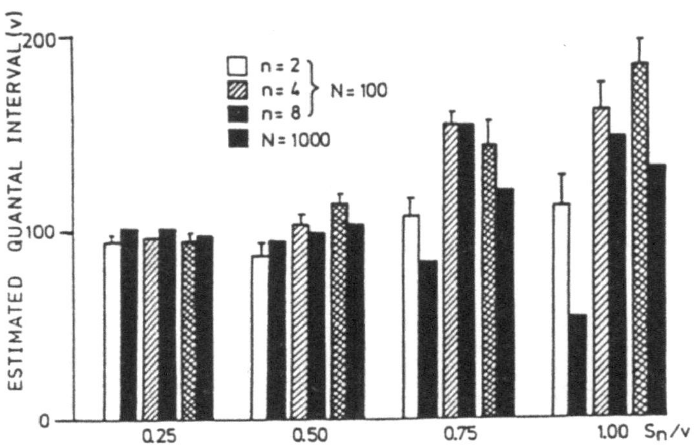

Fig. 15.2. Estimates of v (*vertical axes*) for simulated distributions with various S_n (*horizontal axes*). Results from three distributions are shown (see *inset*): n = 2 (p_1 = 0.3, p_2 = 0.9), n = 4 (p_1 = 0.2, p_2 = 0.5, p_3 = 0.7, p_4 = 0.9), and n = 8 (p = 0.55±0.24, mean±SD). The height of each bar is equal to the mean v (±SEM) calculated from ten samples (N = 100 for each sample). *Black bars* represent v estimates from the corresponding large samples with N = 1000. [1059]

Continuous curves in Fig. 15.1A, C show simulated compound binomial distributions with three different S_n. With S_n = 25, clear regular peaks were visible in the simulated histograms with both large (Fig. 15.1A) and small (Fig. 15.1C) N. Fig. 15.1C1 exemplifies the fact that with small S_n/v (less then 40%), even small N = 100 was usually sufficient for the correct v estimates (within ±5–10% of the simulated v). Similarly satisfactory solutions were often found for simulations with S_n/v = 0.5 when regular peaks were not predicted theoretically (Fig. 15.1A2, dotted curve) and usually were absent in the simulated histograms (Fig. 15.1A2, continuous curve). At 0.6<S_n/v>0.4, single solutions might deviate from the simulated values when small samples (N = 100) and large n (n>2) were simulated (Fig. 15.1A, C), but the average of ten solutions from the small samples usually gave correct values within ±10% of the simulated v (Fig. 15.2, S_n/v = 0.50) similar to estimates from the large (N = 1000) sample (Fig. 15.2, black bars at S_n/v = 0.50).

Figure 15.2 shows that with S_n>0.5v both determinations from large and small N were usually incorrect and shifted to values larger then simulated v (see a recent publication [230] for the similar results for the unconstrained model). Results similar to shown in Fig. 15.2 were obtained in >400 simulations.

15.4 Result of Physiological Experiments

The aim of the first series of our electrophysiological experiments was to evaluate variability of the estimates of m and v and to compare estimates from "small" (N = 100) and "large" (N = 300–1000) samples of EPSP amplitudes. Measurements from the first experimental series of Chap. 13 were used. Figure 15.3 illustrates an example of the selection of a "quasi-stationary" plateau region for a neurone with the largest N considered in Fig. 13.3. The continuous curves in Fig. 15.3C show smoothed noise and amplitude histograms corresponding to the plateau region. The bars which represent the solutions of the deconvolution procedure are 124 μV apart. A slightly larger v was found by the algorithm (Fig. 15.3D, F) for the first "subregion" with N = 100. However, the average v calculated from the estimates for five "subregions" (Fig. 15.3B, AV, full circle) was not significantly different from the estimate based on the large sample (Fig. 15.3B, 1–5 in brackets, full circle). Note that in distinction from the estimates of v, the estimates of p and n differed for small and large samples (Fig. 15.3B, AV, empty circles and triangles).

Similar comparisons for 24 EPSPs showed a high correlations (r = 0.98, P<0.0001) for averages from 3–10 estimates based on N = 100 and the solutions from the respective large samples (N = 300–1000). The respective mean values (±SD) were almost identical: 228±125 and 222±126 μV. Based on our simulations, we might expect that estimates with $S_n/v<0.6$, and especially $S_n/v<0.5$, would be more reliable. However, the results obtained for solutions with different S_n/v were similar to the results of the whole sample. The conclusion was that the average of several estimates from small samples could substitute for the estimate from a large sample.

The evaluation of quantal parameters of hippocampal synapses was made for 11 non-facilitated EPSPs, i.e. EPSPs evoked by first testing stimuli in paired-pulse testing paradigm without tetanization. The data are summarized in Table 15.1.

Altogether, 9 out of 14 neurones recorded for >15 min after tetanization showed a significant increase (LTP) in EPSP amplitude (P<0.01, Student's t-test).

Table 15.1. Quantal parameters of non-facilitated hippocampal EPSPs estimated by the deconvolution procedure based on the quantal model

Set	N	E(μV)	v(μV)	m	n	p	CV_p
All	11	181±141	123±63	1.42±–0.63	4.27±1.05	0.33±0.11	0.58±0.32
		27–545	52–255	0.42–2.37	3–7	0.15–0.53	0.001–1.7
$S_n<0.6v$	7	185±159	133±73	1.30±0.59	4.14±1.25	0.32±0.12	0.70±0.2
		27–545	54–255	0.42–2.14	3–7	0.15–0.53	0.23–1.07
$S_n<0.5v$	3	269±213	184±84	1.20±0.71	4.00±0.00	0.31±0.16	0.92±0.17
		27–545	64–255	0.42–2.14	4–4	0.15–0.53	0.68–1.07

Note. Mean±SD and ranges are given for EPSPs evoked by the first stimulus in the paired-pulse paradigm before tetanic stimulation. Subgroups with two S_n/v ratios (<0.6 and <0.5) are considered separately. CV_p is the coefficient of variation of p_i values.

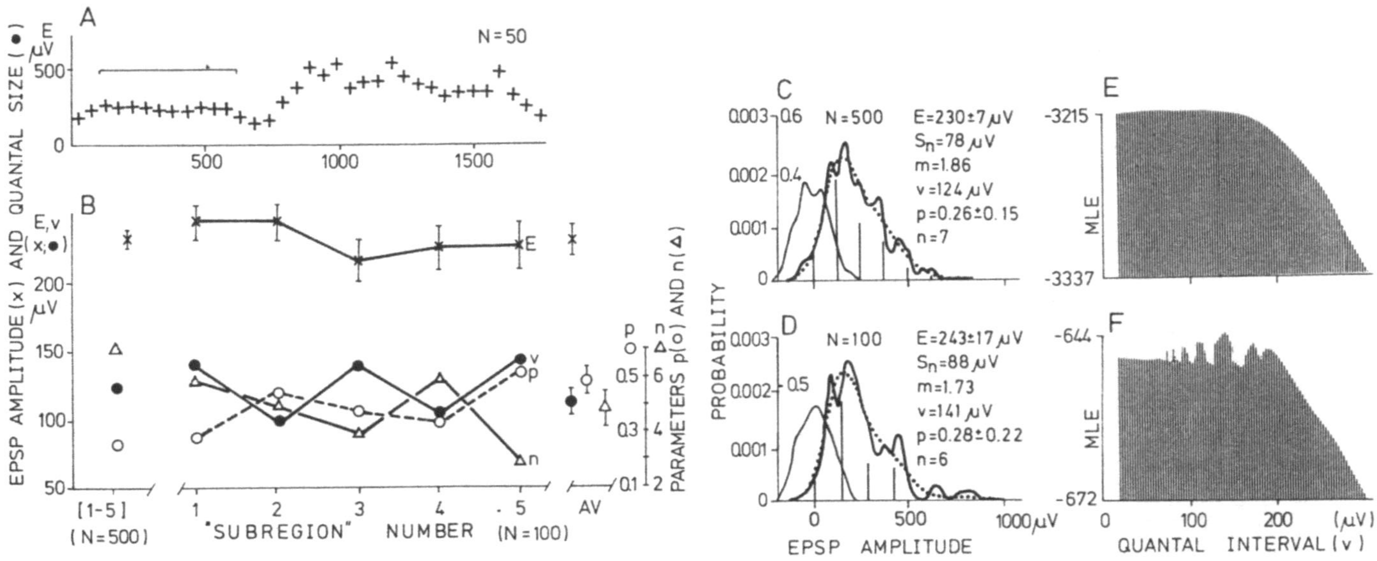

Fig. 15.3A–F. Comparison of estimates of fluctuation parameters of hippocampal EPSPs from large and small samples. **A** Mean EPSP amplitudes (*ordinate*) plotted against response number (*abscissa*). *Vertical bars* represent ±SEM. The *horizontal bar* indicates a plateau region without a significant trend. **B** Mean amplitudes (*crosses*) and quantal parameters (v, p and n, *full circles, empty circles* and *triangles*, respectively) calculated for the plateau region with N = 500 (*1–5 in brackets at horizontal axes*), for single consecutive small samples with N = 100 (*1 to 5 at horizontal axes*) and as averages of the estimates from the small samples (*AV at horizontal axes*). *Left vertical axes* for amplitudes (*crosses* ±SEM) and v (*circles*); *right vertical axes* for p (*dots*) and n (*triangles*). **C, D** Experimental noise and amplitude distributions (*continuous lines*), MLE solutions (bars), and theoretical distributions (*dotted lines*) for the large (**C**) and first small (**D**) samples. **E, F** Maximum likelihood functions plotted against quantal interval. See Fig. 15.1 for more details. [1063]

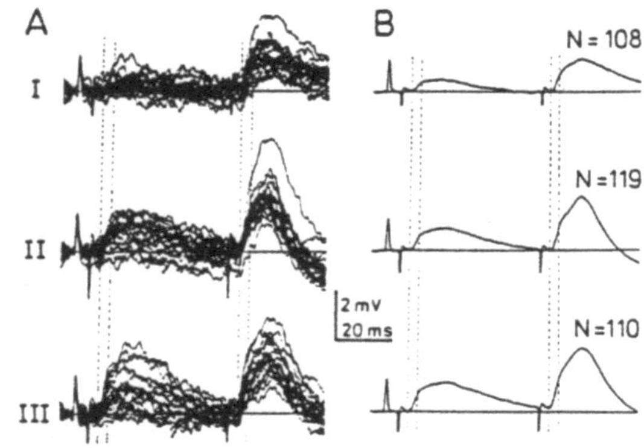

Fig. 15.4A, B. Example of long-term potentiation of hippocampal minimal EPSPs. Single (**A**) and averaged (**B**) EPSPs evoked by paired pulses are shown before (*I*), 4 to 16 min (*II*) and 20 to 35 min (*III*) after tetanic stimulation. *Vertical dotted lines* show the windows chosen for measurements; note calibration pulses (1 mV) in initial parts of sweeps. *N* Number of averaged responses. Membrane potential was –66 mV at the beginning and –67 mV at the end of registration. [1063]

Figure 15.4 shows examples of single (A) and averaged (B) EPSPs before (Fig. 15.4I) and after (Fig. 15.4II, III) tetanization. At least one pretetanic and one post-tetanic plateau region with N>90 could be extracted for nine neurones with ten tetanizations effective for LTP induction. Recordings of more than one EPSP (Fig. 15.4) were analyzed in all neurones: paired- (or triple-) pulse stimulation of one or two pathways were used. Altogether, we estimated fluctuation parameters for 60 regions (Table 15.2, All).

Figure 15.5 illustrates results of the deconvolution analysis for the neurone shown in Fig. 15.4. Only a small post-tetanic increase in v (Fig. 15.5AII,III; CII) or no essential v changes (Fig. 15.5CIII) were found. The lack of changes in v was substantiated by a comparison of visually estimated distances between major peaks in the experimental histograms (Fig. 15.5CI,II,III). In contrast, m increased in all cases (Fig. 15.5AII,III; CII,III) in comparison with the respective pretetanic values (Fig. 15.5AI, CI). In general, m increased in 30 of 37 post-tetanic regions, and the increase was typically close to LTP magnitude: Fig. 15.6A shows that most of the symbols are scattered around the equality line. The changes of the parameters of EPSPs evoked by the first (Fig. 15.6A, dots) and second or third pulses (Fig. 15.6A, triangles) in double- or triple-pulse paradigms were similar. The respective data are summarized in Table 15.2 (see All). There was a significant correlation between the relative m values and LTP magnitude (Fig. 15.6A, continuous line; Table 15.3, r = 0.91).

In contrast, there was no significant correlation for changes in v (Fig. 15.6B, continuous line; Table 15.3, r = –0.17). The mean changes in v were small: the mean post-tetanic relative v was equal to 121% (Table 15.3,W) in comparison with the mean LTP magnitude equal to 225%. However, v increased in the major-

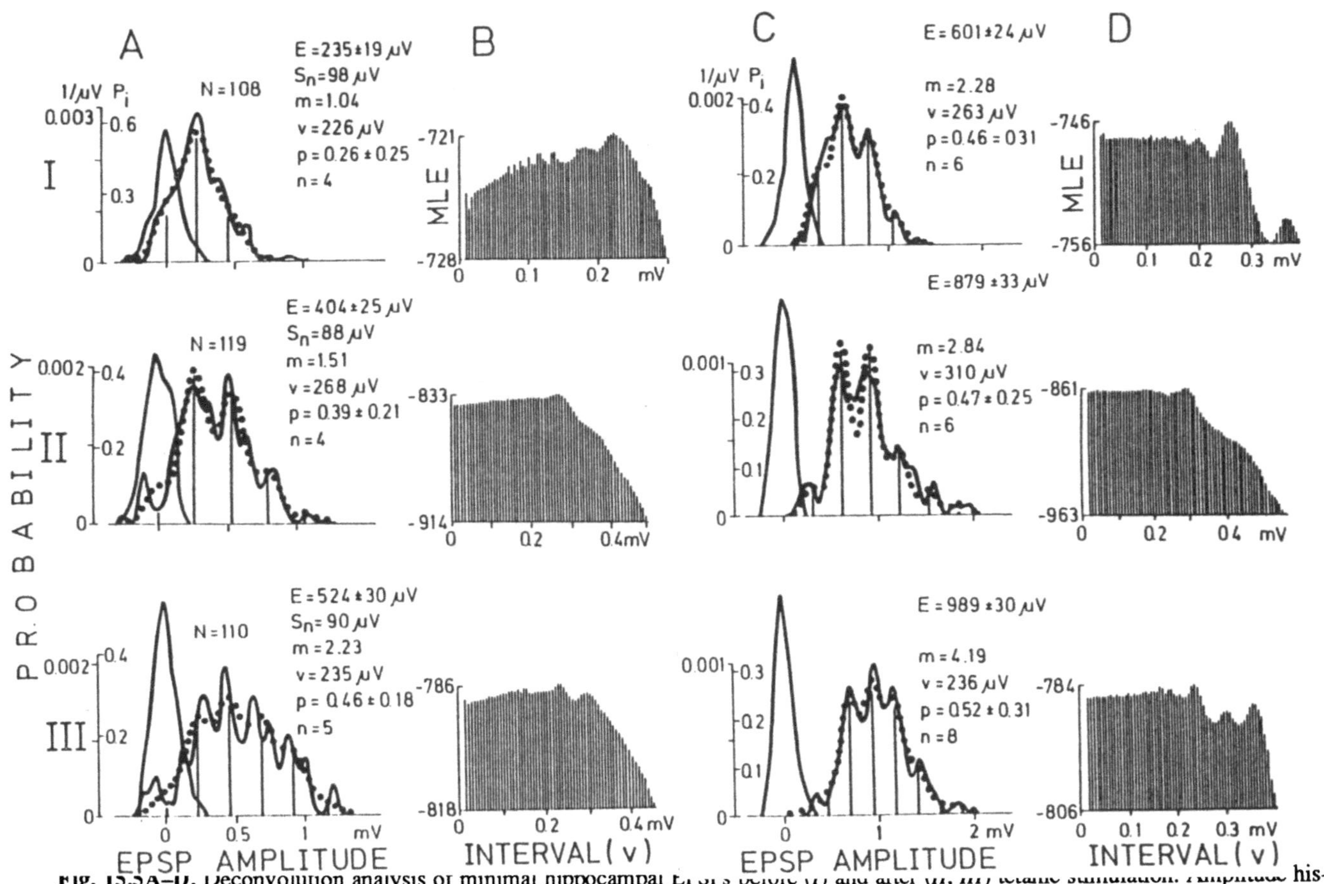

Fig. 15.5A–D. Deconvolution analysis of minimal hippocampal EPSPs before (*I*) and after (*II*, *III*) tetanic stimulation. Amplitude histograms with MLE solutions (**A**, **C**) and MLE (**B**, **D**) are shown for the neurone from Fig. 15.4. See Fig. 15.1 for other explanations. [1063]

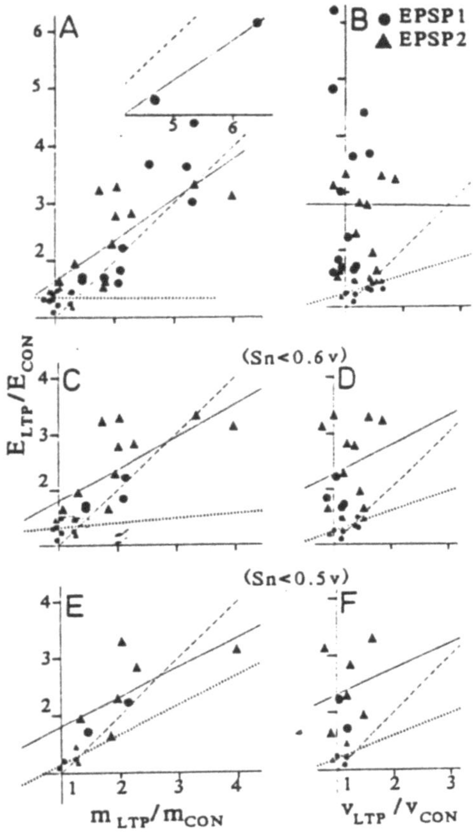

Fig. 15.6A–F. Relative LTP magnitude (E_{LTP}/E_{con}) plotted against changes in the mean quantal content (**A**, **C**, **E**, m_{LTP}/m_{con}) and quantal size (**B**, **D**, **F**, v_{LTP}/v_{con}). **A**, **B** Data for the whole sample. Note the two largest values plotted separately at right upper corner. **C**, **D** Solutions with $S_n/v<0.6$. **E**, **F** Solutions with $S_n/v<0.5$. Cases with LTP magnitude of less than 1.53 are shown by *smaller symbols*. *Dots* represent EPSPs evoked by the first stimulus in the paired- (or triple-) pulse paradigms (EPSP1); *triangles* represent the second (and third) EPSPs (EPSP2). *Thin interrupted line* is the equality line. Linear regressions are shown for cases with large and small LTP as the *continuous* and *dotted lines*, respectively. See Table 15.1 for more details of the correlation analysis. [1063]

ity of the post-tetanic regions (Fig. 15.6B, symbols to the right of the ordinate) and the increase was statistically highly significant (Table 15.2, All; Table 15.3, W). Examples of v increases are illustrated in Fig. 15.5AII.

The calculated S_n/v ratios varied from 0.27 to 0.86. Based on our simulations, we might expect that estimates with $S_n/v>0.6$ would be unreliable. However, their exclusion did not change the general result (Fig. 15.6C, D; Tables 15.2, $S_n<0.6v$ and 15.3, W, $S_n<0.6v$). The analysis of solutions which passed even more stringent criteria ($S_n/v<0.5$ with mean $S_n/v = 0.37$) gave the same result for the pooled post-tetanic regions (Fig. 15.6E, F; Table 15.3, W, $S_n<0.5v$): a significant increase in m which correlated with LTP magnitude, and a small but statistically significant increase in v which was not significantly correlated with EPSP potentiation.

Figure 15.6A suggests only small changes in m for cases with a small LTP magnitude similarly to previous results obtained in the context of the simple binomial model (Chap. 14). To test this suggestion quantitatively, we divided all post-tetanic estimates into two subset with small and large LTP magnitudes. We chose a border at a LTP magnitude equal to 1.53, so that the subset with smaller

Table 15.2. Quantal parameters of hippocampal EPSPs estimated by the deconvolution procedure before (region I) and after tetanic stimulation (regions II and III)

Set	Region	N	E(μV)	v(μV)	m	p	n
All	I	23	247±167	121±62	2.07±1.03	0.41±0.14	4.9±1.5
	II	21	476±345	146±74**	3.23±1.40**	0.51±0.10***	6.2±2.1
	III	16	468±234	139±59***	3.38±0.90**	0.47±0.08*	7.1±1.8**
S_n<0.6v	I	18	257±176	126±67	2.06±1.02	0.42±0.14	4.8±1.4
	II	17	511±368	156±77*	3.27±1.53**	0.54±0.09*	5.9±2.3
	III	12	506±243	154±60	3.32±0.98*	0.50±0.08	6.6±1.6*
S_n<0.5v	I	10	309±195	155±75	1.93±0.78	0.43±0.14	4.5±1.0
	II	15	547±376	164±78	3.38±1.59*	0.55±0.0	6.0±2.3
	III	8	566±223	163±64	3.58±0.78*	0.52±0.05	6.9±1.3*

Notes. Region II and III: 4 to 20 min and 16 to 55 min after tetanization, respectively. *), **) and ***) Significant increases for v, m, p and n in comparison with pretetanic (region I) values, P<0.05, P<0.01 and P<0.001, respectively, Wilcoxon test for matched pairs. Only neurones with statistically significant post-tetanic increase in E were used. See Table 15.1 for other explanations.

LTP magnitudes consisted of slightly more than 1/3 of all post-tetanic estimates. The correlation analysis showed that changes in m in the subset with small LTP were not correlated with LTP magnitude while changes in v were significantly correlated (Table 15.3, All, S; Fig. 15.6A, B, dotted lines). For the EPSP subset with large LTP magnitude, the results were quite similar to these for the whole sample: a strong correlation of the relative m values with LTP magnitude without a correlation for the v (Table 15.3, All, L; Fig. 15.6A, B, continuous lines).

Note that for the subset with small LTP magnitude, the changes in v were accounted for a considerable part of the LTP magnitude (Fig. 15.7A, S). Figure

Table 15.3. Changes in quantal parameters for solutions with different S_n/v ratios and different LTP magnitudes

S_n/v	Set	N	E	m	v(μV)	r(m/E)	r(v/E)
All	W	37	225±118	202±140***	121±29***	0.91***	–0.17
	L	23	280±121	261±149***	114±30***	0.88***	–0.0004
	S	14	135±13	105±16	132±23**	–0.01	0.53*
S_n<0.6v	W	24	195±75	165±76***	123±27***	0.80***	0.17
	L	14	240±68	205±77**	123±31*	0.66*	0.24
	S	10	132±14	109±16	123±20**	0.09	0.54
S_n<0.5v	W	13	195±76	175±80**	114±23*	0.80**	0.30
	L	8	239±63	212±82**	118±28	0.66	0.20
	S	5	124±13	116±14	108±11	0.54	0.32

Notes. Each post-tetanic value of E, m and v was expressed as percentage of the corresponding pretetanic value. Their means±SD are given. W, L and S represent the whole set of measurements (W) and two subsets with large (L) and small (S) LTP. r(m/E) and r(v/E) are the coefficients of correlation of changes in m and v with changes in E. *), **), and ***) Significance levels for coefficients of correlation and for increases in m and v in comparison with pretetanic values. See Table 15.1 for more explanations.

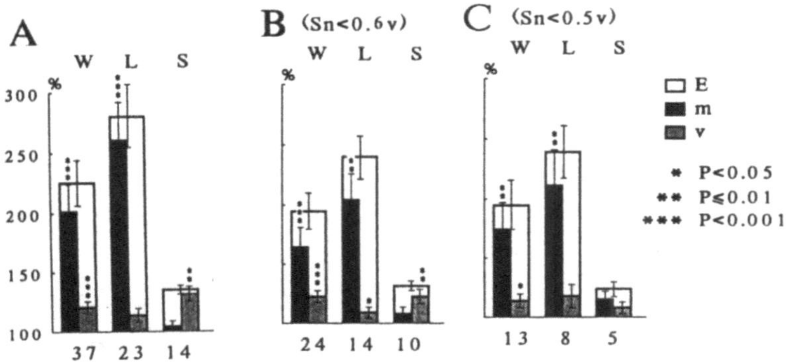

Fig. 15.7A–C. Changes in mean EPSP amplitude (*E*) and quantal parameters (*v* and *m*) for all solutions (**A**), for solutions with $S_n/v<0.6$ (**B**) and $S_n/v<0.5$ (**C**). Values for post-tetanic regions were normalized to pretetanic values. *Bars* represent means (±SEM) for the whole population (*W*) and two subsets of EPSPs: with large (*L*) and small (*S*) LTP magnitude. *Values below horizontal axes* give the number of measurements. *Asterisks* denote significant changes of m and v relative to pretetanic values (Wilcoxon test for matched pairs). [1063]

15.7A, S might even be interpreted as evidence in favour of an exclusive increase in v being responsible for the LTP of small magnitude. However, exclusions of less reliable solutions, firstly with $S_n/v<0.6$ (Fig. 15.7B, Table 15.3), and secondly with $S_n/v<0.5$ (Fig. 15.7C, Table 15.3) diminished the relative contribution of v changes in the LTP magnitude. Therefore, the general conclusion is that changes in both v and m were responsible for the potentiation of EPSP from the subset with small LTP magnitude. Nevertheless, the relative contribution of v changes into LTP was clearly larger for this subset than for the whole set of the solutions (Fig. 15.7).

In Fig. 15.8, changes in EPSP amplitudes and quantal parameters for the whole EPSP set and for two subsets with large and small.LTP magnitudes were compared for two post-tetanic regions. LTP was stable through the whole post-tetanic periods (Fig. 15.8, crosses) with no significant difference in the changes of either m or v (Fig. 15.8W, full and empty circles, regions II, III). There was an apparent tendency to an increase in v during the late post-tetanic period (region III) for solutions with $S_n/v<0.5$ (Fig. 15.8S, thin interrupted line). However, only one case was found for this subset for the late region III (see Sect.15.5.4 for discussion).

To test additionally the reliability of the solutions, we compared quantal parameters in a control group of eight neurones for three subsequent regions with N = 100. Altogether, 121 regions were analyzed. No conditioning tetanus was delivered after the initial 100 stimulations. There was a small (and statistically insignificant trend) towards a decrease in EPSP amplitude but no significant changes in v and m either for the whole set (altogether 121 region) or for regions with the most reliable solutions: with $S_n/v<0.6$ (80 regions) and with $S_n/v<0.5$ (45 regions).

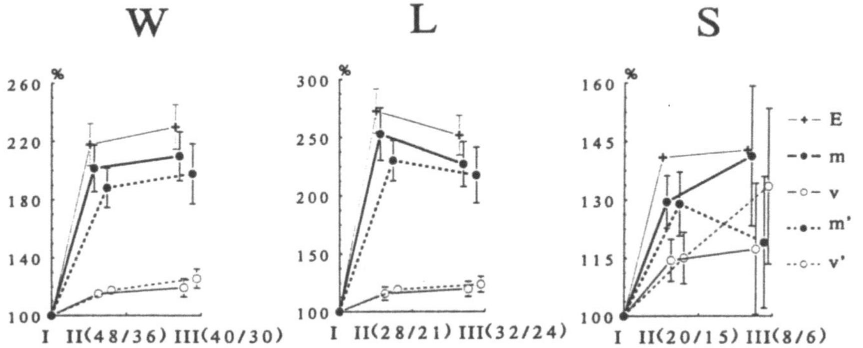

Fig. 15.8. Post-tetanic changes of the mean relative EPSP amplitude (*E, crosses*), and quantal parameters (*m, full circles; v, empty circles*) for the whole EPSP set (*W*) and for two EPSP subsets with large (*L*) and small (*S*) LTP magnitudes. All values calculated for regions *II* and *III* were normalized to pretetanic values (point *I* at horizontal axes) and plotted as a percentage of the latter ±SEM (points *II* and *III* at horizontal axes). *Continuous* and *interrupted lines* represent all data and data for cases with S_n<0.5v, respectively. *Values in parentheses to the left* and *to the right of the slash* show the total number of measurements and the number of measurements for cases with S_n<0.5v, respectively. [1063]

15.5 Discussion

15.5.1 Reliability of the Deconvolution Procedure

The non-unique nature in the fitting of statistical models and uncertainty in confidence intervals for MLE solution have been stressed by others [81, 230, 567, 568, 823, 864, 1112]. We used a variant of the "quantal model" [823, 1068] which suggested equal distances between discrete components of the searched "noise-free" EPSP amplitude distribution. This suggestion is identical to the basic postulate of the quantal hypothesis which was substantiated for the central synapses by many studies (Chap. 7). Specifically for the hippocampal synapses, it was supported by histograms with regular peaks in our previous chapters as well as in the present one (Figs. 15.5A, C). Results obtained with the "unconstrained deconvolution" of hippocampal EPSP [864, 865] also support the quantal nature of the synaptic transmission in area CA1. The "unconstrained quantal" model is a more constrained variant of the deconvolution model than that used by some other authors who studied central synapses [81, 464, 823] including hippocampal ones [864, 865]. It has already been stressed (Chaps. 7, 13) that more restrained models have the advantage of less severe demands on S_n and N to obtain statistically reliable estimates. We were not able to collect large samples of EPSP amplitudes in steady state conditions appropriate for reliable induction of LTP (see Chap. 13 for the discussion of LTP of minimal EPSPs). However, simulation experiments showed that, although with $S_n/v>0.6$, estimates of v might essentially deviate from the true value for both large and small N (see also [230]), a correct solution was obtained for almost every small sample with $S_n/v<0.4$ and $n<10$. With S_n/v in the range of 0.4–0.6, the estimates based on N = 1000 can be substituted by several (about 10) estimates based on small samples (N = 100). It was this approach that we used in physiological experiments.

15.5.2 Post-Tetanic Changes in Quantal Parameters

The deconvolution analysis showed that LTP was accompanyied by an increase in probability of occurrence of larger EPSPs and by a decrease in probability of occurrence of smaller EPSPs. The quantal separation between discrete amplitudes was only slightly increased. This increase accounted for only about 10% of the mean LTP magnitude. The result was consistently reproduced in the majority of our solutions including two groups with smaller estimated S_n/v ratios (<0.6 and <0.5) and more distinct regular peaks (Fig. 15.5). There was no statistically significant correlation between changes in v and LTP magnitude. In contrast, a strong correlation was obtained between LTP magnitude and changes in m. This consistency and correlation cannot be explained by uncertainty of MLE solutions or by a chance hit. In the context of the quantum hypothesis, the basic result of this chapter suggests a presynaptic location of major mechanisms of LTP maintenance (see Sects. 12.7, 14.4 for reservations). Therefore, the present analysis with a more sophisticated model of transmitter release actually led to conclusions similar to those achieved on the basis of the simple binomial model (Chaps. 12, 14). Parameters estimated in the present study for non-facilitated EPSPs were strikingly similar to the respective parameters of the simple binomial model (Table 13.1) if one has in mind that p should be overestimated and n underestimated by the formulae based on the binomial statistics [175]. However, the deviations from true values should not be large when the non-uniformity of p is not large, as might be for the major part of our cases (see CV_p, Table 15.1). Note, however, that the estimates of p, n and CV_p should be considered as much less reliable than v and m (Fig. 15.3B, see also [567]). Nevertheless, the increase in probability of occurrence of larger EPSPs might be interpreted as an increase in release probability for the effective release sites together with appearances of measurable p values for previously noneffective (silent) release sites (see [1054] and discussion related to the silent synapses in Sect. 12.7). It corresponds to post-tetanic increases in both p and n in the context of the present model (non-uniform p and n equal to the maximal number of "effective" release sites).

15.5.3 Differential Changes in Quantal Parameters at Different LTP Magnitudes

In distinction from some other recent studies [100, 662], we used a model which allowed us to describe quantitatively a relative contribution of changes in two basic parameters (v and m). We substantiated here the previous finding of a small post-tetanic increase in v (Chap. 14) and found that the increase was statistically highly significant when a more precise method of the description of EPSP amplitude distributions was used. In addition, we confirmed the differential changes in m and v (Chap. 14, see also [566]) for LTP of different magnitudes, i.e. a relatively larger v increase for cases with small LTP. The initial impression from the present data was that the changes in v might account for all LTP magnitude in cases with small LTP. This conclusion would be in complete agreement with the results of Foster and McNaughton [342, 343], but it would be in variance with our previous data based on the binomial assumption and with results of other authors (Sect. 15.1). However, the consecutive exclusion of less reliable solutions (Fig. 15.7B, C) suggested a contribution of m changes in the EPSP potentiation in the subset with small LTP magnitude as well. Therefore, we keep to our previous suggestion (Chap. 14), namely, that in spite of a larger relative contribution of v changes in cases with small LTP, the absolute net increase in v might be on average at a similar level of about 10 to 30% independent of LTP magnitude.

A tendency towards a larger increase in v for the later post-tetanic regions was noted in Chapter 14. However, a similar tendency was evident here neither in whole EPSP set nor in the EPSP subset with large LTP. Only one EPSP was recorded during both earlier and later post-tetanic periods for the subset with small LTP for the group with the most reliable estimates (S_v<0.5v). A recent analysis of this case with a new method of "noise addition" [1064] did not reveal any additional increase in v during the later post-tetanic period.

15.5.4 Two Types of Synaptic Mechanisms of LTP Maintenance

On the basis of our present and previous observations, two types of synaptic mechanisms of LTP maintenance might be suggested. The first mechanism, based on an increase in m, accounts for the major part of EPSP potentiation. This mechanism is able to support a severalfold increase in EPSP amplitude and it makes a major contribution to moderate and large LTP. The mechanism is localized presynaptically and depends on an increase in p which actually led also to an increase in the number of effective release sites (n). I believe that this is, in fact, analogous to a conversion of previously ineffective (silent) synapses into effective ones. The second mechanism suggested by our data is based on an increase in v. It might be located either pre- or postsynaptically (Sect. 14.4). This mechanism accounts for an essential part of a small LTP and saturates at about 20% above the pretetanic level. It might underlie the maintenance of a recently described smaller LTP component which did not require the activation of N-methyl-D-aspartate (NMDA) receptors [389] and which was frequently dependent. This suggestion is supported by the demonstration [389] that LTP of only a small magnitude (about 10% as judged by the FP slope) was induced when the other (NMDA-sensitive) mechanism was blocked. The suggestion might be tested in experiments with quantal analysis of EPSP amplitudes after conditioning tetani with different frequencies and after application of NMDA receptor antagonists.

15.6 Summary

In previous chapters, quantal analysis assuming the simple binomial model has shown that long-term potentiation (LTP) is accompanied by an increase in both mean quantal content (m) and quantal size (v), whereby the increase in m predominates. In the present computer experiments, "compound" binomial distributions with variable probabilities were convolved with the Gaussian distributions to simulate amplitude histograms of intracellular excitatory postsynaptic potentials (EPSPs). A deconvolution procedure assuming equal "quantal" separation (v) between discrete components, but without assuming the binomial statistics, was applied to the simulated distributions to determine v. It was found that with small ratio of noise standard deviation to v ($S_n/v<0.4$), reliable estimates of v could be usually obtained even for small samples (N = 100). When S_n/v was larger (0.4–0.6), approximate v estimates (within ±10–20% of the simulated v) could be obtained by averaging estimates from about 10 small samples (N = 100). "Minimal" EPSPs were recorded in area CA1 of guinea pig hippocampal slices. Amplitudes of 37 EPSPs (9 neurones) were measured before and up to 55 min after 10 tetanizations of *stratum radiatum*. In accordance with the previous data, the increase in v accounted for only about 10% of the average post-tetanic increase in EPSP amplitude and was not correlated with the latter. However, for an EPSP subset with a small LTP magnitude, an increase in v accounted for an essential part of the LTP magnitude, while an increase in m did not correlate with it. The results are in agreement with the previous data obtained in the context of the simple binomial model and are interpreted as indicating primarily a presynaptic mechanism of LTP maintenance. The results suggest two types of synaptic mechanism of LTP maintenance related to the increases in m and v, respectively. The latter mechanism is saturated at about 10 to 30% increase in EPSP amplitude above the pretetanic level.

General Conclusions

In our studies, a well-pronounced LTP similar to that described by Bliss and his colleagues for the FPs of the DG [143, 145] was demonstrated under new conditions, i.e. at the hippocampus proper, with single neuronal recordings, with intracellular recordings from unanaesthetized animals. Several LTP properties were described in our studies which supported the suggestion that mechanisms similar to those of memory and conditioning underlie this phenomenon. An analysis of LTP mechanisms was performed with recording several electrophysiological indices, including FPs, minimal FPs, single unit spikes and intracellular potentials. Statistical methods based on the quantum hypothesis were used for the first time to investigate a phenomenon of long-lasting plasticity at the vertebrate brain in both the in vivo and in vitro preparations. Our findings suggest that the major mechanism of LTP maintenance consists in changes of synaptic efficacy. Therefore, we support the "synaptic" rather than "membrane" hypothesis to explain major mechanisms of memory trace formation. More specifically, an increase in the efficacy of excitatory synapses was shown without critical participation of changes in inhibitory synapses. Our quantal analysis suggests that the changes in synaptic efficacy are mainly due to presynaptic mechanisms which increase the number of transmitter quanta released by one presynaptic spike (see Sect. 12.7 and below for reservations). Another synaptic mechanism of LTP maintenance is also suggested by our data which might have postsynaptic location (see below), but this mechanism can maintain only a smaller part of LTP. As was noted, our conclusions are in agreement with most literary data obtained, mainly more recently, and almost exclusively in experiments with in vitro preparations (see reviews cited in the Introduction and references in preceding chapters).

It was mentioned that some literature data might also indicate participation of membrane mechanisms such as a decrease in spike threshold to afferent stimulation [32, 221, 910, 955]. The second mechanisms of LTP maintenance suggested by our analysis might, in principle, be related to the postsynaptic membrane changes. However, evidence in the literature in favour of the membrane mechanisms of LTP maintenance requires more precise quantitative evaluation to exclude participation of other factors, e.g. changes in inhibitory inputs. In any case, the general conclusion is that the membrane changes (if any) might constitute only minor additional LTP mechanisms.

The lack of essential changes in quantal size does not reject the possibility of changes in subsynaptic receptor sensitivity under conditions when initially ineffective synapses exist (see Chap. 12). An increase in the number of Glu binding sites following tetanic stimulation in CA1 was initially claimed by Lynch and his colleagues [94, 637, 640] but was not supported by other studies of CA1 [381, 856] and DG [146, 645] (see also [642]). It is pertinent to mention that no learning related changes in Glu receptor binding were found in the hippocampus [734].

First most serious difficulties for the "synaptic" and especially "presynaptic" hypothesis appeared after a demonstration of a blockade of LTP resulting from intracellular injection of a Ca^{2+} chelator [641]. Naturally, this result could be criticized from several aspects, e.g. that it was a negative result observed with application of a harmful agent in experiments based on a statistical approach. These agents might influence the recorded potentials and thus counteract the possible LTP effects. The chelator was applied through the intrasomatic electrode, while the expected changes took place at apical dendrite synapses. Moreover, it is known that Ca^{2+} chelators produce changes in the RMP and R_{in} [553], although Lynch et al. [641] did not reveal such changes. However, these observations [641] were reproduced by others [659] and the involvement of postsynaptic mechanisms was supported by several other lines of evidence. For example, suppression of postsynaptic inhibition was shown to facilitate LTP induction [1094] while intracellular injections of Cs^{+} prevented it [405] (see, however, [659]). Essential evidence for the postsynaptic mechanisms was obtained in experiments with a blockade of subsynaptic receptors [297, 557] (see, however, [276]). More recent studies [235, 237, 422] suggest that it is the activation of a Glu receptor subtype, namely the NMDA receptor that is a prerequisite for the induction of LTP (see reviews cited in the Introduction and [735, 779] for more recent references and [857] for reservations). As a result, a scheme of LTP induction was outlined by Collingridge [235] and Wigström and Gustafsson [1095] (see also [270, 271]) which I shall refer to as the "CWG scheme". It describes LTP induction as a sequence of the following events: tetanic stimulation produces postsynaptic depolarization; the depolarization unblocks the effects of Mg^{2+} on NMDA receptors; the receptors open Ca^{2+} channels; Ca^{2+} enters the cell; elevated intracellular Ca^{2+} triggers further mechanisms which lead to an increase in synaptic efficacy. It should be stressed that NMDA receptor blockers do not influence either control or potentiated low-frequency responses in the hippocampus which indicates that changes of NMDA receptors are not involved in maintenance of hippocampal LTP. It might not be true for LTP-like changes in other structures (e.g. [18, 518, 772]) and for other plasticity phenomena [729, 736]. An essential element of the CWG scheme is a relief of the voltage-dependent Mg^{2+} block of NMDA-activated channels [768] as a result of depolarization during tetanic stimulation. The depolarization is due to EPSP summation and the blockade of inhibition (Chaps. 4, 5). The CWG scheme readily explains LTP cooperativity and associativity and the necessity of Ca^{2+} ions for LTP induction. The role of Ca^{2+} was supported by many LTP studies: see reviews cited in the Introduction, including demonstrations of "Ca^{2+}induced LTP" [984], and more recent data [659] which indicated that an increase in intracellular Ca^{2+} concentration was a necessary and sufficient condition for formation of at least the initial LTP phase (see below). The CWG scheme was further supported by observations showing that postsynaptic hyperpolarizing current during tetanization blocks induction of LTP [501, 661], whereas postsynaptic depolarization facilitates LTP induction after a low intensity tetanization [501]. Moreover, LTP of intracellularly recorded EPSPs could be induced by several conjunctions of single presynaptic volley with Glu iontophoresis [452] or intracellular depolarization [859, 1097]. It should be noted that, in fact, all these observations were a revival of earlier experiments of Bures and his associates, who studied effects of pairings of peripheral stimuli with local activation of single neurones by extracellular current or Glu injection [187]. Increased neuronal responsiveness was found after the pairings in various structures including the hippocampus [366] (see also [70, 909] for similar initial observations at the neocortex). Therefore, cortical modifiable synapses seem to behave in a manner similar to that predicted by Hebb [418] when he proposed that after correlated activation of cells *A* and *B*, "*A*'s efficiency, as one of the cells firing *B*, is increased" ([418], p. 62). The Hebbian condition or Hebb rule was widely used in numerous constructions of modifiable networks (see Sect. 1.4). An obvious contradiction between the data indicating presynaptic LTP mechanisms and the CWG scheme presuming a postsynaptic involvement had been resolved by suggesting a distinction between the localization of mechanisms of LTP *maintenance* (or expression) and the mechanisms (or conditions) of LTP *induction* (or initiation). The presumed presynaptic mechanisms of LTP induction led the necessity of existence of a feedback (retrograde) mechanism from the post- to the presynaptic side. Studies of such retrograde messengers have already started (e.g. [362, 1100, 1123]) and they will undoubtedly be one of the major lines of future studies on LTP mechanisms.

There are some findings which do not fit readily into the widely adopted CWG scheme. I shall mention some of them, which might also constitute new lines of future studies. LTP in the mossy fibre-CA3 circuit was shown not to be blocked by antagonists of NMDA receptors [412]. Respectively, the mossy fibre layers are sparse in NMDA receptors [731]. A different type of LTP without associative properties and induced exclusively by presynaptic activation was proposed (see review [562]) but a postsynaptic involvement in LTP induction at CA3 was recently demonstrated [467, 1099]. NMDA independent LTPs were found also in CA1, firstly, as a component of LTP induced by high-frequency stimulation [389], and secondly, as a potentiation induced by a blocker of K^+ channels [35]. It was suggested that an entry of Ca^{2+} through voltage-dependent channels might induce LTP [35, 389]. Moreover, LTP-like changes were observed in Ca^{2+} free medium after activation of Schaffer collateral-CA1 connections paired with postsynaptic depolarization [685]. The data suggest that transmitter release and activation of subsynaptic receptors might not be necessary for LTP induction, and that a release of intraneuronal Ca^{2+} from storage sites might be sufficient. The same group [860] obtained evidence of glial involvement in LTP induction. They recorded a prolonged depolarization of non-spiking (presumably glial) cells after tetanic stimulation and found persistent EPSP and population spike increases after afferent stimulation paired with current injections into presumed glial cells.

Kuhnt [564] reported LTP-like changes of EPSPs after intracellular tetanic stimulation (5-ms pulses at 100 Hz) without conjunction with presynaptic volley in in vitro CA1 neurones. I shall refer to this phenomenon as the "Kuhnt effect" or "intracellular potentiation" (ICP). The result was especially surprising for me because only short-term (minutes or less) changes had been observed previously after intracellular stimulations of neocortical neurones in experiments of our [1037–1039] and other groups [802, 1118, 1121]. Nevertheless, in our joint experiments performed both in Göttingen and Moscow together with U.Kuhnt, R.N.Davletshin and G.Hess, even more pronounced changes in minimal EPSPs were found after intracellular tetanization of CA1 neurones in slice preparations [1023, 1052, 1054]. Our preliminary quantal analysis indicates an essential increase in m similar to that during usual LTP [1052]. Recently, similar ICP ("non-Hebbian LTP") was described in slices of the entorhinal cortex after longer (100 ms) intracellular pulses delivered at 5 Hz [18]. An NMDA receptor blocker was effective in suppression of the ICP. In distinction from hippocampal LTP, the blocker abolished the potentiated EPSP component. Therefore, both induction and maintenance of the ICP in the entorhinal cortex was dependent on the activation of NMDA receptors. The Kuhnt effect is consistent with involvement of postsynaptic mechanisms in LTP induction, but it is not easy to explain within the CWG scheme. The data disagree with observations of others [397, 1097] that postsynaptic depolarizing pulses not paired to presynaptic activation were not effective in inducing LTP-like changes. There are at least two not mutually exclusive testable explanations of the ICP induction. The first suggests that ICP results from activation of voltage-dependent Ca^{2+} channels. The elevated Ca^{2+} would lead to a cascade usual for LTP. It may be suggested that a relatively small number of unpaired low-frequency depolarizing pulses used by other authors [397, 1097] were not adequate to raise intracellular Ca^{2+} concentration to the magnitude sufficient to induce LTP. An additional (or even major contribution) could give Ca^{2+} release from intracellular storage sites as a result of mechanical damage of charged intracellular organelles when they move relative to the microelectrode under influence of current pulses. However, the first explanation is not supported by the NMDA dependency of the ICP described by Alonso et al. [18]. A critical test would be an evaluation of the NMDA dependency of the hippocampal ICP. A second possible explanation initially proposed by Dr. A.Schütz (pers. comm.) preserves essential features of the CWG scheme and presumes that the intracellular depolarizing pulses are paired with spontaneously released transmitter quanta. The spontaneous release might be different in different pathways, and this explains why ICP was sometimes not equally expressed in both tested inputs in our experiments (unpubl. observ.). Indirect evidence in favour of the second explanation is a presumed significant increase in spontaneous release in testing inputs (see Sect. 13.5.5). The explanation can be tested in experiments with two pathways stimulated before the intracellular tetanization with essentially different frequencies. Although Alonso et al. [18] were able to induce ICP when they hyperpolarized the postsynaptic neurone to prevent spiking, the hyperpolarization could be insufficient to prevent opening of the related channels at remote dendritic spines.

In the course of our study of the ICP, we found that intracellular tetanization of both neurones and non-spiking cells as well as extracellular tetanization with nA currents produced small (about 10%, on average) but statistically significant increase in population spike amplitude [1023, 1055]. These observations seem even more unusual than the ICP and should be reproduced in other laboratories. For the present time, they might be related to several groups of independent observations. One related observation is a widespread increase of optical responses from hippocampal [154] and cortical [532] slices after conjunctions of intracellular stimulation of one neurone with intracellular depolarization of another one. The other related observation is the so-called "field potentiation" revealed after subthreshold (ineffective for FP changes) sinusoidal or long-pulse polarization applied to the whole hippocampal slice [97]. At last, there are demonstrations of various morphological effects (e.g. changes in neurite growth, [795]) induced by currents of the order of nA or fields of <100 μV/cm (see also reviews [321, 905]). Both ICP and this small widespread potentiation of the population spikes might be useful for studies of "retrograde" and other signalling factors of intercellular communication and understanding mechanisms of LTP specificity.

It should be noted that in spite of the results of the quantal analysis and many other related data (Sect. 12.7), there is still no general agreement on the location (pre- or postsynaptic) of major mechanisms of LTP maintenance. Moreover, two sets of recent data were interpreted as strong support of a postsynaptic involvement. Firstly, recent in vitro studies in CA1 [741] and CA3 [1132] confirmed earlier observations obtained in the DG of anaesthetized rats [695] that during LTP no significant changes in the PPF occurred (but see [560]). A lack of changes in PPF magnitude might indicate that LTP is not accompanied by presynaptic changes [695, 741] and particularly not by alterations in the presynaptic Ca^{2+} current associated with transmitter release [741]. However, in these studies [741, 1132], the PPF magnitude in CA1 was derived from FP measurements. We reconsidered the problem of the changes in PPF after LTP induction with intracellular recordings and found that the PPF magnitude of minimal EPSPs significantly decreased during LTP [1033]. We found also a correlation of pretetanic PPF ratio with LTP magnitude. These result are in agreement with the involvement of presynaptic changes in LTP maintenance.

The other set of data which were interpreted as strong support for the postsynaptic hypothesis is a selective post-tetanic increase in EPSP [498] or FP [744] components which are not mediated by NMDA receptors. One way to explain these data in the framework of the presynaptic location of LTP mechanisms is to suggest a spatial segregation between synapses with NMDA and non-NMDA receptors [610, 1033]. This suggestion is in agreement with the known idea of "teacher" and "pupils" synapses [668, 698], the "teacher" being synapses with NMDA receptors. The possibility that different receptor subtypes are segregated and might be activated by different presynaptic endings is supported by experiments on spinal [900] and hippocampal [812] neurones. Observations on cultured rat hippocampus [99] indicate that NMDA and non-NMDA receptors are segregated at about 30% of synapses. Because the cultivation might influence tissue properties, the percentage can be even higher in normal hippocampal slices. Moreover, post-tetanic increases in NMDA components were demonstrated recently under special conditions [86] which totally agree with the presynaptic hypothesis, even without assumption of essential spatial segregation of NMDA and non-NMDA receptors.

In addition to LTP mechanisms, synaptic mechanisms of two simpler plasticity phenomena were under our study, namely FF and LFD. Predominantly presynaptic localization of the underlying mechanisms was indicated as well. On the other hand, essential changes in v were firstly demonstrated under normal conditions without excessive stimulation or suppression of transmitter synthesis. Similarly to our LTP results, these changes in v might indicate participation of not only presynaptic but also subsynaptic and postsynaptic mechanisms. However, as with LTP, these mechanisms seem to be additional to a major presynaptic mechanism which modifies the number of quanta released by every presynaptic spike.

Future studies of LFD is of interest by its own and also in view of its possible relation to LTD which might constitute another line of future studies of long-term plasticities [128, 459, 624, 1088]. Interesting observations are that postsynaptic depolarization of two different levels could induce either LTP or LTD in in vitro visual cortex [45]. There is indirect experimental evidence for action of similar mechanisms in in vivo visual cortex: only synchronous electrical activation of the optic nerves led to maintenance of binocular connections while asynchronous activation led to competition so that only one pathway consolidated and the other became repressed (see [903] for review). A slightly different rule for the "destabilization" of the connections during visual developmental plasticity ("prunning process") was postulated: "the connections destabilize if there is high probability that the presynaptic terminal is inactive at the same time as the postsynaptic target is activated" [903]. It is interesting to note that, similar to LTP in CA1 region, the activation of NMDA receptors is necessary for the disconnection after visual deprivation.

LTP in non-hippocampal structures, especially in the neocortex, became more and more popular for studies of plasticity mechanisms (see reviews cited in the Introduction). Actually, long-lasting changes of neocortical responses under certain conditions were described a long ago (see Introduction to this book). Long-lasting increases in probability of neuronal discharges and amplitudes of EPSPs evoked by single direct cortical stimuli were found on our in vivo studies on waking rabbits after pairings those stimuli with direct cortical trains [1029]. Even more reproducible (but usually not lasting >30 min) changes of neocortical EPSPs were reported with recordings from anaesthetized cats after pairings of thalamic, callosal, peripheral and antidromic stimuli in various combinations including pairings of presynaptic stimuli with intracellular stimulation [68–70]. All such "cellular analogues" [478] of behavioural CRs might be considered now as examples of "associative" LTP (Sect. 1.10 and below). Prolonged (15 min to 1 h) 2 Hz stimulation was found to induce long-lasting enhancement of cortical FPs evoked by the white matter stimulation in slices taken from the developing visual cortex [517]. Post-tetanic changes of neocortical FPs under conditions more similar to these used to induce hippocampal LTP (i.e. with short 200 Hz tetanic trains to the white matter) were reported more recently for in vitro [598, 1020, 1021, 1051] and in vivo [1020, 1021, 1102] preparations. The interpretation of neocortical FPs is less certain than hippocampal ones. However, neocortical responses with more certain interpretation have been also shown to express LTP-like changes: PTRs [543, 544, 1017, 1021], neuronal spikes ([998, 1051] and EPSPs [44, 71, 128, 854]).

LTP-like changes at peripheral synapses (Sects. 7.10.5 and 12.7) hold many advantages for future studies. For example, "pre-" and "post-LTPs" in sympathetic ganglia [562] are promising phenomena to resolve contribution of pre- and postsynaptic mechanisms in the maintenance of long-term changes. Autonomic ganglia [167, 277, 618, 727, 769, 908] might also be convenient for elucidation of the important problem of participation of "modulating" or "reinforcing" structures and substances in long-term plasticities.

Some morphological correlates of hippocampal LTP have already been mentioned (Sect. 6.1). In general, the available morphological data are contradictory. A variety of different modifications is described (e.g. [266, 711]) including pre- and especially postsynaptic ones (see reviews [146, 203, 365, 562]). The changes in dendritic spines (see Sect. 6.1) could be an interesting possibility to explain the relatively small increase in v. However, these changes were not found after afferent tetanization in area CA1 [600], except in one study [41]. Moreover, an increase in the area of spines was reported in a CA3 preparation which failed to develop LTP after tetanization [1080]. It was noted that effects on spine shape disappeared after 2 h, while physiologically demonstrable LTP persisted over the 8-h post-tetanic period [217]. No significant postsynaptic structural changes were found in amygdala and neocortex after many tetanizations in studies [378] of so-called "kindling" phenomenon (see below). A spine swelling in cochlea neurones was shown to be correlated not with response facilitation but rather with an increase in threshold to sound stimulation [831]. Thus, the electron-microscopic data are far from being conclusive, and spine shape changes seem not to be necessary for LTP maintenance.

Intensive biochemical and pharmacological explorations of LTP were initiated later than the related physiological studies, but they generated a large number of challenging and often controversial findings [562, 610, 642, 651]. I shall mention here only some directions related to our data. For example, some studies of transmitter release supported our general conclusion on an increased transmitter release during LTP, but others did not (Sect. 12.7). The CWG scheme stresses the important role of an increase in intracellular Ca^{2+} concentration for LTP induction. There are many biochemical processes which can be activated by Ca^{2+} ions. Among several Ca^{2+}-dependent enzymes, calmodulin-activated and Ca^{2+}/phospholipid-activated protein kinase C (PKC) were most widely discussed candidates (e.g. [562, 610, 651, 751]). Phorbol esters, which are known to activate PKC, produced LTP-like changes of both FPs and EPSPs [658]. Negligible changes in RMP, R_{in} and a depolarization induced by Glu, as well as increases in frequency of spontaneous EPSPs and transmitter release, support presynaptic action of phorbol esters [338, 658] which is compatible with the well-known involvement of PKC and the phosphoinositide (PI) cell-signalling system into regulation of transmitter release in many preparations (e.g. [119, 472]). Increased PI turnover was demonstrated in the DG [67] and in synaptosomes isolated from CA3 area [646]. All these effects are probably related to a sustained increase in amino acid release. On the other hand, Anikszteјn et al. [36] observed only a transient increase in transmitter release after a phorbol ester application, and Hu et al. [442] demonstrated LTP-like changes of EPSPs after intracellular injection of PKC into CA1 neurones. Generally, exact mechanisms of PKC and PI involvement in LTP induction and maintenance are not clear. Although phorbol esters were shown to facilitate induction of "non-decremental" (a few hours) LTP by a weak tetanus [49, 843], the application of phorbol esters alone without the weak tetanus induced a potentiation [658] which was more recently shown to decay to baseline within 30–60 min [663, 743]. Moreover, the occlusion between LTP and effects of phorbol ether reported in the earlier study [658] was found more recently to be only partial [398, 743]. Therefore, the potentiation produced by phorbol esters might not be equivalent to LTP [745]. Nevertheless, all these data suggest that PKC activation might be necessary for the maintenance of at least initial LTP stages.

The studies were mentioned (Sect. 12.7) in which increased phosphorylation of a PKC substrate, a protein called B-50 [66, 373] or F1 [628, 629], was found following LTP and correlated with LTP magnitude. The protein was shown to be identical with other membrane phosphoproteins associated with axonal regeneration (GAP-43 or GAP-48) and axonal growth during development (pp46) (see [114]). The presynaptic localization of the protein favours presynaptic LTP mechanisms [373, 610]. Similarly, a correlation between LTP magnitude and phosphorylation of synaptic vesicle proteins (72 and 55 kdaltons) was found [756].

One important general result of biochemical analysis of LTP is the support of a concept on several LTP stages similar to several memory stages (e.g. [681, 826, 828]). Related physiological data have been already mentioned (Sect. 1.10). For example, Racine et al. [809] demonstrated two time constants of LTP decay (1.4 h and 5.1 days). It is clear now that time courses of different stages might depend on various factors including conditions of LTP induction [657] and the way of LTP testing (Sects. 1.7, 1.10). Nevertheless, both physiological and biochemical studies allow for dividing conventionally LTP maintenance into three following stages [826]: earlier ("decremental") LTP with a return to baseline within about 30–90 min, intermediate (several hours) and late stable (days and weeks) stage. Exact relationships between the stages and their mechanisms await further studies. For example, PKC activation might mediate LTP maintenance for several hours, but it is unlikely to underlie longer LTP stages which might be mediated by increased protein synthesis. Accordingly, an LTP-associated release of newly synthesized proteins was found following tetanic stimulation [219, 292, 332, 626]. Influences of inhibitors of protein synthesis on LTP were reported [255, 285] with some inhibitors showing no effect and others showing LTP blockade only when they were applied before tetanus [930]. Some authors observed the blockade of only late

LTP stages from 4 to 6 h onwards [347, 558, 782] and explained the observations of others by side effects of the inhibitors. Rather surprisingly, predominantly blockade of a slow component of FPs was found in these studies with weaker (if any) effects [782] on LTP of the population spikes. The necessity of intracellular transport for the maintenance of the late LTP stage was suggested by experiments of Frey et al. [348], which demonstrated that LTP lasted only about 4 h in the dendritic region which was separated from the cell bodies of the CA1 area. There are indications of involvement of immediate-early genes in LTP-related processes [233, 285], although activation of several immediate-early genes was noted in the absence of LTP [282, 473, 1108] (see reviews [562, 651] for additional discussion of gene expression and other molecular mechanisms of LTP). One more important result of the development of the biochemical studies is a discovery of a new ("metabotropic") subtype of Glu receptor that stimulates PI hydrolysis in the hippocampus and neocortex [764, 870]. A blocker of metabotropic Glu receptor was shown to block LTP from 2 h onward [826].

Quantal analysis has not yet been performed for periods of >1 h. Analysis of transmitter release [589] and phosphorylation of the presumably presynaptic protein F1 [628] indicated persistence of presynaptic mechanisms at least for 3 days after tetanus while the above-cited iontophoretic studies from the Collingridge group [251, 826] suggested a substitution of presynaptic mechanisms of the initial period by an increase in sensitivity of subsynaptic non-NMDA receptors. Therefore, quantal analysis of later phases of LTP might be one of the important lines of future neurophysiological studies.

A "three-component" scheme of LTP formation was discussed in Section 1.10 with references to the similar hypothesis for CR formation [386]. The scheme suggests that synaptic modification occurs during tetanization due to close sequential activation of excitatory presynaptic fibres, postsynaptic neurone and a "reinforcing" system. By analogy with the three components of this scheme, three peculiarities of the hippocampus were discussed that make this structure especially favourable for studies of long-term modifications [1016]. The more recent studies cited above [501, 859, 1097] seem to prove that, at least for CA1 area, the two-component (Hebbian-like) rather than the three-component scheme is valid to explain LTP induction. In addition to the effects of pairings the presynaptic volley to postsynaptic depolarization [501, 859, 1097], experiments of Sharonova and Khaspekov [890] on cultured hippocampal slices should be mentioned. The authors described LTP under conditions when extrinsic fibres could not survive and therefore extrahippocampal ("reinforcing") inputs were absent. However, numerous data indicate the importance of modulatory influences for LTP induction (see reviews cited in the Introduction). Long-lasting (tens of minutes and hours) LTP-like effects were found following applications of biogenic amines (e.g. [435, 542, 783]). It can be suggested that the Hebbian scheme is applicable only for induction of initial LTP stages in slice preparations, while more complicated schemes are more relevant to explain more robust and more persistent LTP in the intact brain (see e.g., [196, 248]). The importance of noradrenaline (NA) as a "reinforcing" factor is especially probable for LTP induction in mossy fibre input to CA3 neurones and perforant path input to DG neurones. NA was shown to induce LTP-like changes [783] or facilitate LTP induction in CA3 [435] and produced LTP-like changes in the DG [189, 584, 930, 933]. LTP in the DG can be blocked by beta-adrenergic antagonists and chronic NA depletion [930, 931]. However, NA as well as many other receptor agonists and antagonists produced no reproducible effects on LTP in CA1 [657, 930]. Recently, Frey et al. [349] found that dopamine D2 receptor antagonists selectively eliminated LTP from 2 h onward. Therefore, the role of "reinforcing" ("modulatory") systems seems to be an important direction of future studies of LTP mechanisms. The CWG scheme possibly needs revisions. It might be suggested that NMDA, metabotropic and dopamine receptors (and their second messengers) act in a synergistic manner to initiate mechanisms underlying intermediate and stable stages of LTP in CA1 [826]. This suggestion is similar to the three-component scheme. Some intrahippocampal neurones, such as dispersed peptide containing cells [771] or DG granular cells [960] might perform the "reinforcing" function in addition to extrinsic inputs or even substitute for the reinforcement function (e.g., in cultured slices). Studies of the neuronal plasticity in other structures (e.g., in the visual system) also support the view that a simple Hebbian conjunction might not be sufficient for the synaptic modification, and additional "reinforcing" influences are necessary: noradrenergic and cholinergic influences were found to have a permissive role in ocular dominance developmental plasticity [903].

It can be suggested that under conditions of natural (trans-synaptic and humoral) activation of central neurones, the necessary postsynaptic depolarization and the amount of the postsynaptic Ca^{2+} increase can be regulated by conjoint action of several factors. These factors could determine the duration and the sign (potentiation or depression) of the synaptic modification. Activation of two components of the Hebbian scheme might be sufficient only to induce "decremental" LTP traces. Formation of long-term stages can demand participation of the third (reinforcement) component and probably also participation of an additional "motivational" (or "emotional") component [800] as in four-factor schemes [681, 902]. The multiple convergence of several inputs, which seems to be necessary under natural conditions for the formation of long-term memory, can be reduced to a lesser number of factors under artificial experimental conditions when a sufficient depolarization and a sufficient increase of intracellular Ca^{2+} concentration can be achieved due to synchronized activation and also suppression of postsynaptic inhibition during tetanus. The number of sufficient factors can be equal to only two (i.e. presynaptic activation and postsynaptic depolarization) or even one (postsynaptic stimulation when ICP is induced) under these artificial conditions.

A challenging related problem is the role of cholinergic activation in synaptic plasticity. Our own and more recent literature findings of a prominent facilitatory effect of ACh iontophoresis and septal stimulation on hippocampal responses (see Sect. 6.4) indicated a possible participation of cholinergic connections in LTP induction. There is experimental evidence supporting a possible participation of ACh as a "reinforcing" transmitter in long-term neuronal modifications at various brain structures [6, 523, 712, 1121]. It may be suggested that ACh amplifies the postsynaptic depolarization necessary for the induction of LTP due a variety of effects including increased R_{in} (see Sect. 6.4) and facilitation of NMDA receptor-mediated responses recently found [667]. The interest in brain ACh has been intensified with formulation of the cholinergic hypothesis of Alzheimer's disease and geriatric memory disfunction; however, the role of ACh in memory is not clear (see reviews [259, 334]). Similarly, there are contradictions in studies of ACh involvement in hippocampal LTP. Albeit Robinson and Racine [832] found a facilitatory "cooperativity" effect of septal stimulation on LTP in the DG, no effects of muscarinic blockade on LTP was observed. On the other hand, the muscarinic blockade was effective in suppressing LTP induction in CA3 [952] and CA1 [432] (but see [952]), while an anticholinesterase agent enhanced LTP of population spikes in these regions (see review [562]). In apparent contradiction with these observations, a suppression of LTP generation was found in the CA1 area after preliminary muscarine treatment [1098]. Kuba and Kumamoto [562] suggested that the effect might be explained by a desensitization of the muscarinic receptors which resulted in blockade of endogenously released ACh in the latter study [1098]. However, Kichigina and Vinogradova [507] reported more prominent LTP of single neuronal responses in the CA1 region after septo-fimbrial lesions in comparison with intact rabbits. These authors were inclined to explain more intense LTP in the same pathway in in vitro preparations in comparison with intact animals (see Sect. 1.10) by the absence of the septal input in the former. There are only a few LTP studies related to geriatric memory disfunctions [72–74, 586].

Another important problem of potential clinical value is the relationship between LTP and another interesting plasticity phenomenon termed "kindling", which is the appearance of epileptic seizures in response to initially subthreshold brain stimulation after repeated presentation of this stimulation [379]. The available data suggest that LTP is an essential mechanism of kindling [201, 281, 355, 370, 678, 729, 733, 810, 861, 914, 947] but they are not easy to interpret within a simple scheme that the mechanism of kindling is LTP formation in related connections so that further studies are necessary. Quantal analysis of the kindling phenomena might be an interesting line of future studies.

The relationship between mechanisms of LTP, on the one, and memory and behavioural conditioning, on the other side, is an important problem to be solved [194, 592, 701]. We mentioned already that the "unusual" properties of LTP appeared to be similar to the corresponding features of behavioural conditioning, which might suggest common neural mechanisms. Similar influences of mesencephalic reticular stimulation [116, 151] and applications of various drugs (see reviews cited in the Introduction) on LTP and on processes related to memory were reported with various degrees of "similarity". Among the numerous studies of drug effects, these of Morris et al. [734] are especially noteworthy because they demonstrated a selective impairment of learning, which is sensitive to hippocampal damage, by an NMDA receptor antagonist, which blocks hippocampal LTP (see also [591]). Similarly to the above mentioned findings for LTP, recent studies of behavioural conditioning reported increases in Glu release [589, 591] as well as activation of second messengers [591] and of PKC [65, 778].

So-called "associative" LTP (Sect. 1.10) is an especially interesting phenomenon from the point of view of the relationship between LTP and CR mechanisms. In fact, it represents a "neuronal analogue" [478] of behavioural conditioning. However, studies of optimum intervals between the testing (CS) and conditioning (UCS) inputs for the induction of "associative" LTP did not support a straightforward relationship between LTP and CR mechanisms. In fact, for behavioural conditioning, the optimum interval is about 0.5 s with CS ahead of UCS. The "backward" UCS-CS conditioning is usually less effective. For associative LTP, coincidence between stimulations of two inputs was found to be optimal in several publications [178, 394, 604, 657], in agreement with the CWG scheme of LTP induction. However, other authors [594] showed that conditioning ("priming") stimulation is the most effective when delivered about 200 ms before the short burst through the other (testing) input. To resolve the disagreement, Dr. A.M.Kleschevnikov from our Department reconsidered the problem of optimum time intervals for the induction of associative LTP [510, 511, 1023] and found that the coincidence between two inputs is optimal when either sufficiently strong and prolonged conditioning trains are applied as priming stimuli or the experiment is made with the postsynaptic inhibition blocked by picrotoxin. The reverse order with the optimum interval of about 200 ms is more favourable for experiments with short priming trains (or single pulses) similar to those used by Larson and Lynch [594] under normal conditions with preserved postsynaptic inhibition. The result of the priming stimulation is an apparent increase in the amplitude and duration of the testing EPSP [512, 594, 788] due to a suppression of the early IPSP (Sect. 5.8). The prolongation of the EPSP as a result of priming allows short bursts of excitatory synaptic activity to depolarize the postsynaptic neurone sufficiently to induce LTP in agreement with the CWG scheme. These data resolved the disagreement between the data of different authors but, in fact, left open the problem of how the optimum CS-UCS intervals for behavioural conditioning are related to those for LTP.

There are several studies of the influence of LTP induction on CRs. Repeated LTP inductions were found to cause impairment of subsequent maze learning [72] but to increase the rate of acquisition of a discriminative CR [115]. More direct comparison of conditioning performance and LTP was made in rats of different ages. An impairment of both memory and LTP induction was found in aged animals [72, 586] with a correlated decay of spatial learning and LTP [74]. Similarly, positive correlations were found between the animal's learning capacity and the facility to induce LTP [819]. Laroche et al. [283, 590] found that both learning performance and forgetting were linearly correlated with the magnitude of LTP. LTP-like changes of neuronal responses [907, 1078] and FPs [466, 849] were found after behavioural conditioning. However, such a "behavioural LTP" (see [590, 592, 964] for more references and discussion) was not found in a more recent study with elaboration of two different behavioural tasks [411]. The authors noted significant differences in FPs collected during immobility and movements of the animals, and criticized the previous studies which did not control motor activity.

A variant of the same approach is to use electrical stimulation of a monosynaptic pathway as a CS and to record corresponding FPs. Stimulation of the perforant path or entorhinal cortex was utilized as the CS during elaboration of an electrophysiological analogue of CR [786] and behavioural CRs [590, 787, 827]. Reported changes of the FPs represent the most convincing examples of the "behavioural LTP". On the other hand, some authors found no significant changes in early hippocampal responses after behavioural conditioning [254, 627]. An essential decrease in postspike hyperpolarization and an increase in excitability to intracellular current pulses was found in hippocampal neurones after elaboration of a classical CR [258, 627]. Neither of these changes is characteristic for hippocampal LTP.

Besides "synaptic" mechanisms with presynaptic localization, other possible mechanisms of neuronal plasticity deserve future study. Participation of some of these mechanisms had been suggested on the basis of LTP studies, e.g. a blockade of inhibitory systems and an increase in excitability of dendritic spike-triggering zones. I considered them to be additional mechanisms of LTP. It is worth recalling here that the data of Bindman et al. [127] indicated participation of postsynaptic mechanisms in long-lasting increase in PTR after antidromic tetanic stimulation. However, our experiments with recordings of the PTRs [544, 1015, 1017, 1034] support synaptic changes as a basic mechanism of long-term plasticities with the possible additional involvement of postsynaptic changes only during an initial period (<1 h). Nevertheless, the important role of postsynaptic membrane mechanisms in behavioural plasticity is supported by some studies of neuronal CR analogues and behavioural CRs. Changes in spike thresholds of cortical neurones were found in the course of repeated peripheral stimulation in cellular analogues of habituation [391, 978, 979]. However, polysynaptic PSPs were measured, and the claimed threshold shifts could reflect changes in a mixed synaptic input. More direct measurements showed an increase in membrane excitability in snail CNS neurones at initial stages of conditioning [612], but specificity to pairings of the CR studied was not convincingly demonstrated and was denied by other authors [656]. Participation of membrane mechanisms was shown more recently in isolated neurones of the snail CNS on simplified learning analogues [384, 904], in insect motoneurones involved in an operant-conditioning paradigm [440] and especially in mollusc photoreceptors related to a specific type of associative learning [14, 242]. As to the vertebrate CNS, Woody and his colleagues [1118–1120] found a difference in excitability to intracellular currents and in R_{in} between neurones of specific motor projection and other neurones after elaboration of the blink CR in cats. However, comparable (although somewhat smaller) changes could be produced by repeated presentations of unpaired UCS [1118]. In related studies, a long-lasting increase in R_{in} was found in cortical neurones as a consequence of administration of ACh plus current-induced discharge [1118, 1121].

On the other hand, a crucial role of synaptic mechanisms was convincingly demonstrated for behavioural habituation in invertebrates (Sects. 2.5.2, 7.10.3).

Quantal analysis suggested a presynaptic localization [212, 475], but our data demonstrated changes in v and indicated a necessity of additional studies (Chap. 11). Dishabituation and sensitization of defensive behaviour of *Aplysia* were also explained by presynaptic mechanisms (Sects. 2.5.2, 7.10.4). All these studies in invertebrates provided first experimental evidence for the real existence of "modifiable" synapses participating in simple short-term behavioural modifications. Studies of long-term habituation also indicated synaptic localization of the underlying mechanisms [215]. The LFD in the hippocampus (Chaps. 1, 4) might be considered as a cellular analogue of the behavioural habituation. In this case, our data also support the changes in excitatory synaptic action as a basic mechanism of behavioural habituation. Investigations of *Aplysia* defensive behaviour (Sect. 2.5.3) strongly supported the participation of synaptic mechanisms in CRs of simple types, but postsynaptic membrane changes were also found (e.g., [1071]).

In distinction from the studies on invertebrate and LTP studies, the "synaptic" hypothesis has not yet been as strongly supported for CRs in vertebrates. The appearance of short-latency EPSPs and an increase of EPSP amplitudes in neocortical neurones were found in our studies [1008, 1010, 1013, 1019, 1021, 1029] and in experiments of other authors [68–70] on "cellular analogues" of CRs which now can be considered as variants of "associative" LTP in the neocortex. We also found the appearance of short-latency EPSPs in sensorimotor cortex of rabbits with elaborated behavioural CR of a simple type, the LCSR shortly described in Chapter 1 [1010, 1011, 1013, 1019, 1021, 1028]. These data are in agreement with the "synaptic" hypothesis of conditioning but they need future development.

Taking into account the remarkable properties of LTP similar to related properties of behavioural CRs, one might consider the available data on LTP mechanisms as an indirect support of the synaptic hypothesis of conditioning. Moreover, the quantal analysis of LTP shows that synaptic modifications in vertebrate CNS might be based on an increase in transmitter release. I believe that our studies open possibilities for more precise analysis of cellular mechanisms of conditioning and learning at the vertebrate level. Methods of recording and statistical analysis of MPSPs and UPSPs can be applied to study the above-mentioned CR analogues with electrical brain stimulation and intracellular recordings [68–70, 1008, 1010, 1013, 1021, 1029]. Simple short-latency behavioural CRs can be especially useful [258, 272, 317, 538, 862, 969, 980, 1010, 1013, 1021, 1043, 1078, 1118, 1119]. Some of these behavioural paradigms had been under study with intracellular recordings [980, 1010, 1013, 1021, 1028, 1118–1120]. I believe that the LCSR (Sect. 1.9) is an especially convenient paradigm for future studies. Specifically, it is possible to modify the experimental situation so that intracranial electrical stimulation can be used instead of peripheral stimulation as a CS [1021, 1034, 1035]. This should favour precise quantitative analysis and comparison of conditioning mechanisms with mechanisms of neocortical LTP produced under similar experimental conditions [1017, 1020, 1051].

References

1. Abeles M (1982) Local Cortical Circuits. Braitenberg V (ed) Studies of Brain Functions, vol 6. Springer, Berlin Heidelberg New York
2. Abraham WC, Goddard GV (1983) Asymmetric relationships between homosynaptic long-term potentiation and heterosynaptic long-term depression. Nature 305:717–719
3. Abraham WC, Bliss TVP, Goddard GV (1985) Heterosynaptic changes accompany long-term but not short-term potentiation of the perforant path in the anaesthetized rat. J Physiol (Lond) 363:335–349
4. Abraham WC, Gustafsson B, Wigström H (1987) Long-term potentiation involves enhanced synaptic excitation relative of synaptic inhibition in guinea-pig hippocampus. J Physiol (Lond) 394:367–380
5. Abrams TW (1985) Activity-dependent presynaptic facilitation: an associative mechanism in *Aplysia*. Cell Mol Neurobiol 5:123–145
6. Adrianov OS (1986) Sructure-functional basis of the integrative brain activity. In: Batuev AS (ed) Fiziologiya Povedeniya. Rukovodstvo po Fiziologii (Physiology of Behaviour. Handbook of Physiology). Nauka, Leningrad, pp 163–200
7. Agoston DV, Kuhnt U (1986) Increased Ca-uptake of presynaptic terminals during long-term potentiation of hippocampal slices. Exp Brain Res 62:663–668
8. Akert K, Pfenninger K, Sandri C, Moor H (1972) Freeze etching and cytochemistry of vesicles and membrane complexes in synapses of the central nervous system. In: Pappas GD, Purpura DP (eds) Structure and Function of Synapses. Raven, New York, pp 67–86
9. Alger BE, Nicoll RA (1980) Spontaneous inhibitory postsynaptic potentials in hippocampus: mechanism for tonic inhibition. Brain Res 200:195–200
10. Alger BE, Teyler TJ (1976) Long-term and short-term plasticity in the CA1, CA3 and dentate region of the rat hippocampal slice. Brain Res 110:463–480
11. Alger BE, Megela AL, Teyler TJ (1978) Transient heterosynaptic depression in the hippocampal slice. Brain Res Bull 3:181–184
12. Alger BE, Pitler TA, Williamson A (1990) A prolonged post-tetanic hyperpolarization in rat hippocampal pyramidal cells in vitro. Brain Res 521:118–124
13. Aljure E, Day JW, Bennett MVL (1980) Postsynaptic depression of Mauthner cell-mediated startle reflex, a possible contributor to habituation. Brain Res 188:261–268
14. Alkon DL (1987) Memory Traces in the Brain. Oxford University Press, London New York
15. Allen GI, Oshima T, Toyama K (1971) Unitary components in corticopontine activation in the cat. Brain Res 35:245–249
16. Allen GI, Oshima T, Toyama K (1977) The mode of synaptic linkage in the cerebro-ponto-cerebellar pathway investigated with intracellular recording from pontine nucleus cells of the cat. Exp Brain Res 29:123–136
17. Almers W (1990) Exocitosis. Ann Rev Physiol 52: 607–624
18. Alonso A, de Curtis M, Llinas R (1990) Postsynaptic Hebbian and non-Hebbian long-term potentiation of synaptic efficacy in the entorhinal cortex in slices and in the isolated adult guinea pig brain. Proc Natl Acad Sci USA 87:9280–9284
19. Amaral DG, Witter MP (1989) The three-dimensional organization of the hippocampal formation: a review of anatomical data. Neuroscience 31:571–591

20. Amaral DG, Ishizuka N, Claiborne B (1990) Neurons, numbers and the hippocampal network: understanding the brain through the hippocampus. In: Storm-Mathisen J et al. (eds) Progress in Brain Research, vol 83. Elsevier, Amsterdam
21. Amassian VE, Weiner H (1966) Monosynaptic and polysynaptic activation of pyramidal tract neurons by thalamic stimulation. In: Purpura DP, Yahr MD (eds) The Thalamus. Columbia University, New York, pp 255–282
22. Andersen P (1978) Long-lasting facilitation of synaptic transmission. In: Elliot K, Whelau J (eds) Function of the Septo-Hippocampal System. Ciba Foundation Symposium, vol 58. Elsevier, Amsterdam, pp 87–108
23. Andersen P (1987) Long-term potentiation – outstanding problems. In: Changeux JP, Konishi M (eds) The Neural and Molecular Bases of Learning, Wiley, New-York, pp 239–269
24. Andersen P (1990) The main features of long-term potentiation: a model for formation of memory traces. Pontif Acad Sci Scripta 78:369–382
25. Andersen P, Lomo T (1967) Control of hippocampal output by afferent volley frequency. In: Adey RW, Tokizane T (eds) Structure and Function of the Limbic System, Progress in Brain Research, vol 27. Elsevier, Amsterdam, pp 400–412
26. Andersen P, Eccles JC, Loyning Y (1964) Pathway of postsynaptic inhibition in the hippocampus. J Neurophysiol 27: 608–619
27. Andersen P, Holmqvist B, Voorhoeve PE (1966) Entorhinal activation of dentate granule cells. Acta Physiol Scand 66:448–460
28. Andersen P, Bliss TVP, Skrede KK (1971a) Unit analysis of hippocampal population spikes. Exp Brain Res 13:208–221
29. Andersen P, Bliss TVP, Skrede KK (1971b) Laminar organization of hippocampal excitatory pathways. Exp Brain Res 13:222–238
30. Andersen P, Teyler T, Wester K (1973) Long-lasting change of synaptic transmission in a specialized cortical pathway. Acta Physiol Scand Suppl 396, 89:34
31. Andersen P, Sundberg SH, Sveen O, Wigstrom H (1977) Specific long-lasting potentiation of synaptic transmission in hippocampal slices. Nature 266:736–737
32. Andersen P, Sundberg SH, Sveen O, Swann JW, Wigstrom H (1980) Possible mechanisms for long-lasting potentiation of synaptic transmission in hippocampal slices from guinea pigs. J Physiol (Lond) 302:463–482
33. Andersen P, Sayer J, Redman SJ (1985) Fluctuations in the amplitude of excitatory postsynaptic potentials evoked in hippocampal CA1 pyramidal cells in vitro. Proc Australian Physiol Pharmacol Soc 16:100P
34. Andersen P, Raastad M, Storm JF (1990) Excitatory synaptic integration in hippocampal pyramids and dentate granule cells. Cold Spring Harbor Symposium on Quantative Biology, vol 55. Cold Spring Harbor Laboratory, pp 81–86
35. Aniksztejn L, Ben-Ari Y (1991) Blockade of specific K^+ channels produces a Ca^{2+}dependent form of long-lasting potentiation produced in the hippocampus. In Ascher et al. (eds) Glutamate, Cell Death and Memory. Springer, Berlin Heidelberg New York
36. Aniksztejn L, Roisin M-P, Gozlan H, Ben-Ari Y (1987) Long-lasting potentiation produced by phorbol ester in the hippocampus of the anaesthetized rat is not associated with a persistent enhanced release of excitatory amino acids. Neurosci Lett 81:291–295
37. Aniksztejn L, Roisin MP, Amsellem R, Ben-Ari Y (1989) Long-term potentiation in the hippocampus of the anaesthetized rat is not associated with a sustained enhanced release of endogenous excitatory amino acids. Neuroscience 28:387–392
38. Antonov IN, Storozhuk VM, Osipenko ON (1991) Analysis of heterosynaptic facilitation in identified giant neurons from cerebral ganglion of the pond snail *Planorbis corneus*. Comp Biochem Physiol C98:323–327
39. Apostol G, Creutzfeldt OD (1974) Crosscorrelation between the activity of septal units and hippocampal EEG during arousal. Brain Res 67:65–75

40. Appenteng K, O'Donovan MJ, Stephens JA, Taylor A (1977) Synaptic projections of single spindle afferents demonstrated by extracellular spike triggered averaging. J Physiol (Lond) 271:30P–31P
41. Applegate MD, Kerr DS, Landfield PW (1987) Redistribution of synaptic vesicles during long-term potentiation in the hippocampus. Brain Res 401:401–406
42. Arshavskii YuI, Chailakhyan LM (1966) Investigation of the mechanism of influences of direct current on evoked potentials of the cerebellum. Biophysics (Moscow) 11:134–142
43. Artemenko DP (1975) Role of hippocampal neurons in theta-wave generation. Neurophysiology (Kiev) 4:531–539
44. Artola A, Singer W (1987) Long-term potentiation and NMDA receptors in rat visual cortex. Nature 330:649–652
45. Artola A, Bröcher S, Singer W (1990) Different voltage-dependent thresholds for inducing long-term depression and long-term potentiation in slices of rat visual cortex. Nature 347:69–72
46. Arvanov VL, Salanki J, Ayrapetyan SN (1984) Modification of acetylcholine sensitivity of neuronal membrane by presynaptic stimulation. Gen Physiol Biophys 3:483–488
47. Asanuma H, Rosen I (1973) Spread of mono- and polysynaptic connections within cat's motor cortex. Exp Brain Res 16:507–520
48. Asanuma H, Arnold A, Zarzecki P (1976) Further study on the excitation of pyramidal tract cells by intracortical microstimulation. Exp Brain Res 26:443–461
49. Asztely F, Hanse E, Gustafsson B (1990) The early decay of long-term potentiation in the hippocampal CA1 region in vitro is reduced by activators of protein kinase C. Brain Res 521:355–358
50. Atwood HL (1976) Organization and synaptic physiology of crustacean neuromuscular system. Progr Neurobiol 7:291–391
51. Atwood HL, Parnas I (1968) Synaptic transmission in crustacean muscles with dual motor innervation. Comp Biochem Physiol 27:381–404
52. Atwood HL, Tse EV (1988) Changes in binomial parameters of quantal release at crustacean motor axon terminals during presynaptic inhibition. J Physiol (Lond) 402:177–183
53. Atwood HL, Wojtowicz JM (1986) Short-term and long-term plasticity and physiological differentiation of crustacean motor synapses. Int Rev Neurobiol 28:275–362
54. Atwood HL, Swenarchuk LE, Gruenwald CR (1975) Long-lasting synaptic facilitation during sodium accumulation in nerve terminals. Brain Res 100:198–204
55. Auerbach AA (1972) Transmitter release at chemical synapses. In: Pappas GD, Purpura DP (eds) Structure and Function of Synapses. Raven, New York, pp 137–159
56. Auerbach AA, Bennett MVL (1969) Chemically mediated transmission at a giant fiber synapse in the central nervous system of a vertebrate. J Gen Physiol 53:183–193
57. Auerbach AA, Betz W (1971) Does curare affect transmitter release? J Physiol (Lond) 213:691–705
58. Babalian AL, Chmykhova NM (1987) Morphopysiological description of connections between single ventrolateral tract fibers and individual motoneurones in frog spinal cord. Brain Res 407:394–397
59. Babalian AL, Shapovalov AI (1984) Synaptic actions produced by individual ventrolateral tract fibers in frog lumbar motoneurons. Exp Brain Res 54:551–563
60. Bagshaw EV, Evans MH (1976) Measurement of current spread from microelectrodes when stimulating within the nervous system. Exp Brain Res 25:391–400
61. Baimbridge KG, Miller JJ (1981) Calcium uptake and retention during long-term potentiation of neuronal activity in the rat hippocampal slice preparation. Brain Res 221:299–305
62. Baker M (1985) Field potential reversal-point movement after afferent tetanization in the rat dentate gyrus. J Physiol (Lond) 358:46P

63. Balaban PM, Litvinov EG (1977) Command neurons in the unconditioned reflex of the snail (in Russian). Zh Vyssh Nerv Deyat 27:538–544
64. Balanter BI (1977) Veroyatnostnye Modeli v Fiziologii (Probability Models in Physiology). Nauka, Moscow
65. Bank B, DeWeer A, Kuzirian AM, Rasmussen H, Alkon DL (1988) Classical conditioning induces long-term translocation of protein kinase C in rabbit hippocampal CA1 cells. Proc Natl Acad Sci USA 85:1988–1992
66. Bar PR, Tielen AM, Lopes da Silva FH, Zwiers H, Gispen WH (1982) Membrane phosphoproteins of rat hippocampus: Sensitivity to tetanic stimulation and enkephalin. Brain Res 245:69–79
67. Bar PR, Wiegant F, Lopes da Silva FH, Gispen WH (1984) Tetanic stimulation affects the metabolism of phosphoinositides in hippocampal slices. Brain Res 321:381–385
68. Baranyi A, Feher O (1978) Conditioned changes of synaptic transmission in the motor cortex of the cat. Exp Brain Res 33:283–298
69. Baranyi A, Feher O (1981a) Selective facilitation of synapses in the neocortex by heterosynaptic activation. Exp Brain Res 212:164–168
70. Baranyi A, Feher O (1981b) Long-term facilitation of excitatory synaptic transmission in single motor cortical neurons of the cat produced by repetitive pairing of synaptic potentials following intracellular stimulation. Neurosci Lett 23:303–308
71. Baranyi A, Szente MB, Woody CD (1991) Properties of associative long-lasting potentiation induced by cellular conditioning in the motor cortex of conscious cats. Neuroscience 42:321–334
72. Barnes CA (1979) Memory deficits associated with senescence: A neurophysiological and behavioral study in the rat. J Comp Physiol Psychol 93:74–104
73. Barnes CA, McNaughton BL (1980) Physiological compensation for loss of afferent synapses in rat hippocampal granule cells during senescence. J Physiol (Lond) 309:473–485
74. Barnes CA, McNaughton BL (1985) An age comparison of the rates of acquisition and forgetting of spatial information in relation to long-term enhancement of hippocampal synapses. Behav Neurosci 99:1040–1048
75. Barrett JN (1975) Motoneuron dendrites: role in synaptic integration. Fed Proc 34:1398–1407
76. Barrett JN, Crill W (1974) Influence of dendritic location and membrane properties on the effectiveness of synapses on cat motoneurones. J Physiol (Lond) 239:325–345
77. Barrett JN, Graubard K (1970) Fluorescent staining of a cat motoneurons in vivo with bevelled micropipettes. Brain Res 18:558–568
78. Barrionuevo G, Brown TH (1983) Associative long-term potentiation in hippocampal slices. Proc Natl Acad Sci USA 80:7347–7351
79. Barrionuevo G, Schottler F, Lynch G (1980) The effects of repetitive low frequency stimulation on control and 'potentiated' synaptic responses in the hippocampus. Life Sci 27:2385–2391
80. Barrionuevo G, Kelso SR, Johnston D, Brown TH (1986) Conductance mechanism responsible for long-term potentiation in monosynaptic and isolated excitatory synaptic inputs to hippocampus. J Neurophysiol 55:540–550
81. Bart AG, Dityatev AE, Kozhanov VM (1988a) Quantal analysis of the synaptic potentials in the interneuronal synapses: Recovery of a signal from the noise. Neurophysiology (Kiev) 20:350–357
82. Bart AG, Dityatev AE, Kozhanov VM (1988b) Analysis of transmission in the interneuronal synapses by the convolution of two binomial distributions. Neurophysiology (Kiev) 20:487–494
83. Bart AG, Dityatev AE, Kozhanov VM (1989) Analysis of the signal transmission in the interneuronal synapses on the basis of reflection principle. Doklady Biol Sci 306:1503–1507

84. Barton SB, Cohen IS (1977) Are transmitter release statistics meaningful? Nature 268:267–268
85. Barton SB, Cohen IS (1978) Barton and Cohen reply. Nature 271:688
86. Bashir ZI, Alford S, Davies SN, Randall AD, Collingridge GL (1991) Long-term potentiation of NMDA receptor-mediated synaptic transmission in the hippocampus. Nature 349:156–158
87. Baskys A (1979) Analysis of mechanisms of post-tetanic potentiation of synaptic reactions of pyramidal hippocampal neurons. Doklady Biol Sci 250:1268–1271
88. Baskys A, Voronin LL, Derevyagin VI (1978) Analysis of mechanisms of frequency potentiation of synaptic responses in hippocampal neurons. Doklady Biol Sci 239:234–237
89. Baskys A, Voronin LL, Gusev LN, Derevyagin VI, Kesarev VS, Logunov DB, Shidarev DR (1979) Calculations of theoretical amplitude distributions of postsynaptic potentials for quantal analysis. Neurophysiology (Kiev) 11:146–150
90. Baskys A, Voronin LL, Gusev AG (1980) Elementary postsynaptic potentials of hippocampal neurons. Bull Exp Biol Med 90:643–646
91. Baskys A, Carlen PL, Wojtowicz JM (1991) Long-term potentiation of synaptic responses in the rat dentate gyrus is due to increased quantal content. Neurosci Lett 127:169–172
92. Batueva IV, Shapovalov AI (1977) Synaptic effects evoked in lamprey motoneurons by direct stimulation of single presynaptic fibers. Neurophysiology (Kiev) 9:390–396
93. Baudry M, Davis J, eds (1991) Long-Term Potentiation: a Debate of Current Issue. MIT Press, Cambridge, Mass
94. Baudry M, Oliver M, Creager R, Wieraszko A, Lynch G (1980) Increase in glutamate receptors following repetitive electrical stimulation in hippocampal slices. Life Sci 27:325–330
95. Baumgarten R von (1970) Plasticity in the nervous system at the unitary level. In: Schmitt FO (ed) The Neurosciences: Second Study Program. Rockefeller University, New York, pp 260–270
96. Baumgarten RJ von, Jahan-Parwar BD (1967) Time course of repetitive heterosynaptic facilitation in *Aplysia californica*. Brain Res 4:295–297
97. Bawin SM, Sheppard AR, Mahoney MD, Abu-Assal M, Adey WR (1986) Comparison between the effects of extracellular direct and sinusoidal currents on excitability in hippocampal slices. Brain Res 362:350–354
98. Baxter DA, Bittner GD, Brown TH (1985) Quantal mechanism of long-term synaptic potentiation. Proc Natl Acad Sci USA 82:5978–5982
99. Bekkers JM, Stevens CF (1989) NMDA and non-NMDA receptors are co-localized at individual excitatory synapses in cultured rat hippocampus. Nature 341:230–233
100. Bekkers JM, Stevens CF (1990) Presynaptic mechanism for long-term potentiation in the hippocampus. Nature 346:724–729
101. Bekkers JM, Richerson GB, Stevens CF (1990) Origin and variability in quantal size in cultured hippocampal neurons and hippocampal slices. Proc Natl Acad Sci USA 87:5359–5362
102. Belekhova MG (1968) Post-tetanic potentiation (in Russian). Uspekhi Sovr Biol 66:199–225
103. Belluzzi O (1984) Quantal release of neurotransmitter: An iterative method for the automatic computation of binomial distribution parameters. J Neurosci Meth 10:41–50
104. Bennett MR (1983) Development of neuromuscular synapses. Physiol Rev 63:915–1048
105. Bennett MR, Fisher C (1977) The effect of calcium ions on the binomial parameters that control acetylcholine release during trains of nerve impulses at amphibian neuromuscular synapses. J Physiol (Lond) 271:673–698

106. Bennett MR, Florin T (1974) A statistical analysis of the release of acetylcholine at newly formed synapses in striated muscle. J Physiol (Lond) 238:93–107
107. Bennett MR, Lavidis NA (1979) The effect of calcium on the secretion of quanta evoked an impulse at nerve terminal release sites. J Gen Physiol 74:429–456
108. Bennett MR, Lavidis NA (1982) Variation in quantal secretion at different release sites along developing and mature motor terminal branches. Dev Brain Res 5:1–9
109. Bennett MR, Pettigrew AG (1975) The formation of synapses in amphibian striate muscle during development. J Physiol (Lond) 252:203–239
110. Bennett MR, Florin T, Hall K (1975) The effect of calcium ions on the binomial statistic parameters that control acetylcholine release at synapses in striate muscle. J Physiol (Lond) 247:429–445
111. Bennett MR, Florin T, Pettigrew AG (1976) The effect of calcium ions on the binomial statistic parameters that control acetylcholine release at preganglionic nerve terminals. J Physiol (Lond) 257:597–620
112. Bennett MR, Fisher C, Florin T, Quine M, Robinson J (1977) The effect of calcium ions and temperature on the binomial parameters that control acetylcholine release by a nerve impulse at amphibian neuromuscular synapses. J Physiol (Lond) 271:641–672
113. Bennett MVL (1976) Electrical transmission: A functional analysis and comparison to chemical transmission. In: Brookhart JM, Mountcastle VB (eds) Handbook of Physiology. The Nervous System, Sect 1, vol 1 (Kandel ER, ed). Williams and Wilkins, Baltimore
114. Benowitz IJ, Routtenberg A (1987) A membrane phosphoprotein associated with neural development, axonal regeneration phospholipid metabolism and synaptic plasticity. Trends Neurosci 10:527–532
115. Berger TW (1984) Long-term potentiation of hippocampal synaptic transmission affects rate of behavioral learning. Science 224:627–630
116. Bergis O, Bloch V, Laroche S (1990) Enhancement of long-term potentiation in the dentate gyrus two days after associative learning in the rat. Neurosci Res Comm 6:119–128
117. Beritashvili IS (1968) Pamyat Pozvonochnykh Zhivotnykh, Ee Kharakteristika i Proiskhozhdenie (Memory of Vertebrates, Its Characteristic and Origin). Metsniereba, Tbilisi, pp 74–78
118. Beritashvili IS (1972) Characteristic and origin of psycho-nervous memory in vertebrates. In: Beritashvili IS (ed) Gagrskie Besedy (Gagra Talks), vol 6. Metsniereba, Tbilisi, pp 9–36
119. Berridge MJ (1987) Inositol trisphosphate and diacylglycerol: two interacting second messengers. Ann Rev Biochem 56: 159–193
120. Berry MS, Pentreath VW (1976) Criteria for distinguishing between monosynaptic and polysynaptic transmission. Brain Res 105:1–20
121. Beswick FG, Conroy RTL (1965) Optimal tetanic conditioning of heteronymous monosynaptic reflex. J Physiol (Lond) 180:134–146
122. Betz WJ (1970) Depression of transmitter release at the neuromuscular junction of the frog. J Physiol (Lond) 206:629–644
123. Bevan S (1976) Subminiature end-plate potentials at untreated frog neuromuscular junctions. J Physiol (Lond) 258:145–155
124. Bieser A, Wernig A, Zucker H (1984) Different quantal responses within single frog neuromuscular junctions. J Physiol (Lond) 350:401–412
125. Bilkley DK, Goddard GV (1986) Medial septal facilitation of hippocampal granule cell activity is mediated by inhibition of inhibitory interneurones. Brain Res 361:99–106
126. Bindman LJ, Lippold OCJ, Redfearn JWT (1964) The action of brief polarizing currents on the cerebral cortex of the rat (1) during current flow and (2) in the production of long-lasting after-effects. J Physiol (Lond) 172:369–382

127. Bindman LJ, Lippold OCJ, Milne AR (1979) Prolonged changes in excitability of pyramidal tract neurones in the cat: A postsynaptic mechanism. J Physiol (Lond) 286:457–477
128. Bindman LJ, Christofi G, Murphy K, Nowicky A (1991) Long-term potentiation (LTP) and depression (LTD) in the neocortex and hippocampus: an overview. In: Stone TW (ed) Aspects of Synaptic Transmission, vol 1. Taylor and Francis, London, pp 1–16
129. Bird SJ, Aghajanian GK (1975) Denervation supersensitivity in the cholinergic septo-hippocampal pathway: a microiontophoretic study. Brain Res 100:355–370
130. Birks R, MacIntosh FC (1961) Acetylcholine metabolism of a sympathetic ganglion. Can J Biochem Physiol 39:787–827
131. Biscoe TJ, Straughan DV (1966) Micro-electrophoretic studies of neurones in the cat hippocampus. J Physiol (Lond) 183:341–359
132. Bittner GD, Harrison J (1970) A reconsideration of the Poisson hypothesis for transmitter release at the crayfish neuromuscular junction. J Physiol (Lond) 206:1–22
133. Blackman JG, Ginsborg BL, Ray C (1963a) Spontaneous synaptic activity in sympathetic ganglion cells of the frog. J Physiol (Lond) 163:389–401
134. Blackman JG, Ginsborg BL, Ray C (1963b) On the quantal release of transmitter at a sympathetic ganglion synapse. J Physiol (Lond) 167:402–415
135. Blackman JG, Purves RD (1969) Intracellular recordings from ganglia of the thoracic sympathetic chain in the guinea pig. J Physiol (Lond) 203:173–198
136. Blakeley AGH, Cunnane TC, Petersen SA (1982) Local regulation of transmitter release from frog rodent sympathetic nerve terminals. J Physiol (Lond) 325:93–109
137. Bland BH, Kostopoulos GK, Phillis JW (1974) Acetylcholine sensitivity of hippocampal formation neurones. Can J Physiol Pharmacol 52:966–972
138. Blankenship JE, Kuno M (1968) Analysis of spontaneous subthreshold activity in spinal motoneurons of the cat. J Neurophysiol 31:195–209
139. Blanton MG, Lo Turco JJ, Kriegstein AR (1989) Whole cell recording from neurons in slices of reptilian and mammalian cerebral cortex. J Neurosci Meth 30:203–210
140. Bliss TVP (1979) Synaptic plasticity in the hippocampus. Trends Neurosci 2:42–45
141. Bliss TVP, Dolphin AG (1982) What is the mechanism of long-term potentiation in the hippocampus? Trends Neurosci 5:289–290
142. Bliss TVP, Gardner-Medwin AR (1971) Long-lasting increases of synaptic influence in the unanaesthetized hippocampus. J Physiol (Lond) 216:32–33
143. Bliss TVP, Gardner-Medwin AR (1973) Long-lasting potentiation of synaptic transmission in the dentate area of the unanaesthetized rabbit following stimulation of the perforant path. J Physiol (Lond) 232:357–374
144. Bliss TVP, Lomo T (1970) Plasticity in a monosynaptic cortical pathway. J Physiol (Lond) 207:61P
145. Bliss TVP, Lomo T (1973) Long-lasting potentiation of synaptic transmission in the dentate area of the anaesthetized rabbit following stimulation of the perforant path. J Physiol (Lond) 232:331–356
146. Bliss TVP, Lynch MA (1988) Long-term potentiation of synaptic transmission in the hippocampus: Properties and mechanisms. In: Landfield PW, Deadwyler SA (eds) Long-Term Potentiation: From Biophysics to Behavior. Liss, New York, pp 3–72
147. Bliss TVP, Burns, B.D., Uttley, A.M. (1968) Factors affecting the conductivity of pathways in the cerebral cortex. J Physiol. (Lond) 232:331–356.
148. Bliss TVP, Lancaster B, Wheal HV (1983) Long-term potentiation in commissural and Schaffer projections to hippocampal CA1 cells: an in vivo study in the rat. J Physiol 341:617–626
149. Bliss TVP, Douglas RM, Errington ML, Lynch MA (1986) Correlation between long-term potentiation and release endogenous amino acids from dentate gyrus of anaesthetized rats. J Physiol (Lond) 377:391–408

150. Bliss TVP, Errington ML, Laroche S, Lynch MA (1987) Increase in K^+-stimulated Ca^{2+}-dependent release of [3H]glutamate from rat dentate gyrus three days after induction of long-term potentiation. Neurosci Lett 83:107–112
151. Bloch V, Laroche S (1985) Enhancement of long-term potentiation in the rat dentate gyrus by post-trial stimulation of the reticular formation. J Physiol (Lond) 360:215–231
152. Bloedel J, Gage R, Llinas RW, Quastel DM (1966) Transmitter release at the squid giant synapse in the presence of tetrodotoxin. Nature 212:49–50
153. Blokhina GI, Zefirov AL (1984) Electrophysiological and morphological research of synaptic organization of frog's tonic muscular fibers (in Russian). Fiziol Zh SSSR 70:157–165
154. Bonhöffer T, Staike V, Aertsen A (1989) Synaptic plasticity in rat hippocampal slice cultures: local "Hebbian" conjunction of pre- and postsynaptic stimulation leads to distributed synaptic enhancement. Proc Natl Acad Sci USA 86:8113–8117
155. Borisova OV (1990) Cyclic adenosine monophosphate (cAMP)-induced potentiation of synaptic responses in Helix neuron. Cell Mol Neurobiol 10:275–279
156. Bornstein JC (1974) Multiquantal release of acetylcholine in mammalian ganglia. Nature 248:529–531
157. Boyd IA, Martin AR (1956a) Spontaneous subthreshold activity at mammalian neuromuscular junction. J Physiol (Lond) 132:61–73
158. Boyd IA, Martin AR (1956b) The end-plate potential in mammalian muscle. J Physiol (Lond) 132:74–91
159. Bradler JE, Barrionuevo G (1989) Long-term potentiation in hippocampal CA3 neurons: tetanized input regulates heterosynaptic efficacy. Synapse 4:132–142
160. Bragin AG, Otmakhov NA (1979) Comparative characteristics of the direct perforant path influence on the neurons of hippocampal fields CA1 and CA3 in vitro. Neurophysiology (Kiev) 11:303–310
161. Bragin AG, Vinogradova OS (1973) A phenomenon of chronic potentiation of cortical afferent input to CA3 pyramids of the hippocampus. In: Gromova EA (ed) Fiziologicheskie Mekhanizmy Pamyati (Physiological Mechanisms of Memory). Institut of Biophysics, Pushchino, pp 8–24
162. Bragin AG, Vinogradova OS, Otmakhov IA (1977) A study of indirect influence of the perforant pathway on neurons in CA3 hippocampal area in vitro. Doklady Biol Sci 233:249–252
163. Braha O, Dale N, Hochner B, Klein M, Abrams TD, Kandel ER (1990) Second messengers involved in the two processes of presynaptic facilitation that contribute to sensitization and dishabituation in *Aplysia* sensory neurons. Proc Natl Acad Sci USA 87:2040–2044
164. Braitenberg V, Schüz A (1991) Anatomy of the Cortex. Statistics and Geometry. Braitenberg V (ed) Studies of Brain Functions, vol 18. Springer, Berlin Heidelberg New York
165. Branisteanu DD, Miyamoto MD, Volle RL (1975) Quantal release parameters during fall of end-plate potentials. Nannyn-Schmiedeberg's Arch Pharmacol 288:323–527
166. Branisteanu DD, Miyamoto MD, Volle RL (1976) Effects of physiologic alterations on binomial transmitter release at magnesium-depressed neuromuscular junction. J Physiol (Lond) 254:19–38
167. Briggs CA, McAfee DA, McCaman RE (1985) Long-term potentiation of synaptic acetylcholine release in the superior cervical ganglion of the rat. J Physiol (Lond) 363:181–190
168. Brindley GS (1968) The classification of modifiable synapses and their use in models for conditioning. Proc Roy Soc 168:361–376
169. Brink E, Morrison PJ, Jankowska I, McCrea DA, Skoog B (1983) Postsynaptic potentials in a population of motoneurones following activity of single interneurones in the cat. J Physiol (Lond) 343:341–359

170. Brocks J (1976) Suppression of foreign synapses. Nature 260:281
171. Brooks VB, Thies RE (1962) Reduction of quantal content during neuromuscular transmission. J Physiol (Lond) 162:298–310
172. Brown AG (1991) Nerve Cells and Nervous Systems. An Introduction to Neuroscience. Springer, Berlin Heidelberg New York
173. Brown RT, Flaming DG (1974) Bevelling of fine micropipette electrodes by a rapid precision method. Science 185:693–695
174. Brown TH, Johnston D (1983) Voltage-clamp analysis of mossy fiber synaptic input to hippocampal neurons. J Neurophysiol 50:487–507
175. Brown TH, Perkel DH, Feldman MW (1976) Evoked transmitter release: statistical effects of nonuniformity and nonstationarity. Proc Natl Acad Sci USA 73:2913–2917
176. Brown TH, Wong RKS, Prince DA (1979) Spontaneous miniature synaptic potentials in hippocampal neurons. Brain Res 177:194–199
177. Brown TH, Chang VC, Ganong AH, Keenan CL, Kelso SR (1988a) Biophysical properties of dendrites and spines that may control the induction and expression of long-term synaptic potentiation. In: Landfield PW, Deadwyler SA (eds) Long-Term Potentiation: from Biophysics to Behavior. Liss, New York, pp 201–264
178. Brown TH, Ganong AH, Kairiss EW, Keenan CL, Kelso SR (1988b) Long-term potentiation in two synaptic systems in the hippocampal brain slice. In: Byrne JH, Berry WO (eds) Neural Models of Plasticity. Academic, New York, pp 266–306
179. Brown TH, Kairiss EW, Keenan CL (1990) Hebbian synapses: biophysical mechanisms and algorithms. Ann Rev Neurosci 13:475–511
180. Bruner A (1967) Self-stimulation in the rabbit: an anatomical map of stimulation effects. J Comp Neurol 131:615–630
181. Bruner J, Kehoe J (1970) Long-term decrements in the efficacy of synaptic transmission in mollusca and crutaceas. In: Horn G, Hinde RA (eds) Short-term Change in Neural Activity and Behaviour. Cambridge University, Cambridge, pp 323–359
182. Bruner J, Kennedy D (1970) Habituation: occurrence at a neuromuscular junction. Science 169:92–94
183. Bruner J, Tauc L (1966) Long-lasting phenomena in the molluscan nervous system. Sympos Soc Exptl Biol 20:457–475
184. Bryan JS, Atwood HL (1981) Two types of synaptic depression at synapses of a single crustacean motor axon. Mar Behav Physiol 8:99–121
185. Budantsev AYu (1976) Monoaminergicheskie Sistemy Mozga (Monoaminergic Systems of the Brain). Nauka, Moscow
186. Buonomano DV, Byrne JH (1990) Long-term synaptic changes produced by a cellular analog of classical conditioning in *Aplysia*. Science 249:420–422
187. Bures J, Buresova O (1970) Plasticity in single neurones and neural populations. In: Horn G, Hinde RA (eds) Short-term Changes in Neural Activity and Behaviour. Cambridge University Press, Cambridge, pp 363–403
188. Bures J, Buresova O, Huston JP (1983) Techniques and Basic Experiments for the Study of Brain and Behavior. Elsevier, Amsterdam
189. Burgard EC, Decker G, Sarvey JM (1989) NMDA receptor antagonists block norepinephrine-induced long-lasting potentiation and long-term potentiation in rat dentate gyrus. Brain Res 482:351–355
190. Burke RE (1967) Composite nature of the monosynaptic excitatory postsynaptic potential. J Neurophysiol 30:1114–1137
191. Burke RE, Nelson PG (1966) Synaptic activity in motoneurons during natural stimulation of muscle spindles. Science 151:1088–1091
192. Buzsaki G (1980) Long-term potentiation of the commissural path-CA1 pyramidal cell synapse in the hippocampus of freely moving rat. Neuroscience 19:293–296
193. Buzsaki G (1984) Feed-forward inhibition in the hippocampal formation. Progr Neurobiol 22:131–153

194. Buzsaki G (1985) What does the "LTP model of memory" model? In: Will BE et al. (eds) Brain Plasticity, Learning, and Memory. Plenum, New York, pp 157–166
195. Buzsaki G (1989) Two-stage model of memory trace formation: A role for "noisy" brain states. Neuroscience 31:551–570
196. Buzsaki G, Gage FH (1989) Absence of long-term potentiation in the subcortically deafferented dentate gyrus. Brain Res 484:94–101
197. Buzsaki G, Haas HL, Anderson EG (1987) Long-term potentiation induced by physiologically relevant stimulus patterns. Brain Res 435:331–333
198. Byrne JH (1985) Neural and molecular mechanisms underlying information storage in *Aplysia*: implications for learning and memory. Trends Neurosci 8:478–482
199. Byrne JH (1987) Cellular analysis of associative learning. Physiol Rev 76:329–439
200. Byrne JH, Berry WO (eds) (1988) Neural Models of Plasticity. Academic, New York
201. Cain DP (1989) Long-term potentiation and kindling: how similar are the mechanisms? Trends Neurosci 12:6–10
202. Cajal SR (1955) Histologie du Systeme Nerveux de l'Homme et des Vertebres, vol 2. Instituto Ramon y Cajal, Madrid
203. Calverley RKS, Jones DG (1990) Contributions of dendritic spines and perforated synapses to synaptic plasticity. Brain Res Rev 15:215–249
204. Camardo JS, Siegelbaum AS, Kandel ER (1984) Cellular and molecular correlates of sensitization in *Aplysia* and their implications for associative learning. In: Alkon DL, Farley J (eds) Primary Neural Substrates of Learning and Behavioral Change. Cambridge University Press, Cambridge, pp 185–205
205. Capek R, Esplin B (1977) Homosynaptic depression and transmitter turnover in spinal monosynaptic pathway. J Neurophysiol 40:95–105
206. Capek R, Esplin DW, Salehmoghaddam S (1971) Rates of transmitter release at the neuromuscular junction estimated by electrophysiological techniques. J Neurophysiol 34:831–841
207. Carew TJ, Sahley CL (1986) Invertebrate learning and memory: From behavior to molecules. Ann Rev Neurosci 9:435–487
208. Carew TJ, Castellucci VF, Kandel ER (1971) An analysis of dishabituation and sensitization of the gill-withdrawal reflex in *Aplysia*. Int J Neurosci 2:79–98
209. Carew TJ, Walters ET, Kandel ER (1981) Classical conditioning in a simple withdrawal reflex in *Aplysia*. J Neurosci 1:1426–1437
210. Carew TJ, Hawkins RD, Kandel ER (1983) Differential classical conditioning of a defensive withdrawal reflex in *Aplysia*. Science 219:397–400
211. Carew TJ, Abrams TW, Hawkins RD, Kandel ER (1984) The use of simple invertebrate systems to explore physiological issues related to associative learning. In: Alkon DL, Farley J (eds) Primary Neural Substrates of Learning and Behavioral Change. Cambridge University, Cambridge, pp 169–185
212. Castellucci VF, Kandel ER (1974) A quantal analysis of the synaptic depression underlying habituation of the gill-withdrawal reflex in *Aplysia*. Proc Natl Acad Sci USA 71:5004–5008
213. Castellucci VF, Kandel ER (1976) Presynaptic facilitation as a mechanism for behavioral sensitization in *Aplysia*. Science 194:1176–1178
214. Castellucci V, Pinsker H, Kupfermann I, Kandel ER (1970) Neuronal mechanisms of habituation and dishabituation of the gill-withdrawal reflex in *Aplysia*. Science 167:1745–1748
215. Castellucci VF, Carew TJ, Kandel ER (1978) Cellular analysis of long-term habituation of the gill-withdrawal reflex of *Aplysia californica*. Science 202:1306–1308
216. Ceccarelli B, Hurlbut WP, Mauro A (1973) Turnover of transmitter and synaptic vesicles at the frog neuromuscular junction. J Cell Biol 57:499–524

217. Chang F-LF, Greenough WT (1984) Transient and enduring morphological correlates of synaptic activity and efficacy change in the rat hippocampal slice. Brain Res 309:35–46
218. Changeux JP, Konishi M, eds (1987) The Neural and Molecular Bases of Learning. Wiley, Chichester
218a Charlton MP, Smith S, Zucker RS (1982) Role of presynaptic calcium ions and channels in synaptic facilitation and depression at the squid giant synapse. J Physiol (Lond) 323:173–193
219. Charriaut-Marlangue C, Aniksztein L, Roisin MP, Ben-Ari Y (1988) Release of proteins during long-term potentiation in the hippocampus of the anaesthetized rat. Neurosci Lett 91: 308–314
220. Chattarji S, Stanton PK, Sejnowski TJ (1989) Commissural synapses, but not mossy fiber synapses, in hippocampal field CA3 exhibit associative long-term potentiation and depression. Brain Res 495:145–150
221. Chavez-Noriega LE, Halliwell JV, Bliss TVP (1990) A decrease in firing threshold observed after induction of the EPSP-spike (E–S) component of long-term potentiation in rat hippocampal slices. Exp Brain Res 79:633–641
222. Chirwa SS, Sastry BR (1988) Asynchronous synaptic responses in hippocampal CA1 neurons during synaptic long-term potentiation. Neurosci Lett 89:355–360
223. Chmykhova NM, Shapovalov AI, Shiriaev BI, Komissarchik YaJu, Snigirevskaya ES (1987) Ultrastructure analysis of physiological identified and stained by the horse radish peroxidase sensorimotor synapse of the frog. Doklady Biol Sci 293:250–253
224. Christensen BN (1976) Morphological correlates of synaptic transmission in lamprey spinal cord. J Neurophysiol 39:197–212
225. Christensen BN, Martin AR (1970) Estimation of probability of transmitter release at the mammalian neuromuscular junction. J Physiol (Lond) 210:933–945
226. Christoffersen GRJ, Juel C (1979) Synaptic depression in *Helix pomatia* 1. A presynaptic model. Comp Biochem Physiol 62A:611–619
227. Chung SH (1976) Intermittent nerve conduction. Nature 264:313–314
228. Chung SH (1977) Synaptic memory in the hippocampus. Nature 266:677–678
229. Clamann P, Mathies T, Lüscher H-R (1989) Variance analysis of excitatory postsynaptic potentials in cat spinal motoneurons during posttetanic potentiation. J Neurophysiol 61:403–416
230. Clamann P, Rioult-Pedotti MS, Lüscher H-R (1991) The influence of noise on quantal EPSP size obtained by deconvolution in spinal motoneurons in the cat. J Neurophysiol 65:67–75
231. Clements JD, Forsythe ID, Redman SJ (1987) Presynaptic inhibition of synaptic potentials evoked in cat spinal motoneurones by impulses in single group Ia axons. J Physiol (Lond) 383:153–170
232. Cohen I, Van der Kloot W (1985) Calcium and transmitter release. Int Rev Neurobiol 27:299–336
233. Cole AJ, Saffen DW, Baraban JM, Worley PF (1989) Rapid increase of an immediate early gene messenger RNA in hippocampal neurons by synaptic NMDA receptor activation. Nature 340: 474–476
234. Collier TJ, Quirk GJ, Routtenberg A (1987) Separable roles of hippocampal granule cells in forgetting and pyramidal cells in remembering spatial information. Brain Res 409:316–328
235. Collingridge GL (1985) Long-term potentiation in the hippocampus: mechanisms of initiation and modulation by neurotransmitters. Trends Pharmacol Sci 6:407–411
236. Collingridge GL, Bliss TVP (1987) NMDA receptors – their role in long-term potentiation. Trends Neurosci 10:288–293
237. Collingridge GL, Kehl SJ, McLennan H (1983) Excitatory amino acids in synaptic transmission in the Schaffer collateral commissural pathway of the rat hippocampus. J Physiol (Lond) 334:33–46

238. Cooke JD, Quastel DMJ (1973) Transmitter release by mammalian motor nerve terminals in response to focal polarization. J Physiol (Lond) 228:377–405
239. Cope TC, Mendell LM (1982) Distributions of EPSP latency at different group Ia-fiber-alpha-motoneuron connections. J Neurophysiol 47:469–478
240. Covian MR, Antunes-Rodrigues J, O'Flaherty JJ (1964) Effects of stimulation of the septal area upon blood pressure and respiration in the cat. J Neurophysiol 27:394–407
241. Crow TJ (1976) The chemistry of reward. Trends Biochem Sci 1:287–289
242. Crow TJ (1988) Associative learning, memory, and neuromodulation in *Hermissenda*. In: Byrne JH, Berry WO (eds) Neural Models of Plasticity. Academic, New York, pp 1–21
243. Crunelli V, Forda S, Collingridge GL, Kelly JS (1982) Intracellularly recorded synaptic antagonism in the rat dentate gyrus. Nature 300:450–452
244. Csicsaky M, Papadopoulos R, Wiegand H (1985) Detection of sub-miniature endplate potentials by harmonic analysis. J Neurosci Meth 15:113–129
245. Cullheim S, Kellerth JO, Conradi SN (1977) Evidence for direct synaptic interconnections between cat spinal motoneurons via the recurrent axon collaterals: a morphological study using intracellular injection of horseradish peroxidase. Brain Res 132:1–10
246. Cunnane TC (1984) The mechanism of neurotransmitter release from sympathetic nerves. Trends Neurosci 7:248–253
247. Curtis DR, Eccles JC (1960) Synaptic action during and after repetitive stimulation. J Physiol (Lond) 150:374–398
248. Czeh G, Horvath Z, Czopf J (1990) Long-term potentiation in slices from subcortically deafferented hippocampi. Brain Res 518:279–282
249. Dale N, Schacher S, Kandel ER (1988) Long-term facilitation in *Aplysia* involves increase in transmitter release. Science 239:282–285
250. Davies CH, Davies SN, Collingridge GL (1990) Paired-pulse depression of monosynaptic GABA-mediated inhibitory postsynaptic responses in rat hippocampus. J Physiol (Lond) 424:513–531
251. Davies SN, Lester RAJ, Reymann KG, Collingridge GL (1989) Temporally distinct pre- and postsynaptic mechanisms maintain long-term potentiation. Nature 338:500–503
252. Deadwyler SA, Dudek FE, Cotman CW, Lynch G (1975) Intracellular responses of rat dentate granule cells in vitro: post-tetanic potentiation to perforant path stimulation. Brain Res 88:80–85
253. Deadwyler SA, Dunwiddie T, Lynch G (1978) Short-lasting changes in hippocampal neuronal excitability following repetitive synaptic activation. Brain Res 147:884–389
254. Deadwyler SA, West M, Lynch G (1979) Activity of dentate granule cells during learning: differentiation of perforant path input. Brain Res 169:29–43
255. Deadwyler SA, Dunwiddie T, Lynch G. (1987) A critical level of protein synthesis is required for long-term potentiation. Synapse 1: 90–95
256. Deisz RA, Prince DA (1989) Frequency-dependent depression in inhibition in guinea-pig neocortex in vitro by $GABA_B$ receptor feed back on GABA release. J Physiol (Lond) 412:513–541
257. De Jonge M, Racine RJ (1985) The effects of repeated induction of long-term potentiation in the dentate gyrus. Brain Res 328:181–185
258. De Jonge MC, Black J, Deyo RA, Disterhoft JF (1990) Learning-induced afterhyperpolarization reductions in hippocampus are specific for cell type and potassium conductance. Exp Brain Res 80:456–462
259. Dekker JAM, Connor DJ, Thal LJ (1991) The role of cholinergic projections from *nucleus basalis* in memory. Neurosci Biobehav Rev 15:299–317

260. Del Castillo J, Katz B (1954a) The effects of magnesium on the activity of motor nerve endings. J Physiol (Lond) 124:553–559
261. Del Castillo J, Katz B (1954b) Quantal component of the end plate potential. J Physiol (Lond) 124:560–573
262. Del Castillo J, Katz B (1954c) Statistical factors involved in neuromuscular facilitation and depression. J Physiol (Lond) 124:574–585
263. Del Castillo J, Katz B (1958) Localization of active spots within the neuromuscular junction of the frog. J Physiol (Lond) 132:630–649
264. Delcomyn F (1981) Nickel chloride for intracellular staining of neurons in insect. J Neurobiol 12:623–627
265. Dennis MJ, Harris A, Kuffler SW (1971) Synaptic transmission and its duplication by focally applied acetylcholine in parasympathetic neurones in the heart of the frog. Proc Roy Soc B177:509–539
266. Desmond NL, Levy WB (1990) Morphological correlates of long-term potentiation imply the modification of existing synapses, not synaptogenesis, in the hippocampal dentate gyrus. Synapse 5:139–143
267. Deutsch JA (1973) The cholinergic synapse and the site of memory. In: Deutsch JA (ed) The Physiological Basis of Memory. Academic, New York, pp 113–124
268. Diamond I, Yasargil GM (1969) Synaptic function in the fish spinal cord: dendritic integration. Progr Brain Res 31:201–209
269. Ding K (1982) Lack of correlation between physiological and morphological features of regenerating frog neuromuscular junction. Brain Res 253:47–56
270. Dingledine R (1983) N-methyl-D-aspartate activates voltage dependent calcium conductance in rat hippocampal pyramidal cells. J Physiol (Lond) 343:385–495
271. Dingledine R (1986) NMDA receptors: what do they do? Trends Neurosci 9:47–49
272. Disterhoft JF, Golden DJ, Read HR, Coulter DA, Alkon DL (1988) AHP reductions in rabbits hippocampal neurons during conditioning correlate with acquisition of the learned response. Brain Res 462:118–125
273. Dityatev AE (1989) Analysis of the functional heterogeniety of the sensory-motor synapse. Neurophysiology (Kiev) 21:780–788
274. Dodd J, Dingledine R, Kelly JS (1981) The excitatory action of acetylcholine on hippocampal neurones of the guinea pig and rat maintained in vitro. Brain Res 207:109–127
275. Doherty P, Hawgood BJ, Smith ICH (1984) Changes in miniature end-plate potentials after brief nervous stimulation at the frog neuromuscular junction. J Physiol (Lond) 356:349–358
276. Dolphin AC (1983) The excitatory amino acid antagonist gamma-D-glutamil-glycine masks rather than prevents long-term potentiation of the perforant path. Neuroscience 10:377–383
277. Dolphin AC (1985) Long-term potentiation at peripheral synapses. Trends Neurosci 8:376–378
278. Dolphin AC, Errington ML, Bliss TVP (1982) Long-term potentiation of the perforant path in vivo is associated with increased glutamate release. Nature 297:496–498
279. Doty RW (1990) Time and memory. In: McGaugh JL et al. (eds) Brain Organization and Memory: Cells, Systems, and Circuits. Oxford University Press, New York, pp 145–158
280. Douglas RM (1977) Long-lasting synaptic potentiation in the rat dentate gyrus following brief high-frequency stimulation. Brain Res 126:361–365
281. Douglas RM, Goddard GV (1975) Long-term potentiation of the perforant path-granule cell synapse in the rat hippocampus. Brain Res 86:205–217
282. Douglas RM, Dragunow M, Robertson HA (1988) High-frequency discharge of dentate granule cells, but not long-term potentiation, induces c-fos protein. Mol Brain Res 4: 259–262

283. Doyer V, Laroche S (1992) Linear relationship between the maintenance of hippocampal long-term potentiation and retention of an associative memory. Hippocampus 2:461–470
284. Drabkina TM (1976) Estimation of the quantal content of excitatory postsynaptic potentials during facilitation and presynaptic inhibition in crayfish neuromuscular apparatus. In: Grachev II (ed) Nervnaya Sistema (Nervous System), vol 16. Leningrad University, Leningrad, pp 73–78
285. Dragunow M, Currie RW, Faul RLM, Robertson HA, Jansen K (1989) Immediate-early genes, kindling and long-term potentiation. Neurosci Behav Rev 13:301–313
286. Dreyer F, Peper K, Akert K, Sandri C, Moor H (1973) Ultrastructure of the "active zone" in the frog neuromuscular junction. Brain Res 62:373–382
287. Dudek FE, Deadwyler SA, Cotman CW, Lynch G (1976) Intracellular responses from granular cell layer on slices of rat hippocampus perforant path synapse. J Neurophysiol 39:384–393
288. Dudel J, Kuffler SW (1961a) The quantal nature of transmission and spontaneous miniature potentials at the crayfish neuromuscular junction. J Physiol (Lond) 155:514–529
289. Dudel J, Kuffler SW (1961b) Mechanism of facilitation at the crayfish neuromuscular junction. J Physiol (Lond) 155:530–542
290. Dudel J, Kuffler SW (1961c) Presynaptic inhibition at the crayfish neuromuscular junction. J Physiol (Lond) 155:543–562
291. Duffy CJ, Teyler TJ (1978) The development of potentiation in the dentate gyrus of rat: physiology and anatomy. Brain Res Bull 3:425–430
292. Duffy CJ, Teyler TJ, Shashoua VE (1981) Long-term potentiation in the hippocampal slice: evidence for stimulated secretion of newly synthesized proteins. Science 212:1148–1151
293. Dunant Y (1986) On the mechanism of acetylcholine release. Progr Neurobiol 26:55–92
294. Dunin-Barkovskii VL (1978) Informatsionnye Protsessy v Neironnykh Strukturakh (Information Processes in Neuronal Structures). Nauka, Moscow
295. Dunwiddie TV, Lynch GS (1978) Long-term potentiation and depression of synaptic responses in the rat hippocampus: localization and frequency dependency. J Physiol (Lond) 276:353–367
296. Dunwiddie TV, Lynch G (1979) The relationship between extracellular calcium concentrations and the induction of hippocampal long-term potentiation. Brain Res 169:103–110
297. Dunwiddie TV, Madison D, Lynch G (1978) Synaptic transmission is required for the initiation of long-term potentiation. Brain Res 150:413–417
298. Eccles JC (1953) The Neurophysiological Basis of Mind. Clarendon, Oxford
299. Eccles JC (1957) The Physiology of Nerve Cells. Johns Hopkins, Baltimore
300. Eccles JC (1964) The Physiology of Synapses. Springer, Berlin Heidelberg New York
301. Eccles JC (1972) Possible synaptic mechanisms subserving learning. In: Karczmar AG, Eccles JC (eds) Brain and Human Behavior. Springer, Berlin Heidelberg New York, pp 39–61
302. Eccles JC (1983) Calcium in long-term potentiation as a model for memory. Neuroscience 10:1071–1081
303. Eccles JC, Katz B, Kuffler SW (1941) Nature of the "end plate potential" in curarized muscle. J Neurophysiol 4:362–387
304. Eccles JC, Krnjevic K, Miledi R (1959) Delayed effect of peripheral severance of afferent nerve fibres on the efficacy of their central synapses. J Physiol (Lond) 145:204–220
305. Eckert R (1963) Electrical interaction of paired ganglion cells in the leech. J Gen Physiol 46:573–588

306. Edwards FA (1991) LTP is a long-term problem. Nature 350: 271–272
307. Edwards FA, Konnerth A, Sakmann B (1990) Quantal analysis of inhibitory synaptic transmission in the dentate gyrus of rat hippocampal slice: a patch clamp study. J Physiol (Lond) 430:213–249
308. Edwards FR, Redman SJ, Walmsley B (1976a) Statistical fluctuation in charge transfer at Ia synapses on spinal motoneurones. J Physiol (Lond) 259:665–688
309. Edwards FR, Redman SJ, Walmsley B (1976b) Non-quantal fluctuation and transmission failures in charge transfer at Ia synapses on spinal motoneurones. J Physiol (Lond) 259:689–704
310. Edwards FR, Redman SJ, Walmsley B (1976c) The effect of polarizing current on unitary Ia excitatory postsynaptic potentials evoked in spinal motoneurones. J Physiol (Lond) 259:705–723
311. Eide E, Fedina L, Jansen J, Lundberg A, Vyklicky L (1967) Unitary excitatory postsynaptic potential in Clarke's column neurones. Nature 215:1176–1177
312. Elmqvist D, Quastel DMJ (1965a) Presynaptic action of hemicholinum at the neuromuscular junction. J Physiol (Lond) 177:463–482
313. Elmqvist D, Quastel DMJ (1965b) A quantitative study of end-plate potentials in isolated human muscle. J Physiol (Lond) 178:505–529
314. Engberg I, Marshall KG (1979) Reversal potential for Ia excitatory postsynaptic potentials in spinal motoneurones of cats. Neuroscience 4:1583–1591
315. Erxleben C, Kriebel ME (1988) Subunit composition of the spontaneous miniature end-plate currents at the mouse neuromuscular junction. J Physiol (Lond) 400:659–676
316. Esplin DW, Zablocka-Esplin B (1971) Rates of transmitter turnover in spinal monosynaptic pathway investigated by electrophysiological techniques. J Neurophysiol 34:842–859
317. Evarts EV (1973) Motor cortex reflex associated with learned movement. Science 179:501–503
318. Evdokimov SA, Samoilov MO, Semenov DG (1976) Microelectrode recording of responses in visually observed cerebral cortex neurons of cat. Neurophysiology (Kiev) 8:122–125
319. Faber DS, Korn H (1982) Transmission at a central inhibitory synapse. I Magnitude of unitary postsynaptic conductance change and kinetics of channel activation. J Neurophysiol 48:654–678
320. Faber DS, Korn H (1988) Synergism at central synapses due to lateral diffusion of transmitter. Proc Natl Acad Sci USA 85:8708–8712
321. Faber DS, Korn H (1989) Electrical field effects: their relevance in central neural networks. Physiol Rev 69:821–863
322. Faber DS, Lin J-W, Korn H (1991) Silent synaptic connections and their modifiability. Proc Natl Acad Sci USA 627:151–164
323. Faber DS, Korn H (1992) Applicability of the coefficient variation method for analyzing synaptic plasticity. Biophys J 60:1288–1294
324. Faber DS, Funch PG, Korn H (1985) Evidence that receptors mediating central synaptic potentials extend beyond the postsynaptic density. Proc Natl Acad Sci USA 82:3504–3508
325. Fanardjian VV, Gorodnov VL (1984) Postsynaptic processes in rubro-spinal neurons of cat's brain by different corticofugal influences. Neurophysiology (Kiev) 16:67–74
326. Farel PB, Thompson RF (1976) Habituation of a monosynaptic response in frog spinal cord: evidence for a presynaptic mechanism. J Neurophysiol 39:661–666
327. Farel PB, Glanzman DL, Thompson RF (1973) Habituation of a monosynaptic responses in vertebrate central nervous system: lateral column-motoneuron pathway in isolated frog spinal cord. J Neurophysiol 36:1117–1130
328. Farley J, Alkon DL (1985) Cellular mechanisms of learning, memory, and information storage. Ann Rev Psychol 36:419–494

329. Fatt P, Katz B (1950) Some observations on biological noise. Nature 166: 597–598
330. Fatt P, Katz B (1951) An analysis of the end-plate potential recorded with an intracellular electrode. J Physiol (Lond) 115:320–370
331. Fatt P, Katz B (1952) Spontaneous subthreshold activity at motor nerve endings. J Physiol (Lond) 117:109–128
332. Fazeli MS, Errington ML, Dolphin AC, Bliss TVP (1990) Increased efflux of a haemoglobin-like protein and an 80 kDa protease into push-pull perfusates following the induction of long-term potentiation in the dentate gyrus. Brain Res 521:247–253
333. Fessard A (1962) Analysis of closing temporal connections at the neuronal level. In: Smirnov GD (ed) Electroentsefalograficheskoe Issledovanie Vysshei Nervnoi Deyatelnosti (Electroencephalographic Study of Higher Nervous Activity). Academy of Sciences of the USSR, Moscow, pp 147–173
334. Fibiger HC (1991) Cholinergic mechanisms in learning, memory and dementia: a review of recent evidence. Trends Neurosci 14:220–223
335. Fifkova E, Van Harreveld A (1977) Long-lasting morphological changes in dendritic spines of dentate granular cells following stimulation of the entorhinal area. J Neurocytol 6:211–230
336. Fifkova E, Marsala J (1967) Stereotaxic atlases of the cat, rabbit and rat. In: Bures J et al. (eds) Electrophysiological Methods in Biological Research. Academia, Prague, pp 653–746
337. Filimonov IN (1949) Sravnitelnaya Anatomiya Kory Bolshogo Mozga Mlekopitayushchikh. Paleokorteks, Arkhikorteks, Promezhutochnaya Kora (Comparative Anatomy of the Cortex of the Mammalian Brain Paleocortex, Archicortex, Intermediate Cortex). Meditsina, Moscow
338. Finch DM, Jackson MB (1990) Presynaptic enhancement of synaptic transmission in hippocampal cell cultures by phorbol esters. Brain Res 518:269–273
339. Finkel AS, Redman SJ (1983) The synaptic current evoked in cat spinal motoneurones by impulses in single group Ia axons. J Physiol (Lond) 342:615–632
340. Flatman JA, Engberg I, Lanbert JDC (1982) Reversibility of Ia EPSP investigated with intracellularly iontophoresed QX-222. J Neurophysiol 48:419–430
341. Forsythe ID, Clements JD (1990) Presynaptic glutamate receptors depress excitatory monosynaptic transmission between mouse hippocampal neurones. J Physiol (Lond) 429:1–16
342. Foster TC, McNaughton BL (1990) LTE in CA1 is associated with an increase in apparent quantal size. In: Squire LR, Lindenlaub E (eds) Symposia Medica Hoechst 23: The Biology of Memory. Schattauer, Stuttgart, pp 264–272
343. Foster TC, McNaughton BL (1991) Long-term enhancement of CA1 synaptic transmission is due to increased quantal size, not quantal content. Hippocampus 1:79–91
344. Frank E (1973) Matching of facilitation at the neuromuscular junction of the lobster: a possible case for influence of muscle on nerve. J Physiol (Lond) 233:635–658
345. Frank E (1987) The influence of neuronal activity on patterns of synaptic connecting. Trends Neurosci 10:188–189
346. Frazier WT, Kandel ER, Kupfermann I, Waziri R, Coggeshall RE (1967) Morphological and functional properties of identified neurons in the abdominal ganglion of *Aplysia californica*. J Neurophysiol 30:1288–1351
347. Frey U, Krug M, Reymann KG, Matthies H (1988) Anisomycin, an inhibitor of protein synthesis, blocks late phases of LTP phenomena in the hippocampal CA1 region in vitro. 452:57–65
348. Frey U, Krug M, Brödemann R, Reymann K, Matthies H (1989) Long-term potentiation induced in dendrites separated from rat's CA1 pyramidal somata does not establish a late phase. Neurosci Lett 97:135–139
349. Frey U, Schröder H, Matthies H (1990) Dopaminergic antagonists prevent long-term maintenance of post-tetanic LTP in CA1 region of rat hippocampal slices. Brain Res 522:69–75

350. Fricke RA (1984) Development of habituation in the crayfish due to selective weakening of electrical synapses. Brain Res 322:139–143
351. Friedlander MJ, Sayer J, Redman SJ (1990) Evaluation of long-term potentiation of small compound and unitary EPSPs at the hippocampal CA3–CA1 synapse. J Neurosci 10:814–825
352. Friesen WO (1975) Antifacilitation and facilitation in the cardiac ganglion of the spiny lobster *Panulirus interruptus*. J Comp Physiol 101A:207–224
353. Fritz LC, Gardner-Medwin AR (1976) The effects of synaptic activation on the extracellular potassium concentration in the hippocampal dentate area in vitro. Brain Res 112:183–187
354. Frotscher M, Gahwiler BH (1988) Synaptic organization of intracellularly stained CA3 pyramidal neurons in slice cultures of rat hippocampus 24:541–552
355. Fujito Y (1983) Enhancement of EEG spikes and hyperpolarizations of cells in the kindled hippocampus of the rabbit. J Physiol (Lond) 33:227–238
356. Furukawa BT, Matsuura S (1978) Adaptive rounding of excitatory postsynaptic potentials at synapses between hair cells and eighth nerve fibres in the goldfish. J Physiol (Lond) 276:193–209
357. Furukawa BT, Hayashida Y, Matsuura S (1978) Quantal analysis of the size of excitatory postsynaptic potentials at synapses between hair cells and afferent nerve fibres in goldfish. J Physiol (Lond) 276:211–223
358. Furukawa T, Kuno M, Matsuura S (1982) Quantal analysis of a decremental response at hair cell afferent fibre synapses in the goldfish sacculus. J Physiol (Lond) 322:181–195
359. Gage PW (1976) Generation of end-plate potentials. Physiol Rev 56:177–251
360. Gallistel CR (1990) The Organization of Learning. MIT Press, New York
361. Gardner D, Kandel ER (1977) Physiological and kinetic properties of cholinergic receptors activated by multiaction interneurons in buccal ganglia of *Aplysia*. J Neurophysiol 40:333–348
362. Garthwaite J (1991) Glutamate, nitric oxide and cell-to-cell signalling in the nervous system. Trends Neurosci 14:60–67
363. Gasanov UG (1973) The functional properties of the auditory cortical units in a controlled experiment (in Russian). Zh Vyssh Nerv Deyat 23:585–592
364. George JA, Milton PC, Stephen JS (1987) Calcium action in synaptic transmitter release. Ann Rev Neurosci 10:633–693
365. Geinisman Y, de Toledo-Morrell L, Morrell.F (1991) Structural synaptic substrates of kindling and long-term potentiation. In: Morrell F (ed) Kindling and Synaptic Plasticity. Birkhauser, Boston, pp 124–159
366. Gerbrandt LK, Skrebitsky VG, Buresova O, Bures J (1968) Plastic changes of unit activity induced by tactile stimuli followed by electrical stimulation of single hippocampal and reticular neurones. Neuropsychologia 6:3–10
367. German DC, Bowden DM (1974) Catecholamine systems as the neural substrate for intracranial self-stimulation: a hypothesis. Brain Res 73:381–419
368. Gerstein GL (1970) Functional association of neurons: detection and interpretation. In: Schmitt FO (ed) The Neurosciences. Second Study Program. Rockefeller University Press, New York, pp 648–661
369. Ghijsen WEJM, Lopes Da Silva F (1991) Increase in endogeneous amino acid release in long-term potentiation. In: Baudry M, Davis J (eds) Long-Term Potentiation: A Debate of Current Issues. MIT Press, Cambridge, Mass, pp 45–55
370. Gilbert ME, Mack CM (1990) The NMDA antagonist, MK-801, suppresses long-term potentiation, kindling, and kindling-induced potentiation in the perforant path of the unanesthetized rat. Brain Res 519:89–96
371. Gilinskii MA, Dubrovina NI (1976) Laminar analysis of commissural evoked responses of hippocampus by microiontophoretic application of glutamate (in Russian). Fiziol Zh SSSR 62:349–355

372. Gingrich KJ, Byrne JH (1985) Simulation of synaptic depression, post-tetanic potentiation, and presynaptic facilitation of synaptic potentials from sensory neurons mediating gill-withdrawal reflex in *Aplysia*. J Neurophysiol 53:652–669
373. Gispen WH, Leunissen JL, Oestreicher AB, Verkleij AJ, Zwiers H (1985) Presynaptic localization of B-50 phosphoprotein: the (ACTH)-sensitive protein kinase substrate involved in rat brain polyphosphoinositide metabolism. Brain Res 328:381–385
374. Glanzman DL, Thompson RF (1980) Alterations in spontaneous miniature potential activity during habituation of a vertebrate monosynaptic pathway. Brain Res 189:377–390
375. Glanzman DL, Mackey SL, Hawkins RD, Dyke AM, Lloyd PE, Kandel ER (1989) Depletion of serotonin in the nervous system of *Aplysia* reduces the behavioral enhancement of gill withdrawal as well as the heterosynaptic facilitation produced by tail shock. J Neurosci 9:4200–4213
376. Glavinovic MI (1987) Synaptic depression in frog neuromuscular junction. J Neurophysiol 58: 230–246
377. Gnetov AV, Kachalov Yu N, Nozdrachev AD (1986) Steklyannyi Mikroelektrod (Glass Microelectrode). Nauka, Leningrad
378. Goddard GV, Douglas RM (1975) Does the engram of kindling model the engram of normal long-term memory? Canad J Neurol Sci 2:385–394
379. Goddard GV, McIntyre DC, Leech CK (1969) A permanent change in brain function resulting from daily electrical stimulation. Exp Neurol 26:295–330
379a. Godukhin OV, Budantsev AYu, Shchipakina TG, Kondratiev VE (1989) The role of the adenylate cyclase system in cholinergic modulation of the synaptic transmission in the hippocampus. Neurophysiology (Kiev) 21:435–432
380. Goh JW, Sastry BR (1986) On the mechanisms of long-lasting potentiation and depression of synaptically evoked responses in the mammalian hippocampus. In: Buzsaki G, Vanderwolf CH (eds) Electrical Activity of the Archicortex. Academiai Kiado, Budapest, pp 295–317
381. Goh JW, Ho-Asjoe M, Sastry BR (1986) Tetanic stimulation-induced changes in [^{3}H]glutamate binding and uptake in rat hippocampus. Gen Pharmacol 17:537–542
382. Grantyn R, Grantyn A, Schaaf P (1977) Conduction velocity, input resistance and size of cat ocular motoneurons stained with procion yellow. Brain Res 135:167–173
383. Grantyn R, Shapovalov AI, Shiriaev BI (1984) Relation between structural and release parameters at the frog sensory-motor synapse. J Physiol (Lond) 349:459–474
384. Grechenko TN (1984) "Conditioned response" of electrically excitable membrane of isolated neurones of a snail (in Russian). Zh Vyssh Nerv Deyat 34:982–984
385. Green JD (1964) The hippocampus. Physiol Rev 44:561–608
386. Griffith JS (1966) A theory of the nature of memory. Nature 211:1160–1163
387. Griffith JS (1971) Mathematical Neurobiology (An introduction to the mathematics of the nervous system). Academic, London
388. Griffith WH, Brown TH, Johnston D (1986) Voltage-clamp analysis of synaptic inhibition during long-term potentiation in hippocampus. J Neurophysiol 55, 767–775
389. Grover LM, Teyler TJ (1990) Two components of long-term potentiation induced by different patterns of afferent activation. Nature 347:477–479
390. Gruodis Yu, Gutman A, Kuras A, Mildazhis V, Milyukas V, Mitskene V, Mitskis K (1972) EEG-quantum (notion, estimation of the value, registration). Doklady Biol Sci 204:1246–1249
391. Guselnikov VI, Pivovarov AS (1982) Plasticity of cortical neurons during learning: physiological mechanisms. In: Ajmone Marsan C, Matthies H (eds) Neuronal Plasticity and Memory Formation. IBRO Monograph Series, vol 9. Raven, New York, pp 37–62

392. Gusev AG (1985) Organization and plasticity of cortical excitatory and inhibitory connections. In: Adrianov OS (ed) Sinapsy Golovnogo Mozga (Synapses of the Brain). Brain Institute, Moscow, pp 97–102
393. Gustafsson B, Jankowska E (1976) Direct and indirect activation of nerve cells by electrical pulses applied extracellularly. J Physiol (Lond) 258:33–61
394. Gustafsson B, Wigström H (1986) Hippocampal long-lasting potentiation produced by pairing single volleys and brief conditioning tetani evoked in separate afferents. J Neurosci 6:1575–1582
395. Gustafsson B, Wigström H (1988) Physiological mechanisms underlying long-term potentiation. Trends Neurosci 11:156–162
396. Gustafsson B, Wigström H (1990) Long-term potentiation in the hippocampal CA1 region: its induction and early temporal development. Progr Brain Res 83: 223–232
397. Gustafsson B, Wigström H, Abraham W, Huang YY (1987) Long-term potentiation in the hippocampus using depolarizing current pulses as the conditioning stimulus to single volley synaptic potentials. J Neurosci 7:774–780
398. Gustafsson B, Huang Y, Wigström H (1988) Phorbol ester-induced synaptic potentiation differs from LTP in the guinea pig hippocampus in vitro. Neurosci Lett 85:77–81
399. Gutman AM (1971) On the effectiveness of dendrite synapses. Biophysics (Moscow) 16:128–134
400. Gutman AM (1980) Biofizika Vnekletochnykh Tokov Mozga (Biophysics of Extracellular Brain Currents). Nauka, Moscow
401. Gutman AM (1984) Dendrity Nervnykh Kletok (Dendrites of Nerve Cells). Mokslas, Vilnyus
402. Gutman AM, Kuras AV (1974) Investigation of extracellular PSP of single afferent – EEG-quantum in frog *tectum opticum*. Biophysics (Moscow) 19:894–898
403. Gutman AM, Kuras AV (1976) Investigation of retinotectal synapses in the frog by recording of extracellular EPSPs produced by synapses of a single axon. Neurophysiology (Kiev) 8:434–437
404. Haas HL, Rose G (1982) Long-term potentiation of excitatory synaptic transmission in the rat hippocampus: the role of inhibitory processes. J Physiol (Lond) 329:541–552
405. Haas HL, Rose G (1984) The role of inhibitory mechanisms in hippocampal long-term potentiation. Neurosci Lett 47:301–306
406. Hablitz JJ, Lundervold A (1981) Hippocampal excitability and changes in extracellular potassium. Exp Neurol 71:410–420
407. Hackett JT, Cochran SL (1977) Quantal synaptic transmission in frog cerebella in vitro. In: Proc of IUPS, vol 13, 27th Physiol Congr. Paris, p 297
408. Hackett JT, Cochran SL, Greenfield LJJr (1989) Quantal transmission at Mauthner axon target synapses in the goldfish brainstem. Neuroscience 32:49–64
409. Hagiwara S, Tasaki I (1958) A study of the mechanism of impulse transmission across a giant synapse of the squid. J Physiol (Lond) 143:114–137
410. Hamlyn LH (1962) The fine structure of the mossy fibre endings in the hippocampus of the rabbit. J Anat 96:112–120
411. Hargreaves EL, Cain DP, Vanderwolf CH (1990) Learning and behavioral long-term potentiation: importance of controlling for motor activity. J Neurosci 10:1472–1478
412. Harris EW, Cotman CW (1986) Long-term potentiation of guinea pig mossy fiber responses is not blocked by N-methyl-D-aspartate antagonists. Neurosci Lett 70, 132–137
413. Harris EW, Lasher SS, Steward O (1979) Analysis of the habituation-like changes in transmission in the temporo-dentate pathway of the rat. Brain Res 162:21–32
414. Hatt H, Smith DO (1976a) Synaptic depression related to presynaptic axon conduction block. J Physiol (Lond) 259:367–393

415. Hatt H, Smith DO (1976b) Non-uniform probabilities of quantal release at the crayfish neuromuscular junction. J Physiol (Lond) 259:395–404
416. Hawkins RD, Castellucci VF, Kandel ER (1981) Interneurons involved in mediation and modulation of gill-withdrawal reflex in *Aplysia*. II Identified neurons produce heterosynaptic facilitation contributing to behavioral sensitization. J Neurophysiol 45:315–326
417. Hawkins RD, Abrams TW, Carew TJ, Kandel ER (1983) A cellular mechanisms of classical conditioning in *Aplysia*: activity-dependent amplification of presynaptic facilitation. Science 219:400–405
418. Hebb DO (1949) The Organization of Behavior. Wiley, New York
419. Henneman E (1974) Principles governing distribution of sensory input to motor neurones. In: Schmitt FO, Worden FG (eds) The Neurosciences. Third Study Program. MIT Press, Cambridge, Mass, pp 281–291
420. Henneman E, Luescher HR, Mathis J (1984) Simultaneously active and inactive synapses of single Ia fibres on cat spinal motoneurones. J Physiol (Lond) 352:147–161
421. Herrera AA, Grinnell AD (1980) Transmitter release from frog motor nerve terminals depends on motor unit size. Nature 287:649–651
422. Herron CE, Lester RAJ, Coan EJ, Collingridge GL (1986) Frequency-dependent involvement of NMDA receptors in the hippocampus: a novel synaptic mechanism. Nature 322:265–268
423. Herz A, Nacimiento AC (1965)· Ueber die Wirkung von Farmaka an Neurone des Hippocampus nach mikroelectrophoretisscher Verabfogung. Naunnyn-Scmiederb Arch Pharmacol 241:295–314
424. Hess G, Kuhnt U (1989) Low intensity tetanic stimulation induces long-term potentiation in hippocampal neurons. Folia Biol 37:195–202
425. Hess G, Kuhnt U, Voronin LL (1987) Quantal analysis of paired pulse facilitation in guinea pig hippocampal slices. Neurosci Lett 77:187–192
426. Hesse GW, Teyler TJ (1976) Reversible loss of hippocampal long-term potentiation following electroconvulsive seizures. Nature 264:562–564
427. Hestrin S, Nicoll RA, Perkel DJ, Sah P (1990) Analysis of excitatory synaptic action in pyramidal cells using whole-cell recording from rat hippocampal slices. J Physiol (Lond) 422:203–225
428. Heuser JE, Reese TS, Dennis MJ, Jan Y, Jan L, Evans L (1979) Synaptic vesicle exocytosis captured by quick freezing and correlated with quantal transmitter release. J Cell Biol 81:275–300
429. Higashima M, Yamamoto C (1985) Two components of long-term potentiation in mossy fiber-induced excitation in hippocampus. Exp Neurol 90:529–539
430. Higashima M, Sawada S and Yamamoto C (1986) A revised method for generation of unitary postsynaptic potentials for quantal analysis in the hippocampus. Neurosci Lett 68:221–226
431. Hirata K, Sawada S, Yamamoto C (1991) Enhancement of transmitter release accompanying with long-term potentiation in synapses between mossy fibers and CA3 neurons in hippocampus. Neurosci Lett 123:73–76
432. Hirotsu I, Hori N, Katsuda N, Ishihara T (1989) Effect of anticholinergic drug on long-term potentiation in rat hippocampal slices. Brain Res 482:194–197
433. Hirst GDS, Redman SJ, Wong K (1981) Post-tetanic potentiation and facilitation of synaptic potentials evoked in cat spinal motoneurones. J Physiol (Lond) 321:97–109
434. Hochner B, Klein M, Schacher S, Kandel E (1986) Additional component in cellular mechanism of presynaptic facilitation contributes to behavioral dishabituation in *Aplysia*. Proc Natl Acad Sci USA 83: 8794–8798
435. Hopkins WF, Johnston D (1988) Noradrenergic enhancement of long-term potentiation at mossy-fiber synapses in the hippocampus. J Neurophysiol 59:667–687
436. Horel JA (1978) The anatomy of amnesia. Brain 101:403–445
437. Horn G (1967) Neuronal mechanisms of habituation. Nature 215:707–711

438. Horn R, Marty A (1988) Muscarinic activation of ionic currents measured by new whole-cell recording method. J Gen Physiol 92:145–159
439. Hounsgaard J (1978) Inhibition produced by iontophoretically applied acetylcholine in area CA1 of thin hippocampal slice from the rat. Acta Physiol Scand 103:110P
440. Hoyle G (1982) Cellular basis of operant-conditioning of leg position. In: Woody CD (ed) Conditioning: Representation of Involved Neural Functions. Advances in Behavioral Biology, vol 26. Plenum, New York, pp 197–212
441. Hughes JR (1958) Post-tetanic potentiation. Physiol Rev 38: 91–113
442. Hu G-Y, Hvalby O, Walaas SJ, Albert NA, Skjeflo P, Andersen P, Greengard P (1987) Protein kinase C injection into hippocampal pyramidal cells elicits features of LTP. Nature 328:426–429
443. Hubbard JI (1963) Repetitive stimulation at the mammalian neuromuscular junction and the mobilization of transmitter. J Physiol (Lond) 169:641–662
444. Hubbard JI (1973) Microphysiology of vertebrate neuromuscular transmission. Physiol Rev 53:674–723
445. Hubbard JI, Jones SF, Landau EM (1971) The effect of temperature change upon transmitter release, facilitation and post-tetanic potentiation. J Physiol (Lond) 216:591–609
446. Hubbard JI, Llinas R, Quastel DM (1969) Electrophysiological Analysis of Synaptic Transmission. Arnold, London
447. Hughes JR (1958) Post-tetanic potentiation. Physiol Rev 38:91–113
448. Huijbregts GLAM, Schreurs M (1975) Evaluation of amplitude and frequency changes of miniature potentials with a poor signal-to-noise ratio. Comp Biochem Physiol C52:11–16
449. Hurlbut WP, Iezzi N, Fesce R, Ceccarelli B (1990) Correlation between quantal secretion and vesicle loss at the frog neuromuscular junction. J Physiol (Lond) 425:501-526
450. Huston JP (1982) Searching for the neural mechanism of reinforcement of "sampling-in". In: Hoebel BG, Novin D (eds) The Neural Basis of Feeding and Reward. Haer Institute, Brunswick, Ne, pp 75–83
451. Hutter OF (1952) Post-tetanic restoration of neuromuscular transmission blocked by D-tubocurarine. J Physiol (Lond) 118:216–227
452. Hvalby O, Lacaille J-C, Hu G-Y, Andersen P (1986) Postsynaptic long-term potentiation following coupling of dendritic glutamate application and synaptic activation. Experientia 43:599–601
453. Iansek R, Redman SJ (1973) The amplitude, time course and charge of unitary excitatory postsynaptic potentials evoked in spinal motoneurone dendrites. J Physiol (Lond) 234:665–668
454. Iles JF (1977) The speed of passive dendrite conduction of synaptic potentials in a model motoneurone. Proc Roy Soc B197:225–229
455. Ilyuchenok RYu, Dubrovina NI, Gilinskii MA (1974) Septo-hippocampal evoked responses during microiontophoretic application of glutamate and atropine. Neurophysiology (Kiev) 6:434–436
456. Isaacson RL (1974) The Limbic System. Plenum, New York
457. Isaacson RL, Pribram KH, eds (1987) The Hippocampus, vol 3. Plenum, New-York
458. Israel M, Dunant Y, Manaranche R (1979) The present status of the vesicular hypothesis. Progr Neurobiol 13:237–275
459. Ito M (1989) Long-term depression. Ann Rev Neurosci 12:85–102
460. Ito M, Udo M, Mano M, Kawai N (1970) Synaptic action of the fastigiobulbar impulses upon neurones in the medullary reticular formation and vestibular nuclei. Exp Brain Res 11:29–47
461. Jack JJB, Redman SJ (1971) The propagation of transient potentials in some linear cable structures. J Physiol (Lond) 285:283–320

462. Jack JJ, Miller S, Porter R, Redman SJ (1971) The time course of minimal excitatory postsynaptic potentials evoked in spinal motoneurones by group Ia afferent fibres. J Physiol (Lond) 215:353–380
463. Jack JJB, Noble D, Tsien RW (1975) Electrical Current Flow in Excitable Cells. Clarendon, Oxford
464. Jack JJ, Redman SJ, Wong K (1981) The components of synaptic potentials evoked in cat spinal motoneurones by impulses in single group Ia afferents. J Physiol (Lond) 321:65–96
465. Jacklet JW, Rine J (1977) Facilitation at neuromuscular junctions: contribution to habituation and dishabituation of the *Aplysia* gill withdrawal reflex. Proc Natl Acad Sci USA 74:1267–1271
466. Jaffard R, Jeantet Y (1981) Posttraining changes in excitability of the commissural path-CA1 pyramidal cell synapse in the hippocampus of mice. Brain Res 220:167–172
467. Jaffe D, Johnston D (1990) Induction of long-term potentiation at hippocampal mossy-fiber synapses follows a Hebbian rule. J Neurophysiol 64:948–960
468. Jankowska E, Roberts WJ (1972) Synaptic actions of single neurones mediating reciprocal Ia inhibition of motoneurones. J Physiol (Lond) 222:633–642
469. Johnson EW, Wernig A (1971) The binomial nature of transmitter release at the crayfish neuromuscular junction. J Physiol (Lond) 218:757–767
470. Jones SF, Kwanbunbumpen S (1970) Some effects of nerve stimulation and hemicholinium on quantal transmitter release at the mammalian neuromuscular junction. J Physiol (Lond) 207:51–62
471. Juel C, Christoffersen GRJ (1979) Synaptic depression in *Helix pomatia*. II Temperature- and ion-dependency. Comp Biochem Physiol A62:621–626
472. Kaczmareck LK (1987) The role of protein kinase C in the regulation of ion channels and neurotransmitter release. Trends Neurosci 10: 30–31
473. Kaczmareck LK, Siedlecki JA, Danysz W (1988) Proto-oncogene c-fos induction in rat hippocampus. Molec Brain Res 3:183–186.
474. Kamenskaya MA (1972) Modern concepts on the mechanism of the release of transmitter quantum from motor nervous endings of striate muscle (in Russian). Uspekhi Fiziol Nauk 3 (N3): 22–63
475. Kandel ER (1976) Cellular Basis of Behavior. Freeman, San Francisco
476. Kandel ER (1979) Cellular insights into behavior and learning. The Harvey Lectures, Ser 73. Academic, New York, pp 19–92
477. Kandel ER, Schwartz JH (1985) Principles of Neural Science. Elsevier, New York
478. Kandel ER, Spencer WA (1968) Cellular neurophysiological approaches in the study of learning. Physiol Rev 48:65–134
479. Kandel ER, Tauc L (1965) Mechanism of heterosynaptic facilitation in the giant cell of the abdominal ganglion of *Aplysia depilans*. J Physiol (Lond) 181:28–47
480. Kandel ER, Spencer WA, Brinley FJ (1961) Electrophysiology of hippocampal neurons. I Sequential invasion and synaptic organization. J Neurophysiol 24:225–242
481. Kandel ER, Frazier WT, Wachtel H (1969) Organization of inhibition in abdominal ganglion of *Aplysia*. I Role of inhibition and disinhibition in transforming neural activity. J Neurophysiol 32:496–508
482. Karamian OA, Kozhanov VM (1980) Excitatory postsynaptic potentials evoked in rat motoneurons by stimulation of single reticulospinal neurons (in Russian). Fiziol Zh SSSR 66:1333–1343
483. Karamian OA, Kozhanov VM, Chmykova NM (1991) Structural and release parameters at the sensorimotor connection of the rat. Neuroscience 43:577–584
484. Kater SB, Nicholson C, eds (1973) Intracellular Staining in Neurobiology. Springer, Berlin Heidelberg New York
485. Katz B (1966) Nerve, Muscle and Synapse. McGraw-Hill, New York

486. Katz B (1969) The Release of Neural Transmitter Substance. Charles C Thomas, Springfield, Ill
487. Katz B (1978) The release of neuromuscular transmitter and the present state of the hypothesis. In: Porter R (ed) Studies in Neurophysiology. Cambridge University Press, Cambridge, pp 1–21
488. Katz B, Miledi R (1963) A study of spontaneous miniature potentials in spinal motoneurones. J Physiol (Lond) 181:656–670
489. Katz B, Miledi R (1965a) Propagation of electric activity in motor nerve terminals. Proc Roy Soc B161:453–482
490. Katz B, Miledi R (1965b) The measurement of synaptic delay and the time course of acetylcholine release at the neuromuscular junction. Proc Roy Soc B161:483–495
491. Katz B, Miledi R (1970) Further study of the role of calcium in synaptic transmission. J Physiol (Lond) 207:789–801
492. Katz B, Miledi R (1972) The statistical nature of the acetylcholine potential and its molecular components. J Physiol (Lond) 224:665–699
493. Katz B, Miledi R (1973) The binding of acetylcholine to receptors and its removal from the synaptic cleft. J Physiol (Lond) 231:549–574
494. Katz B, Miledi K (1979) Estimates of quantal content during chemical potentiation of transmitter release. Proc Roy Soc B205:365–378
495. Katz B, Thesleff S (1957a) On the factors which determine the amplitude of the "miniature" end-plate potential. J Physiol (Lond) 137:267–278
496. Katz B, Thesleff S (1957b) A study of "desensitization" produced by acetylcholine at the motor end-plate. J Physiol (Lond) 138:63–80
497. Kauer JA, Nicoll RA (1988) An APV-resistant non-associative form of long-term potentiation in the rat hippocampus. In: Haas HL, Buzsaki G (eds) Synaptic Plasticity in the Hippocampus. Springer, Berlin Heidelberg New York, pp 65–66
498. Kauer JA, Malenka RC, Nicoll RA (1988) A persistent postsynaptic modification mediates long-term potentiation in the hippocampus. Neurone 1:911–917
499. Kauer JA, Malenka RC, Nicoll RA (1988) NMDA application potentiate synaptic transmission in the hippocampus. Nature 334:250–252
500. Kauer JA. Schweizer FE, Friel DD, Tsien RW (1991) Quantal analysis of synaptic transmission and potentiation in rat hippocampal slices. Third IBRO World Congr Neurosci, Abstracts. Montreal, p 303
501. Kelso SR, Ganong AH, Brown TH (1986) Hebbian synapses in hippocampus. Proc Natl Acad Sci USA 83:5326–5330
502. Kennedy D, Takeda JK (1965) Reflex control of abdominal flexor muscles in the crayfish. I The twitch system. J Exp Biol 43:211–227
503. Kennedy MB (1989) Regulation of synaptic transmission in the central nervous system: long-term potentiation. Cell 59:777–787
504. Kichigina VF (1974) Role of hippocampal CA3 area in regulation of activity of the CA1 field. In: Livanov MN (ed) Pamyat i Sledovye Protsessy (Memory and Trace Phenomena). Institute of Biophysics, Pushchino, pp 193–195
505. Kichigina VF (1979) Functional characteristics of the Schaffer collaterals-CA1 hippocampal connection in in vivo and in vitro experiments. Neurophysiology (Kiev) 11:208–217
506. Kichigina VF, Bragin AG (1976) Functional characteristics of main internal connection systems of hippocampal formation. Neurophysiology (Kiev) 8:159–266
507. Kichigina VF, Vinogradova OS (1983) Effect of the brain stem-septal input elimination upon long-lasting changes in the hippocampus (in Russian). Zh Vyssh Nerv Deyat 5:918–925
508. Kireev VA (1963) Kratkii Kurs Fizicheskoi Khimii (Short Course of Physical Chemistry). Goskhimizdat, Moscow, Chap 14
509. Kirkwood PA, Sears TA (1980) The measurement of synaptic connections in the mammalian central nervous system by means of spike triggered averaging. In: Spinal

and Supraspinal Mechanisms of Voluntary Motor Control and Locomotion. Basel, pp 44–71

510. Kleschevnikov AM, Ivanova (1990) Influence of picrotoxin and conditioning duration on induction of associative long-term potentiation in the CA1 region of mouse hippocampus in vitro. Neurophysiology (Kiev) 22:215–223
511. Kleschevnikov AM, Voronin LL (1989) Suppression of associative long-term potentiation in the hippocampus after tetanization of the conditioning input. Neurophysiology (Kiev) 21:452–458
512. Kleschevnikov AM, Fedorov NB, Voronin LL (1990) Suppression of the "fast" IPSP as a possible cause of priming in hippocampus of mouse. Neurophysiology (Kiev) 22:730–738
513. Knowles WD, Schwartzkroin PA (1981) Local circuit synaptic interactions in hippocampal brain slices. J Neurosci 1:318–322
514. Knowles WD, Traub RD, Wong RKS, Miles R (1985) Properties of neuronal networks: experimentation and modelling of the epileptic hippocampal slice. Trends Neurosci 8:73–79
515. Koenig J, Pecot-Dechavassine M (1971) Relations entre l'apparition des potentiels miniatures spontanes et l'ultrastructure des plaques motrices en voie de reinnervation et de neoformation chez le Rat. Brain Res 27:43–57
516. Kohonen T (1984) Self-Organization and Associative Memory. Springer, Berlin Heidelberg New York
517. Komatsu Y, Toyama K, Maeda J, Sakaguchi H (1981) Long-term potentiation investigated in a slice preparation of striate cortex of young kittens. Neurosci Lett 26:269–274
518. Komatsu Y, Nakajima S, Toyama K (1991) Induction of long-term potentiation without participation of N-methyl-D-aspartate receptors in kitten visual cortex. J Neurophysiol 65:20–32
519. Konorski J (1948) Conditioned Reflex and Neurone Organization. Cambridge University Press, Cambridge.
520. Konorski J (1967) Integrative Activity of the Brain. An Interdisciplinary Approach. University of Chicago Press, Chicago
521. Kopytova FV, Bogolepova IN (1972) Characteristics of neuronal trace responses in some fields of the rabbit dorsal hippocampus (in Russian). Zh Vyssh Nerv Deyat 22:616–619
522. Kopytova FV, Rabinovich MY (1969) Microelectrode study of the conditioned reflex to time. Neurosci Trans 4:461–468
523. Kopytova FV, Mednikova YuS, Rusinova EV (1981) Analog of a conditioned reflex in sensorimotor cortical units following microinjection of acetylcholine. Neurosci Behav Physiol 11:213–220
524. Korn H, Faber DS (1976) Vertebrate central nervous system: same neurons mediate both electrical and chemical inhibitions. Science 194:1166–1169
525. Korn H, Faber DS (1987) Regulation and significance of probabilistic release mechanisms at central synapses. In: Edelman G et al. (eds) Synaptic Function. Wiley, New York, pp 57–108
526. Korn H, Faber DS (1990) Transmission at a central inhibitory synapse. IV. Quantal structure of synaptic noise. J Neurophysiol 63:198–222
527. Korn H, Mallet A (1984) Transformation of binomial input by the postsynaptic membrane at central synapses. J Neurosci 4:125–130
528. Korn H, Faber DS, Triller A (1982) Transmission at a central inhibitory synapse. II Quantal description of release with a physical correlate for binomial n. J Neurophysiol 48:679–707
529. Korn H, Faber DS, Burnod Y, Triller A (1984) Regulation of efficacy at central synapses. J Neurosci 4:125–140

530. Korn H, Burnod Y, Faber DS (1987) Spontaneous quantal currents in a central neurone match predictions from binomial analysis of evoked responses. Proc Natl Acad Sci USA 84:5981–5985
531. Korn H, Fassnacht C, Faber DS (1991) Is maintenance of LTP presynaptic? Nature 350:282
532. Kossel AT, Bonhoeffer T, Bolz J (1990) Non-Hebbian synapses in rat visual cortex. Neuroreport 1:115–118
533. Kostyuk PG (1960a) Mikroelektrodnaya Tekhnika (Microelectrode Techniques). Academy of Sciences of the USSR, Kiev
534. Kostyuk PG (1960b) On plasticity of synaptic connections. In: Beritashvili IS (ed) Gagrskie Besedy (Gagra Talks), vol 3. Metsniereba, Tbilisi, pp 83–100
535. Kostyuk PG (1972) Synaptic mechanisms of plasticity in the central nervous system. In: Samoregulyatsiya Neirofiziologicheskikh Mekhanizmov Integrativnoi Deyatelnosti. Materialy Simpoziuma (Self-regulation of Neurophysiological Mechanisms of Integrative Activity. Symposium Proceedings). Nauka, Leningrad, pp 25–26
536. Kostyuk PG (1986) Kaltsii i Kletochnnaya Vozbudimost (Calcium and Cellular Excitability). Nauka, Moscow
537. Kotlyar BI (1986) Reorganization of central nervous system in processes of restoration of neural function disorders (in Russian). Biol Nauki 2:23–34
538. Kotlyar BI, Maiorov VI, Timofeeva NO, Shulgovskii VV (1983) Neironnaya Organizatsiya Uslovnoreflektornogo Povedeniya (Neuronal Organization of Conditioned Behaviour). Nauka, Moscow
539. Koyano K, Kuba K, Minota S (1985) Long-term potentiation of transmitter release induced by repetitive presynaptic activities in bull-frog sympathetic ganglia. J Physiol (Lond) 359:219–233
540. Kozhanov VM, Shapovalov AI (1977) Synaptic actions evoked in motoneurons by stimulation of individual propriospinal neurons. Neurophysiology (Kiev) 9:402–407
541. Kozhanov VM, Karamian OA, Chmykhova NM (1989) Morphological and quantal characteristics of sensorimotor synapses in isolated spinal cord of young rats. Neurophysiology (Kiev) 21:522–529
542. Kozhedub RG (1985) Serotonin action on cholinergic septo-hippocampal responses in rabbits (in Russian). Zh Vyssh Nerv Deyat 25:537–544
543. Kozhedub RG (1987) About the duration of trace phenomena in rabbit neocortex (in Russian). Zh Vyssh Nerv Deyat 37: 860–865
544. Kozhedub RG (1989) Protracted changes of synaptic efficiency and cellular excitability at learning (in Russian). Zh Vyssh Nerv Deyat 39: 675–681
545. Kozhedub RG, Kozhechkin SN, Voronin LL (1978) Influence of acetylcholine iontophoresis on cellular and field septo-hippocampal responses (in Russian). Zh Vyssh Nerv Deyat 28:411–414
546. Kozhedub RG, Voronin LL, Kozhechkin SN (1982) Acetylcholine action on hippocampal reactions during long-term potentiation and depression following tetanization of the medial septum and dentate fascia (in Russian). Zh Vyssh Nerv Deyat 32:947–954
547. Krasne FB (1976) Invertebrate systems as a means of gaining insight into the nature of learning and memory. In: Rosenzweig MR, Bennett EL (eds) Neural Mechanisms of Learning and Memory. MIT Press, Cambridge, Mass, pp 401–429
548. Kretz R, Shapiro E, Kandel ER (1982) Post-tetanic potentiation at an identified synapse in *Aplysia* is correlated with a Ca^{2+} accumulation. Proc Natl Acad Sci USA 79:5430–5434
549. Kriebel ME, Llados F, Matteson DR (1976) Spontaneous subminiature end-plate potentials in mouse diaphragm muscle: evidence for synchronous release. J Physiol (Lond) 262:553–581
550. Kriebel ME, Llados F, Matteson DR (1982) Histograms of the unitary evoked potential of the mouse diaphragm show multiple peaks. J Physiol (Lond) 322:211–222

551. Kriebel ME, Vautrin J, Holsapple J (1990) Transmitter release: prepacking and random mechanism of dynamic and deterministic process. Brain Res Rev 15:167–178
552. Krnjevic K, Ropert N (1982) Electrophysiological and pharmacological characteristics of facilitation of hippocampal population spikes by stimulation of the medial septum. Neuroscience 7:2165–2183
553. Krnjevic K, Puil E, Werman R (1978) EGTA and motoneuronal afterpotentials. J Physiol (Lond) 275:199–223
554. Krnjevic K, Reiffenstein RJ, Ropert N (1981) Disinhibitory action of acetylcholine in the rat's hippocampus: extracellular observations. Neuroscience 6:2465–2474
555. Krnjevic K, Morris ME, Reiffenstein RJ (1982) Stimulation-evoked changes in extracellular K^+ and Ca^{2+} in pyramidal layers of the rat's hippocampus. Canad J Physiol Pharmacol 60:1643–1657
556. Krug M, Schmidt J (1975) Langdauernde Erregbarkeistanderungen in sensomotorischen Kortex von Kaninchen nach tetanischer Reizung der Zahnpulpa. Acta Biol Med Germ 34:135–141
557. Krug M, Brödemann R, Ott T (1982) Blockade of LTP in the fascia dentata of freely-moving rats by the glutamic acid antagonist GDEE. Brain Res 249:57–62
558. Krug M, Lössner B, Ott T (1984) Anisomycin blocks the late phase of long-term potentiation in the dentate gyrus of freely moving rats. Brain Res Bull 13:39–42
559. Krug M, Müller-Welde P, Wagner M, Ott T, Matthies H (1985) Functional plasticity in two afferent systems of the granule cells in the rat dentate area: frequency-related changes, long-term potentiation and heterosynaptic depression. Brain Res 360:264–272
560. Krug M, Wagner M, Brödemann R, Matthies H (1989) Habituation of monosynaptic field potentials in the dentate gyrus of freely moving rats interferes with LTP. Brain Res 476:163–166
561. Kruglikov RI, Koshtoyants OKh, Valtsev VB (1977) On some mechanisms of acetylcholine involvement in the processes of formation and fixation of temporary connections (in Russian). Zh Vyssh Nerv Deyat 27:989–996
562. Kuba K, Kumamoto E (1990) Long-term potentiation in vertebrate synapses: a variety of cascades with common subprocesses. Progr Neurobiol 34:197–269
563. Kuffler SW, Yoshikami D (1975) The number of transmitter molecules in a quantum: an estimate from iontophoretic application of acetylcholine at the neuromuscular synapse. J Physiol (Lond) 251: 465–482
564. Kuhnt U (1984) Long-lasting changes of synaptic excitability induced by repetitive intracellular current injections in hippocampal neurons. Neurosci Lett Suppl 18:S 27
565. Kuhnt U, Mihaly A, Siklos L, Joo F (1988) Increased retention of calcium in the dendrites of long-term potentiated CA1 neurons of the hippocampal slice. A combined electrophysiological and electron histochemical study. Histochemistry 90:185–193
566. Kuhnt U, Hess G, Voronin LL (1992) Statistical analysis of long-term potentiation of large postsynaptic potentials recorded in CA1 area of guinea pig hippocampal slices: binomial model. Exp Brain Res 92:265–274
567. Kullmann DM (1989) Application of the expectation maximization algorithm to quantal analysis of postsynaptic potentials. J Neurosci Meth 30: 231–245
568. Kullmann DM, Martin Rl, Redman SJ (1989) Reduction by general anaesthetics of group Ia excitatory postsynaptic potentials and currents in the cat spinal cord. J Physiol (Lond) 412:277–296
569. Kumamoto E, Kuba K (1983) Sustained rise in ACh sensitivity of a sympathetic ganglion cell induced by postsynaptic electrical activities. Nature 305:145–146
570. Kuno M (1964a) Quantal components of excitatory synaptic potentials in spinal motoneurones. J Physiol (Lond) 175:81–99
571. Kuno M (1964b) Mechanism of facilitation and depression of the excitatory synaptic potential in spinal motoneurones. J Physiol (Lond) 175:100–112

572. Kuno M (1971) Quantum aspects of central and ganglionic synaptic transmission in vertebrates. Physiol Rev 51:647–678
573. Kuno M, Llinas R (1970) Alterations of synaptic action in chromatolyzed motoneurones of the cat. J Physiol (Lond) 210:823–838
574. Kuno M, Miyahara JT (1968) Factors responsible for multiple discharge of neurones in Clarke's column. J Neurophysiol 31:624–638
575. Kuno M, Miyahara JT (1969a) Nonlinear summation of unit synaptic potentials in spinal motoneurones in the cat. J Physiol (Lond) 201:464–478
576. Kuno M, Miyahara JT (1969b) Analysis of synaptic efficacy in spinal motoneurones from "quantum" aspects. J Physiol (Lond) 201:479–493
577. Kuno M, Weakly JN (1972a) Facilitation of monosynaptic excitatory synaptic potential in spinal motoneurones of the cat. J Physiol (Lond) 224:271–286
578. Kuno M, Weakly JN (1972b) Quantal components of the inhibitory synaptic potential in spinal motoneurones of the cat. J Physiol (Lond) 224:287–303
579. Kuno M, Turkanis SA, Weakly JN (1971) Correlation between nerve terminal size and transmitter release at the neuromuscular junction of the frog. J Physiol (Lond) 213:545–546
580. Kuras AV, Husainovene NP (1986) Paired-pulse facilitation of mass extracellular synaptic potentials of single retinal afferents in the frog tectum. Neurophysiology (Kiev) 18:45–55
581. Kurchavyi GG (1980) Excitatory postsynaptic currents in the cat motoneurons (in Russian). Fiziol Zh SSSR 66:169–176
582. Kurchavyi GG, Motorina AI, Shapovalov AI (1973) Investigation of distribution of synaptic inputs on analogous model of motoneurons. Neurophysiology (Kiev) 5:289–297
583. Kusano K, Landau EM (1975) Depression and recovery of transmission at the squid giant synapse. J Physiol (Lond) 245:13–32
584. Lacaille J-C, Harley CW (1985) The action of norepinephrine in the dentate gyrus: beta-mediated facilitation of evoked potentials in vitro. Brain Res 358:210–220
585. Landfield PW, Deadwyler SA (eds) (1988) Long-Term Potentiation: from Biophysics to Behavior. Liss, New York
586. Landfield PW, McGaugh J, Lynch G (1978) Impaired synaptic potentiation processes in the hippocampus of aged, memory-deficient rats. Brain Res 150:85–101
587. Langmoen IA, Andersen P (1983) Summation of excitatory postsynaptic potentials in hippocampal pyramidal cells. J Neurophysiol 50:1320–1329
588. Larkman A, Stratford K, Jack J (1991) Quantal analysis of excitatory synaptic action and depression in hippocampal slices. Nature 350:344–347
589. Laroche S, Errington ML, Lynch MA, Bliss TVP (1987) Increase in [^{3}H]glutamate release from slices of dentate gyrus and hippocampus following classical conditioning in the rat. Behav Brain Res 13:39–42
590. Laroche S, Doyere V, Bloch (1989) Linear relation between the magnitude of long-term potentiation in the dentate gyrus and associative learning in the rat: a demonstration using commissural inhibition and local infusion of an N-methyl-D-aspartate receptor antagonist. Neuroscience 28:375–386
591. Laroche S, Redini Del Negro C, Clements MP, Lynch MA (1990) Long-term activation of phosphoinositide turnover associated with increased release of amino acids in the dentate gyrus and hippocampus following classical conditioning in the rat. Eur J Neurosci 2:534–543
592. Laroche S, Bloch V, Doyer V, Redini Del Negro C (1991) The significance of long-term potentiation for learning and memory. In: Morrell F (ed) Kindling and Synaptic Plasticity: Legacy of G. Goddard. Birkhauser, Boston, pp 12–37
593. Larrabee MG, Bronk DW (1947) Prolonged facilitation of synaptic excitation in sympathetic ganglia. J Neurophysiol 10:139–154

594. Larson J, Lynch G (1986) Induction of synaptic potentiation in hippocampus by patterned stimulation involves two events. Science 232:985–988
595. Larson J, Wong D, Lynch G (1986) Patterned stimulation in the theta frequency is optimal for the induction of hippocampal long-term potentiation. Brain Res 368: 347–350
596. Lass Y, Halevi Y, Landau EM, Gitter S (1973) A new model for transmitter mobilization in the frog neuromuscular junction. Pflugers Arch 343:157–163
597. Latash LP (1979) Trace changes in spinal cord and some general questions of neurophysiology of memory. In: Beritashvili IS (ed) Gagrskie Besedy (Gagra Talks), vol 7 Metsniereba, Tbilisi, pp 118–130
598. Lee KS (1982) Sustained enhancement of evoked potentials following brief, high-frequency stimulation of the cerebral cortex in vitro. Brain Res 239:617–625
599. Lee KS (1983) Cooperativity among afferents for the induction of long-term potentiation in the CA1 region of the hippocampus. J Neurosci 3:1369–1372
600. Lee KS, Schottler F, Oliver M, Lynch G (1980) Brief bursts of high frequency stimulation produce two types of structural change in rat hippocampus. J Neurophysiol 44:247–258
601. Letbetter WD, Willis WD, Thompson WM (1968) Monosynaptic excitation of motoneurons by single dorsal root ganglion cells. Fed Proc 27:749
602. Levy WB, Desmond NL (1985) Associative potentiation/depression in the hippocampal dentate gyrus. In: Buzsaki G, Vanderwolf CH (eds) Electrical Activity of the Archicortex. Akademiai Kiado, Budapest, pp 359–371
603. Levy WB, Steward O (1979) Synapses as associative memory elements in the hippocampal formation. Brain Res 175:233–245
604. Levy W, Brassel S, Moore S (1983) Partial quantification of the associative synaptic learning rule of the dentate gyrus. Neuroscience 8:799–808
605. Libet B, Kobayashi H, Tanaka T (1975) Synaptic coupling into the production and storage of memory trace. Nature 258:155–157
606. Liley AW (1956) An investigation of spontaneous activity at the neuromuscular junction of the rat. J Physiol (Lond) 132:650–666
607. Liley AW (1957) Spontaneous release of transmitter substance in multiquantal units. J Physiol (Lond) 136:595–605
608. Liley AW, North KAK (1953) An electrical investigation of effects of repetitive stimulation on mammalian neuromuscular junction. J Neurophysiol 16:509–527
609. Lin J-W, Faber DS (1988) Synaptic transmission mediated by single club endings on the goldfish Mauthner cell. II. Plasticity of excitatory postsynaptic potentials. J Neurosci 8:1313–1325
610. Linden DJ, Routtenberg A (1989) The role of protein kinase C in long-term potentiation: a testable model. Brain Res Rev 14:279–296
611. Linder T (1973) Calcium and facilitation at two classes of crustacean neuromuscular synapses. J Gen Physiol 61:56–73
612. Litvinov EG, Logunov DB (1979) Changes in excitability of command neurone in the initial period of conditioning in the snail (in Russian). Zh Vyssh Nerv Deyat 29:284–293
613. Livanov MN (1965) Neurokinetics. In: Chernigovskii VN (ed) Problemy Sovremennoi Neirofiziologii (Problems of Modern Neurophysiology). Nauka, Moscow, pp 37–72
614. Livingston RB (1985) Neurophysiology. In: West J (ed) Physiological Basis of Medical Practice, Sect 9. Williams and Wilkins, Baltimore, pp 970–1297
615. Llinas R (1975) Electrical synaptic transmission in the mammalian central nervous system. In: Santini M (ed) Golgi Centennial Symposium. Raven, New York, pp 374–386
616. Llinas R, Nicholson C (1976) Reversal properties of climbing fiber potential in Purkinje cells: an example of a distributed synapse. J Neurophysiol 39:311–323

617. Lloyd DPC (1949) Post-tetanic potentiation of response in monosynaptic reflex pathways of the spinal cord. J Gen Physiol 33:147–170
618. Loewy AD, Spyer M, eds (1990) Central Regulation of Autonomic Functions. Oxford University Press, New York
619. Logunov DB (1984) Homosynaptic post-tetanic potentiation of identified synapses in central nervous system of snail. Doklady Biol Sci 274:475–478
620. Logunov DB, Balaban PM (1978) Monosynaptic connection between identified neurons in snail. Doklady Biol Sci 240:237–240
621. Logunov DB, Voronin LL, Derevyagin VI (1980) Quantal analysis of low-frequency depression in the central nervous system of the snail. Doklady Biol Sci 253:1005–1009
622. Lomo T (1966) Frequency potentiation of excitatory synaptic activity in the dentate area of the hippocampal formation. Acta Physiol Scand 68 (Suppl 277):128
623. Lopes da Silva FH, Arnolds DEAT (1978) Physiology of the hippocampus and related structures. Ann Rev Physiol 40:185–216
624. Lopez HS, Burger B, Dickstein R, Desmond NL, Levy WB (1990) Associative synaptic potentiation and depression: quantification of dissociable modifications in the hippocampal dentate gyrus favors a particular class of synaptic modification equations. Synapse 5:33–47
625. Lorente de No R (1934) Studies on the structure of the cerebral cortex. II Continuation of the study of the ammonic system. J Psychol Neurol 46:113–177
626. Lössner B, Schweigert Ch, Krug M, Matthies HK (1990) Post-tetanic long-term potentiation in the dentate gyrus of freely moving rats is accompanied by an increase in protein synthesis. Biomed Biochem Acta 49:385–392
627. LoTurco JJ, Coulter DA, Alkon DL (1988) Enhancement of synaptic potentials in rabbit CA1 pyramidal neurons following classical conditioning. Proc Natl Acad Sci USA 85:1672–1676
628. Lovinger DM, Akers RF, Nelson RB, Barnes CA, McNaughton BL, Routtenberg A (1985) A selective increase in phosphorylation of protein F1, a protein kinase C substrate, directly related to three day growth of long-term synapse enhancement. Brain Res 343:137–143
629. Lovinger DM, Colley PA, Akers RF, Nelson RB, Routtenberg A (1986) Direct relation of long-term synaptic potentiation to phosphorylation membrane protein F1, a substrate for membrane protein kinase C. Brain Res 399:205–211
630. Low WC, Bement SL (1980) Enhancement of afferent fiber activity in hippocampal slices. Brain Res 198:472–477
631. Lucas SM, Cope TC, Binder MD (1984) Analysis of individual Ia-afferent EPSPs in a homonymous motoneuron pool with respect to muscle topography. J Neurophysiol 51:64–74
632. Lundberg A, Quilisch H (1953) Presynaptic potentiation and depression of neuromuscular transmission in frog and rat. Acta Physiol Scand 30:111–120
633. Lüscher HR, Ruenzel P, Henneman E (1979) How the size of motoneurones determines their susceptibility to discharge. Nature 282:859–861
634. Lüscher HR, Ruenzel P, Henneman E (1983) Composite EPSPs in motoneurons of different sizes before and during PTP: implications for transmission failure and its relief in Ia projections. J Neurophysiol 49:269–289
635. Lustig C, Parnas H, Segel LA (1986) On the quantal hypothesis of neurotransmitter release: an explanation for calcium dependence of the binomial parameters. J Theor Biol 120:205–213
636. Lynch G (1986) Synapses, Circuits, and the Beginnings of Memory. MIT Press, Cambridge, Mass
637. Lynch GS, Baudry M (1984) The biochemistry of memory: a new and specific hypothesis. Science 224:1057–1063

638. Lynch GS, Gribkoff VK, Deadwyler SA (1976) Long-term potentiation is accompanied by a reduction in dendritic responsiveness to glutamic acid. Nature 263, 151–153
639. Lynch GS, Dunwiddie T, Gribkoff V (1977) Heterosynaptic depression: a postsynaptic correlate to long-term potentiation. Nature 266:737–739
640. Lynch G, Halpain S, Baudry M (1982) Effects of high-frequency synaptic stimulation on glutamate receptor binding studied with a modified in vitro hippocampal slice preparation. Brain Res 244:101–111
641. Lynch G, Larson J, Kelso S, Barrionuevo G, Schottler F (1983) Intracellular injections of EGTA block the induction of hippocampal long-term potentiation. Nature 305:719–721
642. Lynch G, Muller D, Seubert P, Larson J (1988) Long-term potentiation: persistent problems and recent results. Brain Res Bull 21:363–372
643. Lynch G, Kessler M, Arai A, Larson J (1990) The nature and causes of hippocampal long-term potentiation. In: Storm-Mathisen J et al. (eds) Progress in Brain Research, vol 83. Elsevier, Amsterdam, pp 233–250
644. Lynch MA, Voss KL (1991) Presynaptic changes in long-term potentiation: elevated synaptosomal calcium concentration and basal phosphoinositide turnover in dentate gyrus. J Neurochem 56:113–118
645. Lynch MA, Errington ML, Bliss TVP (1985) Long-term potentiation of synaptic transmission in the dentate gyrus: increased release of glutamate without increase in receptor binding. Neurosci Lett 62:123–129
646. Lynch MA, Clements MP, Errington ML, Bliss TVP (1988) Increased hydrolyses of phosphatidylinositol-4, 5-bisphosphate in long-term potentiation. Neurosci Lett 84:291–296
647. Mackey SL, Kandel ER, Hawkins RD (1989) Identified serotonergic neurons LCB1 and RCB1 in the cerebral ganglia of *Aplysia* produce presynaptic facilitation of siphon sensory neurons. J Neurosci 9:4227–4235
648. MacVicar BA, Dudek FE (1982) Electrotonic coupling between granule cells of rat dentate gyrus: physiological and anatomical evidence. J Neurophysiol 47:579–592
649. Madison DV, Nicoll RA (1988) Enkephaline hyperpolarizes interneurones in the rat hippocampus. J Physiol (Lond) 398:123–130
650. Madison DV, Lancaster B, Nicoll RA (1987) Voltage clamp analysis of cholinergic action in the hippocampus. J Neurosci 7:733–741
651. Madison DV, Malenka RC, Nicoll RA (1991) Mechanisms underlying long-term potentiation of synaptic transmission. Ann Rev Neurosci 14:379–397
652. Maeno T (1969) Analysis of mobilization and demobilization processes in neuromuscular transmission in the frog. J Neurophysiol 32:793–800
653. Maeno T, Edwards C (1969) Neuromuscular facilitation with low-frequency stimulation and effects of some drugs. J Neurophysiol 32:785–792
654. Magleby KL (1973) The effect of tetanic and post-tetanic potentiation on facilitation of transmitter release at the frog neuromuscular junction. J Physiol (Lond) 234:353–371
655. Magleby KL, Miller DC (1981) Is the quantum of transmitter release composed of subunits? A critical analysis in the mouse and frog. J Physiol (Lond) 311:267–287
656. Maksimova OA, Balaban PM (1983) Neironnye Mekhanizmy Plastichnosti Povedeniya (Neuron Mechanisms of Behaviour Plasticity). Nauka, Moscow
657. Malenka RC (1991) Postsynaptic factors control the duration of synaptic enhancement in area CA1 of the hippocampus. Neuron 6:53–60
658. Malenka RC, Ayoub GA, Nicoll RA (1987) Phorbol esters enhance transmitter release in rat hippocampal slices. Brain Res 403:198–203
659. Malenka RS, Kauer JA, Zucker RS, Nicoll RA (1988) Postsynaptic calcium is sufficient for potentiation of hippocampal synaptic transmission. Science 242:81–84

660. Malinow R (1991) Transmission between pairs of hippocampal slice neurons: quantal levels, oscillations, and LTP. Science 252:722–724
661. Malinow R, Miller JP (1986) Postsynaptic hyperpolarization during conditioning reversibly blocks induction of long-term potentiation. Nature 320:529–530
662. Malinow R, Tsien RW (1990) Presynaptic enhancement shown by whole-cell recordings of long-term potentiation in hippocampal slices. Nature 346:177–180
663. Malinow R, Madison DV, Tsien RW (1988) Persistent protein kinase activity underlying long-term potentiation. Nature 335:820–824
664. Mallart A, Martin AN (1968) The relation between quantum content and facilitation at the neuromuscular junction of the frog. J Physiol (Lond) 196:593–604
665. Maloshevskii SG, Lenkov DN, Bragina TA (1976) About changes in synaptic vesicles of motor-cortex axon terminals after stimulation in cats (in Russian). Neurophysiology (Kiev) 7:639–646
666. Marchbanks RM (1981) The vesicular hypothesis questioned. Trends Neurosci 1:83–84
667. Markram H, Segal M (1990) Long-lasting facilitation of excitatory postsynaptic potentials in the rat hippocampus by acetylcholine. J Physiol (Lond) 427:381–393
668. Marr D (1971) Simple memory: a theory for archicortex. Phil Trans Roy Soc B262:23–81
669. Martin AR (1955) A further study of the statistical composition of the end-plate potential. J Physiol (Lond) 130:114–122
670. Martin AR (1966) Quantal nature of synaptic transmission. Physiol Rev 46:51–66
671. Martin AR (1976) The effect of membrane capacitance on nonlinear summation of synaptic potentials. J Theor Biol 59:179–187
672. Martin AR (1977) Junctional transmission. II Presynaptic mechanisms. In: Brookhart JM, Mountcastle VB (eds) Handbook of Physiology. The Nervous System, sect 1, vol 1 (Kandel ER, ed). Williams and Wilkins, Baltimore, pp 329–355
673. Martin AR, Pilar G (1964a) Quantal components of the synaptic potential in the ciliary ganglion of the chick. J Physiol (Lond) 175:1–16
674. Martin AR, Pilar G (1964b) Presynaptic and postsynaptic events during post-tetanic potentiation and facilitation in the avian ciliary ganglion. J Physiol (Lond) 175:17–30
675. Martin AR, Ringham GL (1975) Synaptic transfer at a vertebrate central nervous synapse. J Physiol (Lond) 251:409–426
676. Martinez-Soler R, Shekhter ED (1975) Chemical sensitivity and plasticity of the neuron. In: Sokolov EN (ed) Peismekernyi Potentsial Neirona (Pacemaker Potential of Neuron). Metsniereba, Tbilisi, pp 109–129
677. Marty A, Neher E (1983) Tight-seal whole-cell recording. In: Sakman B, Neher E (eds) Single-Channel Recording. Plenum, New York, pp 107–122
678. Maru E, Tatsuno J, Okamoto J, Ashida H (1982) Development and reduction of synaptic potentiation induced by perforant path kindling. Exp Neurol 78:409–424
679. Matteson DR, Kriebel ME, Llados PA (1979) A statistical model supports the subunit hypothesis of transmitter release. Neurosci Lett 15:147–152
680. Matteson DR, Kriebel ME, Llados PA (1981) A statistical model indicates that miniature end-plate potentials and unitary evoked end-plate potentials are composed of subunits. J Theor Biol 90:337–363
681. Matthies H (1989) In search of cellular mechanisms of memory. Progr Neurobiol 32: 277–349
682. Matyushkin DP (1989) Obratnye Sviazi v Sinapse (Feedback connections in the synapse). Nauka, Moscow
683. Matyushkin DP, Drabkina TM (1973) To evaluation of a quantal structure of excitatory postsynaptic potentials in neuromuscular synapses of a complex type (in Russian). Fiziol Zh SSSR 59:541–546

684. Matyushkin DP, Drabkina TM, Shabunova IA (1980) Quantitative estimation of function of presynaptic apparatus in single and multiple synapses (in Russian). Uspekhi Fiziol Nauk 11:49–70
685. May PBY, Goh JW, Sastry BR (1987) Induction of hippocampal long-term potentiation in the absence of extracellular Ca^{2+}. Synapse 1:273–278
686. McCanolles DL, Zabloska-Esplin B, Esplin DW (1971) Rates of transmitter turnover in the cat superior cervical ganglion estimated by electrophysiological techniques. J Neurophysiol 34:817–830
687. McCarren M, Alger BE (1985) Use-dependent depression of IPSPs in rat hippocampal pyramidal cells in vitro. J Neurophysiol 53:557–571
688. McGaugh JL (1989) Involvement of hormonal and neuromodulatory systems in the regulation of memory storage. Ann Rev Neurosci 12:255–287
689. McGaugh L, Weinberger NM, Lynch G, eds (1990) Brain Organization and Memory: Cells, Systems, and Circuits, Oxford University Press, New York
690. McIlwain JT, Creutzfeldt OD (1967) Microelectrode study of synaptic excitation and inhibition in the lateral geniculate nucleus of the cat. J Neurophysiol 30:1–12
691. McIntyre AK, Holman ME, Veale J (1967) Cortical responses to impulses from single Pacinian corpuscles in the cat's hind limb. Exp Brain Res 4:243–255
692. McLachlan EM (1975) An analysis of the release of acetylcholine from preganglionic nerve terminals. J Physiol (Lond) 245:447–466
693. McLachlan EM (1978) The statistics of transmitter release at chemical synapses. In: Porter R (ed) International Review of Physiology. Neurophysiology III, vol 17. University Park, Baltimore, MD, pp 49–117
694. McLachlan EM, Martin AR (1981) Non-linear summation of end-plate potentials in the frog and mouse. J Physiol (Lond) 311:307–324
695. McNaughton BL (1982) Long-term synaptic enhancement and short-term potentiation in rat fascia dentata act through different mechanisms. J Physiol (Lond) 324:249–262
696. McNaughton BL, Barnes CA (1977) Physiological identification and analysis of the medial and lateral perforant pathways in the rat. J Comp Neurol 175:439–453
697. McNaughton BL, Foster TC (1990) The cellular basis of memory. Science 249:148
698. McNaughton BL, Morris RGH (1987) Hippocampal synaptic enhancement and information storage within a distributed memory system. Trends Neurosci 10:408–415
699. McNaughton BL, Douglas RM, Goddard GV (1978) Synaptic enhancement in fascia dentata: cooperativity among coactive afferents. Brain Res 157:277–293
700. McNaughton BL, Barnes CA, Andersen P (1981) Synaptic efficacy and EPSP summation in granule cells of rat fascia dentata studied in vitro. J Neurophysiol 46:952–966
701. McNaughton BL, Barnes CA, Rao G, Baldwin J, Rasmussen M (1986) Long-term enhancement of hippocampal synaptic transmission and the acquisition of spatial information. J Neurosci 6:653–571
702. McNaughton N, Miller JJ (1984) Medial septal projections to the dentate gyrus of the rat: Electrophysiological analysis of distribution and plasticity. Exp Brain Res 37:360–363
703. Meiss DE, Govind CK (1979) Multiterminal innervation: non-uniform density along single lobster muscle fibers. Brain Res 160:163–169
704. Melchers BPC, Waldman WJ, Zijps LJ, Lopes da Silva FH (1988) Chloride-mediated feedforward inhibition is not involved in long-term potentiation. In: Woody CD et al. (eds) Cellular Mechanisms of Conditioning and Behavioral Plasticity. Plenum, New York, pp 57–63
705. Melkonyan DS (1986) Perekhodnye Protsessy v Neironnykh Sistemakh (Transitional Processes in Neuronal Systems). Academy of Sciences of ArmSSR, Erevan
706. Melkonyan DS, Mkrtchan HH, Fanardjian VV (1982) Simulation of learning processes in neuronal networks of the cerebellum. Biol Cybern 45:79–88

707. Mendell LM (1984) Modifiability of spinal synapses. Physiol Rev 64:260–324
708. Mendell LM, Henneman E (1971) Terminals of single Ia fibers: location, density and distribution within a pool of 300 homonymous motoneurons. J Neurophysiol 34:171–187
709. Mendell LM, Weiner K (1976) Analysis of pairs of individual Ia-EPSPs in single motoneurones. J Physiol (Lond) 255:81–104
710. Mendell LM, Munson B, Scott JG (1976) Alterations of synapses of axotomized motoneurones. J Physiol (Lond) 255:67–79
711. Meshul CK, Hopkins WF (1990) Presynaptic ultrastructure correlates of long-term potentiation in the CA1 subfield of the hippocampus. Brain Res 514:310–319
712. Metherate R, Weinberger NM (1989) Acetylcholine produces stimulus-specific receptive field alterations in cat auditory cortex. Brain Res 480:372–377
713. Migunova EYa, Kudryashov IE, Busnyuk MM, Voronin LL (1981) Activity of enzymes of transmitter metabolism in rabbit hippocampus during long-term post-tetanic potentiation (in Russian). Zh Vyssh Nerv Deyat 31:278–286
714. Miledi R (1967) Spontaneous synaptic potentials and quantal release of transmitter in the stellate ganglion of the squid. J Physiol (Lond) 192:379–406
715. Miles R (1988) Plasticity of recurrent excitatory synapses between CA3 hippocampal pyramidal cells. Soc Neurosci Abstr 14:19.
716. Miles R (1990) Variation in strength of inhibitory synapses in CA3 region of guinea-pig hippocampus in vitro. J Physiol (Lond) 431:659–676
717. Miles R, Wong RKS (1984) Unitary inhibitory synaptic potentials in the guinea-pig hippocampus in vitro. J Physiol (Lond) 356:97–113
718. Miles R, Wong RKS (1986) Excitatory synaptic interactions between CA3 neurones in the guineapig hippocampus. J Physiol (Lond) 373:397–418
719. Miles R, Wong RKS (1987a) Latent synaptic pathways revealed after tetanic stimulation in the hippocampus. Nature 329:724–726
720. Miles R, Wong RKS (1987b) Inhibitory control of local excitatory circuits in the guinea pig hippocampus. J Physiol (Lond) 388:611–619
721. Miles R, Wong RKS, Traub RKS (1991) Recurrent excitatory synapses between CA3 neurons in the hippocampus. In: Thomson A, Wheal H (eds) Excitatory Amino Acids and Synaptic Function. Academic, London, pp 141–154
722. Miller DC, Weinsbock MM, Magleby KL (1978) Is the quantum of transmitter release composed of subunits? Nature 274:388–390.
723. Misgeld U, Sarvey JM, Klee MK (1979) Heterosynaptic postactivation potentiation in hippocampal CA3 neurons. Long-term changes of postsynaptic potentials. Exp Brain Res 37:217–230
724. Mitlener J, Strejckova A (1977) Ultrasonic grinding of capillary microelectrode tips. Physiol Bohemosl 26:95–96
725. Miyamoto MD (1975) Binomial analysis of quantal transmitter release at glycerol-treated frog neuromuscular junctions. J Physiol (Lond) 250:121–142
726. Miyamoto MD (1986) Probability of quantal transmitter release from nerve terminals: theoretical considerations in the determination of spatial variation. J Theoret Biol 123:289–304
727. Mochida S, Libet B (1985) Synaptic long-term enhancement (LTE) induced by a heterosynaptic neural input. Brain Res 329:360–363
728. Model MG, Highstein SM, Bennett MVL (1975) Depletion of vesicles and fatigue of transmission at a vertebrate central synapse. Brain Res 98:209–228
729. Mody I, Stanton PK, Heinemann U (1988) Activation of N-methyl-D-aspartate (NMDA) receptors parallels changes in cellular and synaptic properties of dentate gyrus granule cells after kindling. J Neurophysiol 59:133–1054
730. Molgo J, Pecot-Dechavassine M (1988) Effects of carbonyl cyanide m-chlorophenylhydrazone on quantal transmitter release and ultrastructure of frog motor nerve terminals. Neuroscience 24:695–708

731. Monaghan DT, Cotman (1985) Distribution of N-methyl-D-aspartate-sensitive L-[^{3}H]glutamate-binding sites in rat brain. J Neurosci 5:2909–2919
732. Morrell F (1962) Effect of anodic surface polarization on motor response and on the nature of discharges from individual cortical cells (in Russian). Fiziol Zh SSSR 48:251–263
733. Morrell F, ed (1991) Kindling and Synaptic Plasticity. Birkhauser, Boston
734. Morris RGM, Anderson E, Lynch GS, Baudry M (1986) Selective impairment of learning and blockade of long-term potentiation by an N-methyl-D-aspartate receptor antagonist, AP5. Nature 319: 774–776
735. Morris RGM, Davis S, Butcher SP (1990) Hippocampal synaptic plasticity and NMDA receptors: a role in information storage. Phil Trans Roy Soc B329:187–204
736. Morrisett RA, Chow C, Nadler JV, McNamara JO (1989) Biochemical evidence for enhanced sensitivity to N-methyl-D-aspartate in the hippocampal formation of kindled rats. Brain Res 496:25–28
737. Moshkov DA, Petrovskaya LL, Bragin AG (1977) Post-tetanic changes in ultrastructure of giant spine synapses of hippocampal CA3 field. Doklady Biol Sci 237:1525–1528
738. Moshkov DA, Pavlik LL, Otmakhov NA (1988) A study of the synaptic actin in hippocampal slices at long-lasting potentiation. Cytology (Leningrad) 30:259–267
739. Müller CM (1992) A role for glial cells in activity-dependent central nervous plasticity? A review and hypothesis. Int Rev Neurobiol 34:215–280
740. Muller D, Dunant Y (1985) Subminiature electroplaque potentials are present in *Torpedo* electric organ. Experientia 41:824
741. Muller D, Lynch G (1989) Evidence that changes in presynaptic currents are not responsible for long-term potentiation in hippocampus. Brain Res 479:290–299
742. Muller D, Lynch G (1990) Synaptic modulation of N-methyl-D-aspartate receptor mediated responses in hippocampus. Synapse 5:94–103
743. Muller D, Turnbull J, Baudry M, Lynch G (1988) Phorbol-ester induced synaptic facilitation is different than long-term potentiation. Proc Natl Acad Sci USA 85: 6997–7000
744. Muller D, Larson J, Lynch G (1989) The NMDA receptor-mediated components of responses evoked by patterned stimulation is not increased by long-term potentiation. Brain Res 477:396–399
745. Muller D, Buchs P-A, Dunant Y, Lynch G (1990) Protein kinase C activity is not responsible for the expression of long-term potentiation in hippocampus. Proc Natl Acad Sci USA 87:4073–4077
746. Muller KJ, Nicholls JG (1974) Different properties of synapses between a single sensory neurone and two different motor cells in the leech CNS. J Physiol (Lond) 238:357–369
747. Mumiak CG, Kriebel ME, Carlson CG (1982) Changes in MEPP and amplitude distributions in the mouse diaphragm during synapse formation and degeneration. Dev Brain Res 5:123–138
748. Munson JB, Sypert GW (1979) Properties of single fibre excitatory postsynaptic potentials in triceps surae motoneurones. J Physiol (Lond) 296:329–342
749. Murakami F, Fujito Y, Tsukahara N (1977) Analysis of unitary EPSPs mediated by the newly-formed cortico-rubral synapses after lesion of the nucleus interpositus of the cerebellum. Exp Brain Res 30:245–258
750. Murakami F, Katsumaru H, Saito K, Tsukahara N (1982) A quantitative study of synaptic reorganization in red nucleus neurons after lesion of the nucleus interpositus of the cat: an electron microscopic study involving intracellular injection of horseradish peroxidase. Brain Res 242:41–53
751. Murzina GB (1991) Mathematical model of Ca^{2+}/calmodulin-dependent protein kinase role for a long-term increase of synaptic plasticity. Biophysics (Moscow) 36:157–161

752. Nakajima Y, Nakajima S, Leonard RJ, Yamaguchi K (1986) Acetylcholine rises excitability by inhibiting the fast transient potassium current in cultured hippocampal neurons. Proc Natl Acad Sci USA 83:3022–3026
753. Nathan T, Jensen MS, Lambert JDC (1990) $GABA_B$ receptors play a major role in paired pulse facilitation in area CA1 of the rat hippocampus. Brain Res 531:55–65
754. Neale EA, Nelson PG, Macdonald RL, Christian CN, Bowers LM (1983) Synaptic transmission between mammalian central neurons in cell culture. III Morphophysiological correlates of quantal synaptic transmission. J Neurophysiol 49:1459–1468
755. Nelson PG, Marshall KC, Pun RYK, Christian CN, Sheriff WHJr, Macdonald RL, Neale EA (1983) Synaptic interactions between mammalian central neurons in cell culture. II Quantal analysis of EPSPs. J Neurophysiol 49:1442–1458
756. Nelson PG, Yu C, Fields RD, Neale EA (1989) Synaptic connections in vitro: modulation of number and efficacy by electrical activity. Science 244:585–587
757. Nelson RB, Linden DJ, Routtenberg A (1989) Phosphoproteins localized to presynaptic terminal linked to persistence of long-term potentiation (LTP): quantitative analysis of two-dimensional gels. Brain Res 497: 30–42
758. Nelson SG, Frank K (1967) Anomalous rectification in cat spinal motoneurons and effect of polarizing currents on excitatory postsynaptic potential. J Neurophysiol 30:1097–1113
759. Nelson SG, Collatos TC, Niechaj A, Mendell LM (1979) Immediate increase in Ia-motoneuron synaptic transmission caudal to spinal cord transaction. J Neurophysiol 42:655–664
760. Newlon PG, Goldberg SJ, Hayes RL (1991) High-frequency septal stimulation suppresses long-term potentiation (LTP) in the CA1 region of rat hippocampus. Brain Res 544:320–324
761. Nicholls JG, Kuffler DP (1990) Quantal release of serotonin from postsynaptic nerve terminals. Neurochem Int 17:157–163
762. Nicholls JG, Purves D (1970) Monosynaptic chemical and electrical connections between sensory and motor cells in the central nervous system of the leech. J Physiol (Lond) 209:647–667
763. Nicholls J, Wallace BG (1978) Quantal analysis of transmitter release at an inhibitory synapse in the central nervous system of the leech. J Physiol (Lond) 281:171–185
764. Nicoletti F, Meek JL, Iadarola MJ, Chuang DM, Roth BL, Costa E (1986) Coupling of inositol phospholipid metabolism with excitatory amino acid recognition sites in rat hippocampus. J Neurochem 46: 40–46
765. Nicoll RA (1985) The septo-hippocampal projection: a model cholinergic pathway. Trends Neurosci 8:533–536
766. Nicoll RA (1988) The coupling of neurotransmitter receptors to ion channels in the brain. Science 241:545–551
767. Nishi S, Soeda H, Koketsu K (1965) Studies on sympathetic B and C neurons and patterns of preganglionic innervation. J Cell Compar Physiol 66:19–32
768. Nowak L, Bregestovski P, Ascher P, Herbert A, Prochiantz A (1985) Magnesium gates glutamate-activated channels in mouse central neurones. Nature 307:462–465
769. Nozdrachev AD (1983) Fiziologiya Vegetativnoi Nervnoi Sistemy (Physiology of Vegetative Nervous System). Meditsina, Leningrad
770. Numann RE, Wong RKS (1984) Voltage-clamp study on GABA response desensitization in single pyramidal cells dissociated from the hippocampus of adult guinea pigs. Neurosci Lett 47: 289–294
771. Nunzi MG, Gorio A, Milan F, Freund TF, Somogyi P, Smith AD (1985) Cholecystokinin-immunoreactive cells form symmetrical synaptic contacts with pyramidal and nonpyramidal neurons in the hippocampus. J Comp Neurol 237:485–505
772. Okada Y, Miyamoto T (1989) Formation of long-term potentiation in superior colliculus slices from the guinea pig. Neurosci Lett 96:108–113

773. Okamoto K, Kimura H, Sakai G (1984) Miniature synaptic potential recorded intracellularly from Purkinje cell dendrites in guinea-pig cerebellar slices. Brain Res 311:281–282
774. O'Keefe J, Nadel L (1978) The Hippocampus as a Cognitive Map. Clarendon, Oxford
775. Okujava VM (1963) Aktivnost Verkhushechnykh Dendritov v Kore Bolshikh Polusharii (Activity of the Apical Dendrites in Cerebral Cortex). Academy of Sciences of GruzSSR, Tbilisi
776. Olds J (1972) Learning and the hippocampus. Rev Canad Biol Suppl 31:215–238
777. Olds J (1977) Drives and Rewards: Behavior Studies of Hypothalamic Functions. Raven, New York
778. Olds JL, Anderson ML, McPhie DL, Staten LD, Alkon DL (1990) Imaging of memory-specific changes in the distribution of protein kinase C in the hippocampus. Science 245:866–870
779. Oliver MW, Larson J, Lynch G (1990) Activation of the glycine site associated with the NMDA receptor is required for induction of LTP in neonatal hippocampus. Int J Dev Neurosci 8:417–424
780. Olton DS (1983) Memory functions and the hippocampus. In: Seifert W (ed) Neurobiology of the Hippocampus. Academic, London, pp 335–373
781. Oniani TN (1980) Integrativnye Funktsii Limbicheskoi Sistemy. (Integrative Functions of Limbic System). Metsniereba, Tbilisi
782. Otani S, Marshall CJ, Tate WP, Goddard GV, Abraham WC (1989) Maintenance of long-term potentiation in rat dentate gyrus requires protein synthesis but not messenger RNA synthesis immediately post-tetanization. Neuroscience 28: 519–526
783. Otmakhov NA (1982) Effect of norepinephrine and 5-hydroxytryptamine on responses of CA3 neurons in guinea-pig hippocampal slices to electrical stimulation of mossy fibres. Neurophysiology (Kiev) 14:410–417
784. Otmakhov NA (1985) Multiple acetylcholine action on hippocampal pyramidal cells: phenomenology, ionic mechanisms (in Russian). Uspekhi Sovr Biol 99:450–462
785. Otsuka M, Endo M, Nomomura Y (1962) Presynaptic nature of neuromuscular depression. Jap J Physiol 12:573–584
786. Ott T, Bures J (1975) Behavioural and pharmacological modification of fascia dentata responses to stimulation of entorhinal cortex in rats. Physiol Bohemoslov 24:458–459
787. Ott T, Ruethrich H, Reymann K, Lindenau L, Matthies H (1982) Direct evidence for the participation of changes in synaptic efficacy in the development of behavioral plasticity. In: Ajmone Marsan C, Matthies H (eds) Neuronal Plasticity and Memory Formation. IBRO Monograph Series, vol 9. Raven, New York pp 441–452
788. Pacelli GJ, Su W, Kelso SR (1991) Activity-induced decrease in early and late inhibitory synaptic conductances in hippocampus. Synapse 7:1–13
789. Pakula A, Sokolov EN (1975) Habituation in gastropoda: behavioral, interneuronal and endoneuronal aspects. In: Peeke HVS, Herz MJ (eds) Habituation, vol 2. Physiological Substrates. Academic, New York, pp 35–107
790. Palay SL, Chan-Palay J (1974) Cerebellar Cortex. Cytology and Organization. Springer, Berlin Heidelberg New York
791. Palm G (1990) Cell assemblies as a guideline for brain research. Concept Neurosci 1: 133–147
792. Parnas H, Segel LA (1981) A theoretical study of calcium entry in nerve terminals with application to neurotransmitter release. J Theor Biol 91:125–169
793. Parnas I, Armstrong E, Strumwasser F (1974) Prolonged excitatory and inhibitory synaptic modulation of a bursting pacemaker neuron. J Neurophysiol 97:594–608
794. Parsons RL, Hoffman WW, Feigen GA (1965) Presynaptic effects of potassium ions on the mammalian neuromuscular junction. Nature 208:590–591

795. Patel N, Poo M (1984) Perturbation of the direction of neurite growth by pulses of focal electric fields. J Neurosci 4:2939–2947
796. Patterson MM, Steinmetz JE (1984) Spinal fixation: alteration of spinal reflex excitability after sensory nerve stimulation. J Amer Osteopath Assoc 84:221–222
797. Pavlov IP (1951) Polnoe Sobranie Sochinenii (Complete Works), vols 4 and 5. Academy of Sciences of the USSR, Moscow
798. Pawson PA, Chase R (1988) The development of transmission at an identified molluscian synapse. II A quantal analysis of transmission. J Neurophysiol 60:2211–2322
799. Penfield W, Milner B (1958) Memory deficit produced by bilateral lesions in the hippocampal zone. Arch Neurol Psychiat 79:475–497
800. Pfaff DW, ed (1982) The Physiological Mechanisms of Motivation. Springer, Berlin Heidelberg New York
801. Pirogov AA, Lenkov DN (1972) Habituation of postsynaptic responses of motor cortex neurons during different modality stimulation in the cat. Neurophysiology (Kiev) 4:545–553
802. Pivovarov AS, Tsagareli MG (1982) Excitability of the electroexcitable membrane of cortical neuron after intracellular tetanization with hyperpolarizing current (in Russian). Zh Vyssh Nerv Deyat 32:773–775
803. Pockett S, Pygott V (1982) Chronic nerve stimulation increases quantal output in rat muscles. Exp Neurol 78:477–480
804. Porter R (1970) Early facilitation at corticomotoneuronal synapses. J Physiol (Lond) 207:733–745
805. Porter R, Hore J (1969) The time course of minimal corticomotoneuronal excitatory postsynaptic potentials in lumbar motoneurons of the monkey. J Neurophysiol 32:443–451
806. Pun RYK, Neale EA, Guthrie PB, Nelson PG (1986) Active and inactive central synapses in cell culture. J Neurophysiol 56:1242–1256
807. Purves RD (1981) Microelectrode Methods for Intracellular Recording and Ionophoresis. Academic, London
808. Quastel DMJ (1979) Correction of end-plate potentials and currents for nonlinear summation. Can J Physiol Pharmacol 57:702–709
809. Racine RJ, Milgram NW, Hafner S (1983) Long-term potentiation phenomena in the rat limbic forebrain. Brain Res 260:217–231
810. Racine RJ, Moore KA, Wicks S (1991) Activation of the NMDA receptor: a correlate in the dentate gyrus field potential and its relationship to long-term potentiation and kindling. Brain Res 556:226–239
811. Radicheva NI, Tamarova ZA, Shapovalov AI, Shiriaev BI (1980) Electrical activity of the motoneurons in the kitten spinal cord under interarterial perfusion (in Russian). Fiziol Zh SSSR 66:489–496
812. Radpour S, Thomson AM (1990) Coactivation of local circuit connections may be adequate for induction of long-term potentiation (LTP) of minimal Schaffer collateral EPSPs in the CA1 region of the rat hippocampus in vitro. J Physiol (Lond) 425:88P
813. Rahamimoff R, Yaari Y (1973) Delayed release of transmitter at the frog neuromuscular junction. J Physiol (Lond) 228:241–257
814. Rahamimoff R, Lew-Tov A, Nussinovitch I, Meiri H (1981) Calcium and presynaptic factors in synaptic plasticity. In: Feher O, Joo F (eds) Cellular Analogues of Conditioning and Neural Plasticity. Adv Physiol Sci, vol 36. Akademiai Kiado, Pergamon, Budapest, Oxford, pp 87–96
815. Rahmann H, ed (1989) Fundamentals of Memory Formation: Neuronal Plasticity and Brain Function, Progress in Zoology, vol 37. Gustav Fisher, Stuttgart
816. Rahmann H, Wiegandt (1992) Gangliosides and modulation of neuronal functions. In: Neuhoff V, Friend J (eds) NATO ASI Series, Vol H51, Cell to Cell Signals in Plants and Animals. Springer, Berlin Heidelberg New York, pp 212–232

817. Rall W (1967) Distinguishing theoretical synaptic potentials computed for different soma-dendritic distribution of synaptic input. J Neurophysiol 30:1138–1168
818. Rall W, Burke RF, Smith TG, Nelson PG, Frank K (1967) Dendritic location of synapses and possible mechanism for the monosynaptic EPSP in motoneurons. J Neurophysiol 30:1169–1193
819. Ramirez OA, Carrer HF (1989) Correlation between threshold to induce long-term potentiation in the hippocampus and performance in a shuttle box avoidance response in rats. Neurosci Lett 104:152–156
820. Ranck JB (1975) Which elements are excited in electrical stimulation of mammalian central nervous system: a review. Brain Res 98:417–440
821. Rapoport S, Susswein A, Uchino Y, Wilson VI (1977) Synaptic actions of individual vestibular neurones on cat neck motoneurones. J Physiol (Lond) 272:367–382
822. Redman SJ (1973) The attenuation of passively propagating dendritic potentials in a motoneurone cable model. J Physiol (Lond) 234:637–664
823. Redman S (1990) Quantal analysis of synaptic potentials in neurons of the central nervous system. Physiol Rev 70:165–198
824. Redman S, Walmsley B (1981) The synaptic basis of the monosynaptic stretch reflex. Trends Neurosci 4:248–251
825. Redman S, Walmsley B (1983) The time course of synaptic potentials evoked in cat spinal motoneurones at identified group Ia synapses. J Physiol (Lond) 343:117–133
826. Reymann K (1991) Postsynaptic mechanisms of hippocampal long-term potentiation. The involvement of glutamate receptors and protein kinases in its late main. In: Meldrum BS et al. (eds) Excitatory Amino Acids. Raven, New York, pp 475–485
827. Reymann KG, Ruethrich H, Lindenau L, Ott T, Matthies H (1982) Monosynaptic activation of the hippocampus as a conditioned stimulus: behavioral effects. Physiol Behav 29:1007–1012
828. Reymann KG, Matthies HK, Schulzeck, Matthies H (1989) N-methyl-D-aspartate receptor activation is required for the induction of both early and late phases of long-term potentiation in rat hippocampal slices. Neurosci Lett 96:96–101
829. Richards CD (1972) Potentiation and depression of synaptic transmission in the olfactory cortex of the guinea pig. J Physiol (Lond) 222:209–231
830. Ringham GL (1975) Localization and electrical characteristics of a giant synapse in the spinal cord of the lamprey. J Physiol (Lond) 251:395–407
831. Robertson D (1982) Functional significance of dendrite swelling in the guinea pig cochlea. Proc Austral Physiol Pharmacol Soc 13:48P
832. Robinson GB, Racine RJ (1982) Heterosynaptic interactions between septal and entorhinal inputs to the dentate gyrus: long-term potentiation effects. Brain Res 249:162–166
833. Robinson J (1976) Estimation of parameters for a model of transmitter release at synapses. Biometrics 32:61–68
834. Robitaille R, Tremblay JP (1987) Non-uniform release at frog neuromuscular junction: evidence of morphological and physiological plasticity. Brain Res Rev 12: 95–116
835. Robitaille R, Tremblay JP (1991) Non-uniform responses to Ca^{2+} along the frog neuromuscular junction: effects on the probability of spontaneous and evoked transmitter release. J Neurosci 40:571–585
836. Roitbak AI (1953) Oscillographic investigation of centres of increased sensitivity in cortex (in Russian). Trudy Inst Fiziol AN GruzSSR 2:97–131
837. Roitbak AI, Kashakashvili RP, Labakhua TSh (1976) Influence of temperature on post-tetanic potentiation of dendritic potentials. In: Antelidze RF (ed) Sovremennye Problemy Deyatelnosti i Stroeniya Tsentralnoi Nervnoi Sistemy (Modern Problems of Activity and Structure of the Central Nervous System). Metsniereba, Tbilisi, pp 109–116

838. Ropert N (1985) Modulation of inhibition in the hippocampus in vivo. Can J Physiol Pharmacol 63:838–842
839. Ropert N, Miles R, Korn H (1990) Characteristics of miniature inhibitory postsynaptic currents in CA1 pyramidal neurons of rat hippocampus. J Physiol (London) 428:707–722
840. Rose AM, Hattori T, Fibiger HC (1976) Analysis of the septo-hippocampal pathway by light and electron microscopic autoradiography. Brain Res 108:170–174
841. Rose GM, Dunwiddie TV (1986) Induction of hippocampal long-term potentiation using physiologically patterned stimulation. Neurosci Lett 69:244–248
842. Rosenthal J (1969) Post-tetanic potentiation at the neuromuscular junction of the frog. J Physiol (Lond) 203:123–133
843. Routtenberg A, Colley P, Linden D, Lovinger D, Murakami K, Sheu FS (1986) Phorbol ester promotes growth of synaptic plasticity. Brain Res 378: 374–378
844. Rovainen CM (1967) Physiological and anatomical studies on large neurons of central nervous system of the sea lamprey (*Petromyzon marinus*). II Dorsal cells and giant interneurons. J Neurophysiol 30:1024–1042
845. Rovainen CM (1974) Synaptic interactions of identified nerve cells in the spinal cord of sea lamprey. J Compar Neurol 154:189–206
846. Rudomin P (1990) Presynaptic inhibition of muscle spindle and tendon organ afferent in the mammalian spinal cord. Trends Neurosci 13:499–505
847. Rusakov DA, Skibo GG (1989) Geometrical characteristics of mitochondria and presynaptic active zones in the dorsal horn of the cat spinal cord. Neurophysiology (Kiev) 21:741–747
848. Rusakov DA, Shugurov OA (1987) Changes of the numeric characteristics of the structure of the transmission efficacy in synapses of the cat spinal cord. during repeated activation of cutaneous afferents. Neurophysiology (Kiev) 19:491–497
849. Rüthrich H, Matthies H, Ott T (1982) Long-term changes in synaptic excitability of hippocampal cell populations as a result of training. In: Ajmone Marsan C, Matthies H (eds) Neuronal Plasticity and Memory Formation. IBRO Monograph Series, vol 9. Raven, New York, pp 589–594
850. Sacchi O, Perri V (1970) Quantal mechanism of transmitter release during progressive depletion of the presynaptic stores at a ganglion synapse. The action of hemicholinium and thiamine deprivation. J Gen Physiol 61:342–360
851. Sacchi O, Perri V (1971) Quantal release of acetylcholine from the nerve endings of the guinea pig superior cervical ganglion. Pfluegers Arch 329:207–219
852. Sacchi O, Perri V (1976) Some properties of the transmitter release mechanism at the rat ganglionic synapse during potassium stimulation. Brain Res 107:275–289
853. Sahgal A (1980) Functions of the hippocampal system. Trends Neurosci 3:116–119
854. Sakamoto T, Porter LL, Asanuma H (1987) Long-lasting potentiation of synaptic potentials in the motor cortex produced by stimulation of the sensory cortex in the cat: a basis of motor learning. Brain Res 413:360–364
855. Sakharov DA (1974) Genealogiya Neirona (Genealogy of the Neurone). Nauka, Moscow
856. Sastry BR, Goh JW (1984) Long-lasting potentiation in hippocampus is not due to an increase in glutamate receptors. Life Sci 34:1497–1501
857. Sastry BR, Goh JW, Pandanaboina MM (1984) Verapamil counteracts the masking of long-lasting potentiation of hippocampal population spike produced by 2-amino-5-phosphonovalerate. Life Sci 34:323–329
858. Sastry BR, Murali Mohan P, Goh JW (1985) A transient increase in the activity of the CA3 neurons induced a long-lasting reduction in the excitability of Schaffer collateral terminals in rat hippocampus. Neurosci Lett 53:51–56
859. Sastry BR, Goh JW, Auyeung A (1986) Associative induction of posttetanic and long-term potentiation of rat hippocampus. Science 232: 988–990

860. Sastry BR, Goh JW, May PBY, Chirwa SS (1988) The involvement of nonspiking cells in long-term potentiation of synaptic transmission in the hippocampus. Can J Physiol Pharmacol 66:841–844
861. Sato M, Racine RJ, McIntyre DC (1990) Kindling: basic mechanisms and clinical validity. Electroencephal Clin Neurophysiol 76:459–472
862. Savchenko EI, Maiorov VI, Kotlyar BI (1976) Neuronal activity in the cat motor cortex during conditioned reaction of placing the forepaw on a support (in Russian). Zh Vyssh Nerv Deyat 26:65–72
863. Sawada S, Kamiya H, Yamamoto C (1989) Simultaneous recording of presynaptic spikes and excitatory postsynaptic potentials from monosynaptically connected hippocampal neurons. Neurosci Lett 103:34–38
864. Sayer RJ, Redman SJ, Andersen P (1989) Amplitude fluctuations in small EPSPs recorded from CA1 pyramidal cells in the guinea pig hippocampal slices. J Neurosci 9:840–850
865. Sayer RJ, Friedlander MJ, Redman SJ (1990) The time course and amplitude of EPSPs evoked at synapses between pairs of CA3/CA1 neurons in the hippocampal slice. J Neurosci 10:826–836
866. Schlapfer WT, Woodson PBJ, Tremblay JP, Barondes SH (1974) Depression and frequency facilitation at a synapse in *Aplysia californica*: evidence for regulation by availability of transmitter. Brain Res 76:267–280
867. Schlapfer WT, Tremblay JP, Woodson J, Barondes S (1976) Frequency facilitation and post-tetanic potentiation of a unitary synaptic potential in *Aplysia californica* are limited by different processes. Brain Res 109:1–20
868. Schmajuk NA (1990) Role of the hippocampus in temporal and spatial navigation: an adaptive neural network. Behav Brain Res 39:205–229
869. Schmidt J, Kamerer E, Votrich H (1974) Changes in evoked activity and hemosensitivity of cortical neurons with elaboration of conditioned reflexes. In: Sudakov KV (ed) Sistemnyi Analiz Integrativnoi Deyatelnosti Neirona (System Analysis of Integrative Activity of the Neuron). Nauka, Moscow, pp 109–114
870. Schoepp D, Bockaert J, Sladeczek F (1990) Pharmacological and functional characteristics of metabotropic excitatory amino acid receptors. Trends Pharmacol Sci 11:508–515
871. Schwartzkroin PA, Wester K (1975) Long-lasting facilitation at a synaptic potential following tetanization of the in vitro hippocampal slice. Brain Res 89:107–119
872. Scott JG, Mendell LM (1976) Individual EPSPs produced by single triceps surea Ia afferent fibers in homonymous motoneurons. J Neurophysiol 39:679–692
873. Segal JR, Ceccarelli B, Fesce R, Hurlbut WP (1985) Miniature endplate potential frequency and amplitude determined by an extension of Campbell's theorem. Biophys J 47:183–202
874. Segal M (1977) Supersensitivity of hippocampal neurons to acidic amino acids in decommissurated rats. Brain Res 119:476–479
875. Segal M (1979) A potent inhibitory monosynaptic hypotalamo-hippocampal connection. Brain Res 162:137–141
876. Segal M (1982) Multiple actions of acetylcholine at muscarinic receptor studied in the rat hippocampal slices. Brain Res 246:77–87
877. Segal M (1983) Rat hippocampal neurons in culture: responses to electrical and chemical stimuli. J Neurophysiol 50:1249–1264
878. Segal M (1988) Synaptic activation of a cholinergic receptor in rat hippocampus. Brain Res 452:79–86
879. Segal M, Bjorklund A, Gage FH (1985) Transplanted septal neurons make viable cholinergic synapses with a host hippocampus. Brain Res 336:302–307
880. Segawa M, Saito H, Nishiyma N (1990) Alterations in acetylcholine content and muscarinic cholinergic binding of various brain areas after the acquisition of active avoidance tasks in mice. Biogenic Amines 7:181–190

881. Segundo JP, Bell CC (1970) Habituation of single nerve cell in the vertebrate nervous system. In: Horn G, Hinde RA (eds) Short-term Changes in Neural Activity and Behavior. Cambridge University, Cambridge, pp 977–994
882. Sejnowski TJ, Tesauro G (1988) The Hebb rule for synaptic plasticity: algorithms and implementations. In: Byrne JH, Berry WO (eds) Neural Models of Plasticity. Academic, San Diego, pp 94–103
883. Serkov FN (1985) Nature and synaptic mechanisms of inhibition in neurons of the cerebral cortex (in Russian). Fiziol Zh (Kiev) 31:513–525
884. Serkov FN, Yanovskii ESh, Talnov AN (1975) Monosynaptic inhibitory postsynaptic potentials in cortical neurons. Neurophysiology (Kiev) 7:458–467
885. Shapovalov AI (1980) Interneuronal synapses with electrical, dual and chemical mode of transmission in vertebrates. Neuroscience 5:1113–1124
886. Shapovalov AI, Shiriaev BI (1979) Single-fibre EPSPs in amphibian motoneurons. Brain Res 160:519–523
887. Shapovalov AI, Shiriaev BI (1980) Dual mode of junctional transmission at synapses between single primary afferent fibres and motoneurones in the amphibian. J Physiol (Lond) 306:1–15
888. Shapovalov AI, Shiriaev BI (1987) Peredacha Signalov v Mezhneironnykh Sinapsakh (Transmission of Signals in Interneuronal Synapses). Nauka, Leningrad
889. Shapovalov AI, Karamjan OA, Tamarova ZA, Kurchavyi GG (1971) Synaptic actions evoked from the red nucleus on the spinal alpha-motoneurons in the rhesus monkey. Brain Res 32:325–348
890. Sharonova IN, Khaspekov LG (1982) Long-lasting post-tetanic potentiation in the neurons of cultivated hippocampal tissue. Bull Exp Biol Med 93:14–16
891. Sharonova IN, Voronin LL, Skrebitsky VG (1976) Long-lasting post-tetanic potentiation of hippocampal evoked potentials to the stimulation of Schaffer collaterals (in Russian). Zh Vyssh Nerv Deyat 26:214–217
892. Sherman RG, Atwood HL (1971) Synaptic facilitation: long-term neuromuscular facilitation in crustaceans. Science 171:1248–1250
893. Shik ML, Yagodnitsyn AS (1973) A study of connections of neurons in dorso-caudal part of mesencephalic tegmentum in cats with microstimulation technique. Neurophysiology (Kiev) 5:593–601
894. Shimahara T (1983) Presynaptic modulation of transmitter release by the early outward potassium current in *Aplysia*. Brain Res 263:51–56
895. Shimahara T, Tauc L (1975) Heterosynaptic facilitation in the giant cell of *Aplysia*. J Physiol (Lond) 247:321–341
896. Shimahara T, Tauc L (1976) Identification of a neuron inducing heterosynaptic facilitation on a specific synapse in *Aplysia*. Brain Res 118:142–146
897. Shiriaev BI (1983) Synaptic actions evoked in frog spinal motoneurones by adequate muscle spindle stimulation (in Russian). Fiziol Zh SSSR 69:1157–1163
898. Shramm VA, Naletov IA, Osipov BS (1975) Heterosynaptic facilitation in the mollusc *Lumnaea stagnalis*. In: Osipov BS (ed) Teoreticheskaya i Eksperimentalnaya Biofizika (Theoretical and Experimental Biophysics), vol 5. Kaliningrad University, Kaliningrad, pp 157–161
899. Shute CCD, Lewis PR (1963) Cholinesterase-containing systems of the brain of the rat. Nature 199:1160–1164
900. Sillar KT, Roberts A (1991) Segregation of NMDA and non-NMDA receptors at separate synaptic contacts: evidence from spontaneous EPSPs in *Xenopus* embryo spinal neurons. Brain Res 545:24–32
901. Simonneau M, Tauc L, Baux G (1980) Quantal release of acetylcholine examined by current fluctuation analysis at an identified neuro-neuronal synapse of *Aplysia*. Proc Natl Acad Sci USA 77:1661–1665
902. Simonov PV (1981) Emotsionalnyi Mozg (The Emotional Brain). Nauka, Moscow

903. Singer W (1990) Search for coherence: a basic principle of cortical self-organization. Concepts Neurosci 1:1–26
904. Sinz R, Grechenko TH, Sokolov EN (1982) Plasticity in isolated neurons. In: Ajmone Marsan C, Matthies H (eds) Neuronal Plasticity and Memory Formation. IBRO Monograph Series, vol 9. Raven, New York, pp 285–290
905. Sisken BF (1988) Effects of electromagnetic fields on nerve regeneration. In: Marino AA (ed) Modern Bioelectricity, Marcel Dekker, New York, Basel
906. Skelton RW, Miller JJ Phillips AG (1983) Low-frequency stimulation of the perforant path produces long-term potentiation in the dentate gyrus of unanesthetized rats. Can J Physiol Pharmacol 61:1156–1161
907. Skelton R, Scarth A, Wikie D, Miller J, Phillips (1987) Long-term increases in dentate granule cell responsivity accompany operant conditioning. J Neurosci 7:3080–3087
908. Skok VI, Ivanov AJa (1986) Natural Activity of Vegetative Ganglia. Naukova Dumka, Kiev
909. Skrebitsky VG, Chepkova AN (1973) The role of time intervals between "conditioned" and "unconditioned" stimuli in elaboration of neuronal analogs of conditioned reflexes (in Russian). Zh Vyssh Nerv Deyat 23:576–583
910. Skrebitsky VG, Vorobyev VS (1979) A study of synaptic plasticity in hippocampal slices. Acta Neurobiol Exp 39:633–642
911. Skrebitsky VG, Voronin LL (1966) Intracellular records of the single unit activity of the visual cortex in unanaesthetized rabbits (in Russian). Zh Vyssh Nerv Deyat 16:864–865
912. Skrebitsky VG, Voronin LL, Vorobyev VS, Kudryashov IE, Saakyan SA, Sharonova IN (1978) Frequency and post-tetanic potentiation in hippocampal pathways (in Russian). Zh Vyssh Nerv Deyat 28:650–659
913. Skrede KK, Malthe-Sorenssen D (1981) Increased resting and evoked release of transmitter following repetitive electrical tetanization in the hippocampus: a biochemical correlate to long-lasting synaptic potentiation. Brain Res 208:436–441
914. Slater NT, Stelzer A, Galvan M (1985) Kindling-like stimulus patterns induce epileptiform discharges in the guinea pig in vitro hippocampus. Neurosci Lett 60:25–31
915. Smith DO (1983) Variable activation of synaptic release sites at the muscular junction. Exp Neurol 80:520–528
916. Smith DO, Hatt H (1976) Axon conduction block in a region of dense connective tissue in crayfish. J Neurophysiol 39:794–801
917. Smith S (1987) Progress on LTP at hippocampal synapses: a post-synaptic Ca^{2+} trigger for memory storage? Trend Neurosci. 10:142–144
918. Sokolov EN (1981) Neironnye Mekhanizmy Pamyati i Obucheniya (Neuronal Mechanism of Memory and Learning). Nauka, Moscow
919. Sokolov EN, Ter-Margaryan AG (1984) Long-term synaptic habituation in neurones LPa3 and RPa3 of a snail (in Russian). Zh Vyssh Nerv Deyat 34:985–987
920. Solodkin M, Jimenez I, Collins WF, Mendell LM, Rudomin P (1991) Interaction of baseline synaptic noise and Ia EPSPs: Evidence for appreciable negative correlation under physiological conditions. J Neurophysiol 65:927–945
921. Spencer WA, April RS (1970) Plastic properties of monosynaptic pathways in mammals. In: Horn G, Hinde RA (eds) Short-term Changes in Neural Activity and Behaviour. Cambridge University Press, Cambridge, pp 433–474
922. Spencer WA, Kandel ER (1961) Electrophysiology of hippocampal neurons. III Firing level and time constant. J Neurophysiol 24:260–271
923. Spencer WA, Wigdor R (1965) Ultra-late PTP of monosynaptic reflex responses in cat. Physiologist 8:278
924. Spencer WA, Thompson RF, Neilson DR (1966) Decrements of ventral root electrotonus and intracellularly recorded PSPs produced by iterated cutaneous afferent volleys. J Neurophysiol 29:253–274

925. Spray DC, Bennett MVL (1985) Physiology and pharmacology of gap junctions. Ann Rev Physiol 47:281–303
926. Squire LR (1986) Mechanisms of memory. Science 232:1612, 1986
927. Squire LR, Butters N (1984) Neuropsychology of Memory. Guilford, New York
928. Standaert FG (1982) Release of transmitter at the neuromuscular junction. Brit J Anaesth 54:131–145
929. Stanley JC, France JF de, Marchand JE (1980) Characteristics of tetanic and post-tetanic potentiation in the septo-hippocampal commissural system in the acute rabbit. J Neurobiol 11:193–208
930. Stanton PK, Sarvey JM (1985) Blockade of norepinephrine-induced long-lasting potentiation in the hippocampal dentate gyrus by an inhibitor of protein synthesis. Brain Res 361:276–283
931. Stanton PK, Sarvey JM (1987) Norepinephrine regulate long-term potentiation of both the population spike and dendrite EPSP in hippocampal dentate gyrus. Brain Res Bull 18:115–119
932. Stanton PK, Sejnowski TJ (1989) Associative long-term depression in the hippocampus induced by Hebbian covariance. Nature 339:215–218
933. Stanton PK, Mody I, Heinemann U (1989) A role for N-methyl-D-aspartate receptors in norepinephrine-induced long-lasting potentiation in the dentate gyrus. Exp Brain Res 77:517–530
934. Staubli U, Lynch G (1987) Stable hippocampal long-term potentiation elicited by "theta" pattern stimulation. Brain Res 435:227–234
935. Stefanis C (1964) Hippocampal neurons: their responsiveness to microelectrophoretically administered endogenous amines. Pharmacologist 6:171
936. Steiner FA (1968) Influence of microelectrophoretically applied acetylcholine on the responsiveness of hippocampal and lateral geniculate neurones. Pflugers Arch 303:173–180
937. Stellar JR, Stellar E (1985) The Neurobiology of Motivation and Reward. Springer, Berlin Heidelberg New York
938. Stephens CL (1973) Relative contribution of synaptic and non-synaptic influences to response decrements in a postsynaptic neurone. J Exp Biol 59:315–321
939. Stevens CF (1976) A comment on Martin's relation. BiophysJ 16:891–895
940. Steward O, White WF, Cotman CW (1977) Potentiation of the excitatory synaptic action of commissural, associational and entorhinal afferents to dentate granule cells. Brain Res 134:551–560
941. Stjarne L, Astrand P (1984) Discrete events measure single quanta of adenosine 5'-triphosphate secreted from sympathetic nerves of guinea-pig and mouse *vas deferens*. Neuroscience 13:21–28
942. Stoney SD, Thompson WD, Asanuma H (1968) Excitation of pyramidal tract cells by intracortical microstimulation: effective extent of stimulating current. J Neurophysiol 31:659–669
943. Storm-Mathisen J, Zimmer J, Ottersen OP, eds (1990) Understanding the Brain through the Hippocampus. Progr Brain Res, vol 83. Elsevier Amsterdam
944. Storozhuk VM, Antonov IN (1980) Peculiarities of heterosynaptic facilitation in the giant neurons of the mollusc cerebral ganglion. Neurophysiology (Kiev) 12:498–507
945. Sutherland RJ, Donald RJ (1990) Hippocampus, amygdala and memory deficits in rats. Behav Brain Res 37:57–79
946. Sutor B, Hablitz JJ (1989) EPSPs in rat neocortical neurons in vitro. II Involvement of N-methyl-D-aspartate receptors in the generation of EPSPs. J Neurophysiol 61:621–634
947. Sutula T, Stavard O (1985) Quantative analyses of synaptic potentiation during kindling of the perforant path. J Neurophysiol 56:732–746

948. Swanson LW (1978) The anatomical organization of septo-hippocampal projections. In: Elliot K, Whelan J (eds) Functions of the Septo-Hippocampal System, Ciba Foundations Symposium. Elsevier, Amsterdam, pp 25–43
949. Swanson LW, Teyler TJ, Thompson RF (1982) Hippocampal long-term potentiation: mechanisms and implication for memory. Neurosci Res Progr Bull 20:613–769
950. Takeuchi A (1958) The long-lasting depression in neuromuscular transmission of frog. Jap J Physiol 8:102–113
951. Takeuchi A, Takeuchi N (1962) Electrical changes in pre- and postsynaptic axons of the giant synapse of *Loligo*. J Gen Physiol 45:1181–1193
952. Tanaka Y, Sakurai M, Hayashi S (1989) Effect of scopolamine and HP 029, a cholinesterase inhibitor, on long-term potentiation in hippocampal slices of the guinea pig. Neurosci Lett 98:179–183.
953. Tamarova ZA, Shapovalov AI, Shiriaev BI (1978) Action of magnesium and manganese ions and calcium deficiency on the synaptic transmission in the isolated perfused rat spinal cord. Neurophysiology (Kiev) 10:530–533
954. Tashiro N, Gallacher JP, Nishi S (1976) Facilitation and depression of synaptic transmission in amphibian sympathetic ganglia. Brain Res 118:45–62
955. Taube JS, Schwartzkroin PA (1988) Mechanisms of long-term potentiation: EPSP/spike dissociation, intradendritic recordings, and glutamate sensitivity. J Neurosci 8:1632–1644
956. Tauc L (1982) Nonvesicular release of neurotransmitter. Physiol Rev 62:857–893
957. Tauc L, Epstein R (1967) Heterosynaptic facilitation as a distinct mechanism in *Aplysia*. Nature 214:724–725
958. Tauc L, Gerschenfeld HM (1961) Cholinergic transmission mechanisms for both excitation and inhibition in molluscan central synapses. Nature 192:366–367
959. Taylor A, Stephens JA, Somjen G, Appenteng K, O'Donovan M (1978) Extracellular spike triggered averaging for plotting synaptic projections. Brain Res 140:344–348
960. Terrian DM, Gannon RL, Rea MA (1990) Glutamate is the endogenous amino acid selectively released by rat hippocampal mossy fiber synaptosomes concomitantly with prodynorphin-derived peptides. Neurochem Res 15:1–5
961. Teyler TJ, Alger BF (1976) Monosynaptic habituation in the vertebrate forebrain: the dentate gyrus examined in vitro. Brain Res 115:413–425
962. Teyler TJ, Alger BE (1978) Plasticity in the vertebrate central nervous system. In: Teyler TJ (ed) Brain and Learning. Reidel, Dordrecht, pp 33–50
963. Teyler TJ, DiScenna P (1984) Long-term potentiation as a candidate mnemonic device. Brain Res Rev 7:15–28
964. Teyler TJ, DiScenna P (1987) Long-term potentiation. Ann Rev Neurosci 10:131–161
965. Thesleff S (1959) Motor end-plate "desensitization" by repetitive nerve stimuli. J Physiol (Lond) 148:659–664
966. Thies RE (1965) Neuromuscular depression and the apparent depletion of transmitter in mammalian muscles. J Neurophysiol 28:427–442
967. Thompson SM, Gahwiler BH (1989) Activity-dependent disinhibition. I. Repetitive stimulation reduces IPSP driving force and conductance in the hippocamus in vitro. J Neurophysiol 61:501–516
968. Thompson RF, Spencer WA (1966) Habituation: a model phenomenon for the study of neuronal substrates of behavior. Physiol Rev 173:16–43
969. Thompson RF, Berger TW, Madden J (1983) Cellular processes of learning and memory in the mammalian CNS. Ann Rev Neurosci 6:447–449
970. Thomson AM, Girdlestone D, West DC (1988) Voltage-dependent currents prolong single-axon postsynaptic potentials in layer III pyramidal neurons in rat neocortical slices. J Neurophysiol 60:1896–1907
971. Tonge DA (1974) Physiological characteristics of reinnervation of skeletal muscle in the mouse. J Physiol (Lond) 241:141–153

972. Toyama K, Tsukahara N, Kosaka K, Matsunami K (1970) Synaptic excitation of red nucleus neurones by fibres from interpositus nucleus. Exp Brain Res 11:187–188
973. Tracey DJ, Walmsley B (1984) Synaptic input identified muscle afferents to neurones of the dorsal spinocerebellar tract in the cat. J Physiol (Lond) 350:599–614
974. Tremblay JP, Woodson PBJ, Schlapfer WT, Barondes SH (1976) Dopamine, serotonin and related depression, frequency facilitation and post-tetanic potentiation at a synapse in *Aplysia californica*. Brain Res 109:61–81
975. Tremblay JP, Laurie RE, Colonnier M (1983) Is the MEPP due to the release of one vesicle or to the simultaneous release of several vesicles at the active zone? Brain Res Rev 6:299–314
976. Triller A, Korn H (1982) Transmission at a central inhibitory synapse. III Ultrastructure of physiologically identified and stained terminal. J Neurophysiol 48:708–736
977. Trimble WS, Linial M, Scheller RH (1991) Cellular and molecular biology of the presynaptic terminal. Ann Rev Neurosci 14:93–122
978. Tsitolovskii LE (1986) Integrative function of neurons in record of memory trace (in Russian). Uspekhi Fiziol Nauk 17:83–103
979. Tsitolovskii LE, Pivovarov AS (1976) Changes of postsynaptic and spike potentials in the cortex neuron with habituation. Neurophysiology (Kiev) 8:22–29
980. Tsukahara N (1981) Synaptic plasticity in the mammalian central nervous system. Ann Rev Neurosci 4:351–379
981. Tsukahara N, Kosaka K (1968) The mode of cerebral excitation of red nucleus neurons. Exp Brain Res 5:102–117
982. Tsukahara N, Bando T, Murakami F, Oda J (1983) Properties of cerebello-precerebellar reverberating circuits. Brain Res 274:249–259
983. Turner DA (1988) Waveform and amplitude characteristics of evoked responses to dendritic stimulation of CA1 guinea-pig pyramidal cells. J Physiol (Lond) 395:419–439
984. Turner RW, Baimbridge KG, Miller JJ (1982) Calcium-induced long-term potentiation in the hippocampus. Neuroscience 7:1411–1416
985. Turpaev TM (1952) Dynamics of processes of acetylcholine release and dissociation (in Russian). Izv AN SSSR, Ser Biol N4, 120–132
986. Turpaev TM (1962) Mediatornaya Funktsiya Atsetilkholina i Priroda Kholinoretseptora (Transmitter Function of Acetylcholine and Nature of Cholinoreceptor). Academy of Sciences of the USSR, Moscow
987. Usherwood PNR (1972) Transmitter release from insect excitatory motor nerve terminals. J Physiol (Lond) 227:527–552
988. Usherwood PNR (1973) Release of transmitter from degenerating locust motoneurones. J Exp Biol 59:1–16
989. Valentino RJ, Dingledine R (1981) Acetylcholine inhibits synaptic transmission in the rat hippocampal slice. Fed Proc 40:321
990. Van der Kivy D, Fibiger HC, Phillips AG (1977) Monoamine involvement in hippocampal self-stimulation. Brain Res 136:119–130
991. Van der Kloot W (1977) Quantal acetylcholine release: vesicles or gated channels? Gen Pharmacol 8:21–25
992. Van der Kloot W (1988) Acetylcholine quanta are released from vesicles by exocitosis (and why some think not). Neuroscience 24:1–8
993. Van der Kloot W (1989) Statistical and graphical methods for testing the hypothesis that quanta are made up of subunits. J Neurosci Meth 27:81–89
994. Van der Kloot W, Cohen IS (1984) End-plate potentials in a model muscle fiber. Corrections for the effects of membrane potential on currents and on channel lifetimes. Biophys J 45:905–911
995. Van der Kloot W, Van der Kloot TE (1985) Activity increases quantal size at the frog neuromuscular junction. Experientia 41:47–48

996. Vautrin J (1986) Subunits in quantal transmission at the mouse neuromuscular junction: tests of peak intervals in amplitude distributions. J Theor Biol 120:363–370
997. Veber NV (1976) Post-tetanic facilitation of evoked electrical responses of sensorimotor cortex (in Russian). Zh Vyssh Nerv Deyat 26:381–389
998. Veber NV, Rapoport SSh, Silkis IG (1985) Long-term homo- and heterosynaptic post-tetanic changes in responses of neurons of somatosensory cortex. Doklady Biol Sci 281:486–489
999. Vere-Jones D (1966) Simple stochastic models for the release of quanta of transmitter from a nerve terminal. Aust J Statistics 8:53–63
1000. Vinogradova OS (1975) Gippokamp i Pamyat (Hippocampus and Memory). Nauka, Moscow
1001. Vinogradova OS (1984) Contemporary notions about general characteristics and plastic phenomenon in hippocampal neurons (in Russian). Uspekhi Fiziol Nauk 15:28–54
1002. Vinogradova OS, Konovalov VF, Semenova TP (1969) Neuronal correlates of orienting reflex extinction in different brain structures (in Russian). Vopr Psikhol 1:46–59
1003. Vizi ES (1984) Physiological role of cytoplasmic and non-synaptic release of transmitter. Neurochem Int 6:435–440
1004. Vorobyev VS, Skrebitsky VG (1981) Synaptic plasticity of neurones in hippocampal slices (in Russian). Zh Vyssh Nerv Deyat 31:395–402
1005. Voronin LL (1967) Postsynaptic potentials in motor cortex of waking rabbit (in Russian). Fiziol Zh SSSR 53:623–631
1006. Voronin LL (1968) Some statistical relationships between spontaneous and evoked activity of cortical neurons. In: Narushevichus E (ed) Statisticheskaya Elektrofiziologiya (Statistical Electrophysiology). Vilnius Medical Institute, Vilnius, pp 100–114
1007. Voronin LL (1969) Changes in the excitability of individual neurons after electrical stimulation of the cortical surface. Doklady Biol Sci 186:474–477
1008. Voronin LL (1970) Microelectrode investigations of cellular analogs of learning. Soviet Neurol Psychiatry 4:99–125
1009. Voronin LL (1971) Microelectrode study of a cellular analog of a conditioned reflex to time (in Russian). Zh Vyssh Nerv Deyat 21:1283–1246
1010. Voronin LL (1976) Microelectrode Study of Neurophysiological Mechanisms of Conditioning. Soviet Research Reports, vol 2 (Woody CD, ed). Brain Information Service, Brain Research Institute, UCLA, Los Angeles
1011. Voronin LL (1977) Involvement of cortical neurones in conditioned and unconditioned startle reflex. Neuroscience 3:133–137
1012. Voronin LL (1979) Quantal analysis of postsynaptic potentials. Neurophysiology (Kiev) 11:366–379
1013. Voronin LL (1980) Microelectrode analysis of the cellular mechanisms of conditioned reflex in rabbits. Acta Neurobiol Exp 40:335–370
1014. Voronin LL (1981) Cellular mechanisms of long-term post-tetanic potentiation in the hippocampus. In: Feher O, Joo F (eds) Cellular Analogues of Conditioning and Neural Plasticity. Adv Physiol Sci, vol 36. Akademiai Kiado, Pergamon, Budapest, London, pp 165–174
1015. Voronin LL (1982) Analiz Plasticheskikh Svoistv Tsentralnoi Nervnoi Sistemy (Analysis of Plastic Properties of the Central Nervous System). Metsniereba, Tbilisi, 1982
1016. Voronin LL (1983) Long-term potentiation in the hippocampus. Neuroscience 10:1051–1069
1017. Voronin LL (1984) A study of synaptic plasticity at archicortical and neocortical levels. Neurophysiology (Kiev) 16:651–665

1018. Voronin LL (1985) Plasticity of hippocampal responses. In: Buzsaki G, Vanderwolf CH (eds) Electrical Activity of the Archicortex. Akademiai Kiado, Budapest, pp 295–317
1019. Voronin LL (1986a) Synaptic and membrane mechanisms of temporary connections. In: Batuev AS (ed) Fiziologiya Povedeniya. Rukovodstvo po Fiziologii (Physiology of Behavior. Handbook of Physiology). Nauka, Leningrad, pp 524–580
1020. Voronin LL (1986b) Long-term potentiation at neocortical level. In: Matthies H (ed) Learning and Memory: Mechanisms of Information Storage in the Nervous System. Pergamon, New-York, pp 13–24
1021. Voronin LL (1988a) Investigations of neurophysiological mechanisms of plasticity and learning. In: Turpaev TM (ed) Soviet Science Reviews, Ser F, Physiol Gen Biol, vol 2. Harwood Academic, London, pp 529–596
1022. Voronin LL (1988b) Quantal analysis of long-term potentiation. In: Haas HL, Buzsaki G (eds) Synaptic Plasticity in the Hippocampus. Springer, Berlin Heidelberg New York, pp 27–30
1023. Voronin LL (1989) Principles of long-term potentiation in mammalian brain. In: Rahmann H (ed) Fundamentals of Memory Formation: Neuronal Plasticity and Brain Function. Progress in Zoology, vol 37. Fisher, Stuttgart, pp 175–188
1024. Voronin LL (1991) Quantal analysis of long-term potentiation. Neurosci Facts 2(N20):4
1025. Voronin LL, Derevyagin VI (1984) The model of transmitter depletion and analysis of synaptic plasticity (in Russian). Uspekhi Fiziol Nauk 15:86–110
1026. Voronin LL, Ezrokhi VL (1971) Multisensory convergence on neurons of the sensorimotor cortex of unanaesthetized cat. Neurophysiology (Kiev) 3:563–573
1027. Voronin LL, Gusev AG (1984) Intracortical interneuronal connections in the motor cortex of the alert rabbit. Int J Neurosci 22:190–191
1028. Voronin LL, Ioffe SV (1974) Changes in unit postsynaptic responses at sensorimotor cortex with conditioning in rabbits. Acta Neurobiol Exp 34:505–513
1029. Voronin LL, Kozhedub RG (1971) Analysis of changes in postsynaptic potentials on a cellular analogue of a conditioned reflex (in Russian). Zh Vyssh Nerv Deyat 21:997–1005
1030. Voronin LL, Kudryashov IE (1977) Long-lasting hippocampal post-tetanic potentiation with particular reference to mechanisms of conditioned reflex. In: Proc of IUPS, vol 13, 27th Physiol Congr. Paris, p 794
1031. Voronin LL, Kudryashov IE (1978) Excitatory postsynaptic potentials of hippocampal neurons and their habituation with repetitive stimulation. Neurophysiology (Kiev) 10:3–12
1032. Voronin LL, Kudryashov IE (1979) Unit responses in hippocampus during long-lasting post-tetanic potentiation (in Russian). Zh Vyssh Nerv Deyat 29:141–145
1033. Voronin LL, Kuhnt U (1990) Long-term potentiation affects facilitation ratio of EPSPs recorded from CA1 pyramidal cells in the guinea pig hippocampal slice. Neurosci Res Comm 6:149–155
1034. Voronin LL, Markevich VA (1980) Conditioned reflex analogue with registration of pyramidal tract response to direct cortical stimulation. Doklady Biol Sci 253:1005–1009
1035. Voronin LL, Markevich VA (1982) A neural analog of conditioning: modifications of pyramidal tract response. In: Woody C (ed) Conditioning: Representation of Involved Neural Functions. Advances in Behavioral Biology, vol 26. Plenum, New York, pp 663–676
1036. Voronin LL, Skrebitsky VG (1965) An intracellular investigation of the neurons of the cerebral cortex of the unanaesthetized rabbit. Bull Exp Biol Med 59:473–477 (Translated in Fed Proc 25:574T–576T)
1037. Voronin LL, Solntseva EI (1969a) Analyse intracellulaire des effets consecutifs a la micropolarisation des neurones corticaux. CR Acad Sci D268:1430–1433

1038. Voronin LL, Solntseva EI (1969b) After-effects of intracellular polarization of single cortical neurons. Neurosci Transl 13:65–74

1039. Voronin LL, Solntseva EI, Ezrokhi VL (1968) Mechanism of after-effects following polarization of single neurons. Biophysics (Moscow) 13:656–661

1040. Voronin LL, Zakharov VF, Rabinovich MYa (1969) Effect of electrodermal stimulation of a limb on unit responses in the motor cortex. Neurosci Transl 10:75–82

1041. Voronin LL, Kopytova FV, Rabinovich MY (1972) Prolonged aftereffects in cortical neuronal activity as a possible neurophysiological basis of memory. In: Beritashvili IS (ed) Gagrskie Besedy (Gagra Talks), vol 6. Metsniereba, Tbilisi pp 234–256

1042. Voronin LL, Kudryashov IE, Ioffe SV (1974) Post-tetanic potentiation in hippocampus and its relation with mechanisms of conditioned reflex. Doklady Biol Sci 217:1453–1456

1043. Voronin LL, Gerstein GL, Kudryashov IE, Ioffe SV (1975) Elaboration of a conditioned reflex in single experiment with simultaneous recording of neural activity. Brain Res 92:385–403

1044. Voronin LL, Kudryashov IE, Skrebitsky VG, Uvarov VG, Sharonova IN (1977) "Minimal" evoked potentials in the hippocampus and their post-tetanic changes. Neurophysiology (Kiev) 9:124–134

1045. Voronin LL, Derevyagin VI, Kudryashov IE (1978) Analysis of postsynaptic potentials of the hippocampus and their low-frequency depression. In: Kostyuk PG (ed) Sovremennye Problemy Obshchei Fiziologii Vozbudimykh Obrazovanii (Contemporary Problems of General Physiology of Excitable Structures). Naukova Dumka, Kiev, pp 129–139

1046. Voronin LL, Baskys AV, Derevyagin VI, Kudryashov IE (1979) Plasticity of hippocampal synaptic connections. In: Beritashvili IS (ed) Gagrskie Besedy (Gagra Talks), vol 7. Metsniereba, Tbilisi, pp 37–57

1047. Voronin LL, Baskys A, Gusev AG, Kudryashov IE, Derevyagin VI, Logunov DB (1980) Quantal analysis of postsynaptic potentials in central nervous system (CNS). In: Proc of IUPS, vol 14, 28th Physiol Congr. Budapest, p 776

1048. Voronin LL, Kozhedub RG, Kozhechkin SN (1980) Changes in sensitivity of septo-hippocampal responses to acetylcholine during long-term post-tetanic potentiation. Bull Exp Biol Med 89:515–518

1049. Voronin LL, Derevyagin VI, Logunov DB (1981) Parameters of depletion model of central synapses of the snail. Neurophysiology (Kiev) 13:88–97

1050. Voronin LL, Baskys AV, Gusev AG, Derevyagin VI, Logunov DB (1983) Comparative analysis of statistical parameters of the synaptic transmission in the central neurones of the snail *Helix lucorum* and rabbit. Evol Biochem Physiol 19:359–368

1051. Voronin LL, Baskys AV, Buldakova SL, Skrebitsky VG (1984) Long-term potentiation of evoked reactions in the neocortex of intact brain and cortical slices (in Russian). Fiziol Zh SSSR 70:1167–1177

1052. Voronin LL, Kuhnt U, Hess G, Ehrenreich L (1987) Statistical analysis of hippocampal long-term potentiation evoked by afferent and intracellular tetanization in vitro. Neuroscience Suppl 22: 41S

1053. Voronin LL, Kuhnt U, Davletshin RG, Hess G (1988a) Ineffective synapses in in vitro hippocampal slices. Doklady Biol Sci 302:746–749

1054. Voronin LL, Kuhnt U, Mitiyshkin MV, Hess G (1988b) Mechanisms of long-lasting synaptic plasticity in the hippocampus (in Russian). In: Adrianov OS (ed) Integrativnaya Deyatelnost Mozga (Integrative Activity of the Brain). Moscow, Brain Research Institute, pp 53–55

1055. Voronin LL, Davletshin RN, Ivanova EI, Kleschevnikov AM, Kudryashov IE, Fedorov NB (1989) Long-term facilitation after microstimulation in mouse hippocampal slices. Doklady Biol Sci 305:1015–1019

1056. Voronin LL, Gusev AG, Mityushkin MV (1990a) Statistical methods of quantal analysis in computerized compared with physiological experiments. Neurophysiology (Kiev) 22:162–169

1057. Voronin LL, Kuhnt U, Hess G (1990b) Quantal analysis of long-term potentiation of compound postsynaptic potentials of neurones in vitro hippocampal slices. Neurophysiology (Kiev) 22:465–472

1058. Voronin LL, Kuhnt U, Hess G (1990c) Quantal analysis of long-term post-tetanic modifications of minimal excitatory postsynaptic potentials of in vitro hippocampal slices. Neurophysiology (Kiev) 22:561–570

1059. Voronin LL, Kuhnt U, Gusev AG (1991a) Analysis of EPSP fluctuations indicates increased presynaptic release during long-term potentiation in area CA1 of hippocampal slices. Neurosci Res Comm 8:87–94

1060. Voronin LL, Kuhnt U, Gusev AG (1991b) Changes in quantal parameters of CA1 synaptic transmission are dependent on the magnitude of long-term potentiation. Neurosci Res Comm 9:1–7

1061. Voronin LL, Kuhnt U, Hess G, Gusev A, Roschin V (1992a) Quantal parameters of "minimal" excitatory postsynaptic potentials in guinea pig hippocampal slices: binomial approach. Exp Brain Res 89:248–264

1062. Voronin LL, Kuhnt U, Gusev AG, Hess G (1992b) Quantal analysis of long-term potentiation of "minimal" excitatory postsynaptic potentials in guinea pig hippocampal slices: binomial approach. Exp Brain Res 89:275–287

1063. Voronin LL, Kuhnt U, Gusev AG (1992c) Analysis of fluctuations of "minimal" excitatory postsynaptic potentials during long-term potentiation in guinea pig hippocampal slices. Exp Brain Res 89:288–299

1064. Voronin LL, Kuhnt U, Gusev AG, Ivanov NV (1992d) Reliability of quantal parameter estimates and their changes during long-term potentiation in guinea pig hippocampal slices. Neurosci Lett 146:111–114

1065. Wachtel H, Kandel ER (1971) Conversion of synaptic excitation to inhibition at a dual chemical synapse. J Neurophysiol 34:56–68

1066. Wall PD (1970) Habituation and post-tetanic potentiation in the spinal cord. In: Horn G, Hinde R (eds) Short-Term Changes in Neural Activity and Behaviour. Cambridge University Press, Cambridge, pp 181–210

1067. Walmsley B, Edwards FR, Tracey DJ (1987) The probabilistic nature of synaptic transmission at a mammalian excitatory central synapse. J Neurosci 7:1037–1046

1068. Walmsley B, Edwards FR, Tracey DJ (1988) Nonuniform release probabilities underlie quantal synaptic transmission at a mammalian excitatory central synapse. J Neurophysiol 60:889–908

1069. Walsh JW, Houk JC, Atluri RL, Mugnaini E (1972) Synaptic transmission at single glomeruli in the turtle cerebellum. Science 178:881–883

1070. Walters ET, Byrne JH (1983a) Associative conditioning of single sensory neurons suggests a cellular mechanism for learning. Science 219:405–408

1071. Walters ET, Byrne JH (1983b) Slow depolarization produced by associative conditioning of *Aplysia* sensory neurons may enhance Ca^{2+} entry. Brain Res 280:165–168

1072. Walters ET, Byrne JH (1984) Post-tetanic potentiation in *Aplysia* sensory neurons. Brain Res 293:377–380

1073. Watanabe A, Grundfest H (1961) Impulse propagation at the septal and commissural junction of crayfish lateral giant axons. J Gen Physiol 45:267–308

1074. Watanabe S, Creutzfeldt OD (1966) Spontane postsynaptische Potentiale von Nervenzellen des motorischen Cortex der Katze. Exp Brain Res 1:48–64

1075. Watt DGD, Staufer EK, Taylor A, Reinking RM, Stuart DG (1976) Analysis of muscle receptor connections by spike-triggered averaging. I Spindle primary and tendon organ afferents. J Neurophysiol 39:1375–1392

1076. Waxman SG (1975) Integrative properties and design principles of axons. Int Rev Neurobiol 18:1–40
1077. Weakly GN (1969) Effect of barbiturates on "quantal" synaptic transmission in spinal motoneurones. J Physiol (Lond) 204:63–77
1078. Weisz DJ, Clark GA, Thompson RF (1984) Increased responsivity of dentate granule cells during nictitating membrane response conditioning in rabbit. Behav Brain Res 12:145–154
1079. Wenzel J, Bogolepov NN (1976) Electronenmikroskopische Untersuchungen zur Synaptologie des Hippocampus der Katze. Jf Hirnforsch 17:399–448
1080. Wenzel J, Bogolepov NN, Skrebitsky VG (1978) Structurelle Aspekte der Plastizitat in Synapsen des Hippocampus. In: Adrianov OS (ed) Lokalizatsiya i Organizatsiya Funktsii Mozga (Localization and Organization of the Cerebral Functions). Brain Institute, Moscow, pp 211–212
1081. Werman R, Carlen PL (1976) Unusual behavior of the Ia EPSP in cat spinal motoneurons. Brain Res 112:395–401
1082. Wernig A (1972a) Changes in statistical parameters during facilitation at the crayfish neuromuscular junction. J Physiol (Lond) 226:751–759
1083. Wernig A (1972b) The effects of calcium and magnesium on statistical release parameters at the crayfish neuromuscular junction. J Physiol (Lond) 226:761–768
1084. Wernig A (1975) Estimates of statistical release parameters from crayfish and frog neuromuscular junctions. J Physiol (Lond) 244:207–221
1085. Wernig A (1976) Localization of active sites in the neuromuscular junction of the frog. Brain Res 118:63–72
1086. Wernig A, Motelical-Heino I (1978) On the presynaptic nature of the quantal subunit. Neurosci Lett. 8:231–234
1087. Wernig A, Stirner H (1977) Quantum amplitude distributions point to functional unity of the synaptic active zone. Nature 269:820–822
1088. White G, Levy WB, Steward O (1990) Spatial overlap between populations of synapses determines the extent of their associative interaction during the induction of long-term potentiation and depression. J Neurophysiol 64:1186–1198
1089. Wickens J (1988) Electrically coupled but chemically isolated synapses: dendritic spines and calcium in a rule for synaptic modification. Progr Neurobiol 31:507–528
1090. Wieckens J (1988) Electrically coupled but chemically isolated synapses: dendritic spines and calcium in a rule for synaptic modification. Progr Neurobiol 31:507–528
1091. Wieraszko A (1983) Uptake of D-[^{3}H] aspartic acid by hippocampal slices: the influence of low and high frequency activation of nerve endings. Brain Res 259:324–326
1092. Wigström H, Gustafsson B (1981) Increased excitability of hippocampal unmyelinated fibres following conditioning stimulation. Brain Res 229:507–513
1093. Wigström H, Gustafsson B (1983a) Heterosynaptic modulation of homosynaptic long-lasting potentiation in the hippocampal slice. Acta Physiol Scand 119:455–458
1094. Wigström H, Gustafsson B (1983b) Large long-lasting potentiation in the dentate gyrus in vitro during blockade of inhibition. Brain Res 275:153–158
1095. Wigström H, Gustafsson B (1985) On long-lasting potentiation in the hippocampus: a proposed mechanism for its dependence on coincident pre- and postsynaptic activity. Acta Physiol Scand 123:519–522
1096. Wigström H, Swann JW, Andersen P (1979) Calcium dependency of synaptic long-lasting potentiation in the hippocampal slice. Acta Physiol Scand 105:126–128
1097. Wigström H, Gustafsson B, Huang Y-Y (1986) Hippocampal long-term potentiation is induced by pairing single afferent volleys with intracellularly injected depolarizing current pulses. Acta Physiol Scand 126:317–319
1098. Williams S, Johnston D (1988) Muscarinic depression of long-term potentiation in CA3 hippocampal neurons. Science 242: 84–87

1099. Williams SH, Johnston D (1989) Hippocampal CA3 mossy fiber LTP is blocked by postsynaptic calcium chelators. Soc Neurosci Abstr 15:399

1100. Williams JH, Errington ML, Lynch MA, Bliss TVP (1989) Arachidonic acid induces a long-term activity-dependent enhancement of synaptic transmission in the hippocampus. Nature 341: 739–742

1101. Willis WD., Letbetter WD, Thompson WM (1968) A small system of neurons in the mammalian spinal cord. Brain Res 9:152–155

1102. Wilson DA, Racine RJ (1983) The postnatal development of postactivation in the rat neocortex. Dev Brain Res 7:271–276

1103. Wilson DF (1977) Estimates of quantal release and binomial statistical release parameters at rat neuromuscular junction. Amer J Physiol 233:157–163

1104. Wilson DF, Skirboll LR (1974) Basis for post-tetanic potentiation at the mammalian neuromuscular junction. Amer J Physiol 227:92–95

1105. Wilson R, Levy WB, Steward O (1979) Functional effects of lesion-induced plasticity: long-term potentiation in the normal and lesion-induced temporodentate circuits. Brain Res 176:65–78

1106. Wilson RC, Levy WB, Steward O (1981) Changes in translation of synaptic excitation to dentate granule cell discharge accompanying long-term potentiation. II An evaluation of mechanisms utilizing dentate gyrus dually innervated by surviving ipsilateral and sprouted crossed temporodentate inputs. J Neurophysiol 46:339–355

1107. Wise RA, Bozarth H (1984) Brain reward circuitry: four circuit elements "wired" in apparent series. Brain Res Bull 12:203–208

1108. Wisden W, Errington ML, Williams S, Dunnet SB, Waters C et al. (1990) Differential expression of immediate early genes in the hippocampus and spinal cord. Neuron 4:603–614

1109. Wojtowicz JM, Atwood HL (1985) Correlation of presynaptic and postsynaptic events during establishment of long-term facilitation at crayfish neuromuscular junction. J Neurophysiol 54:220–230

1110. Wojtowicz JM, Atwood HL (1986) Long-term facilitation alters transmitter releasing properties at the crayfish neuromuscular junction. J Neurophysiol 55:484–498

1111. Wojtowicz JM, Parnas I, Parnas H, Atwood HL (1988) Long-term facilitation of synaptic transmission demonstrated with macro-patch recording at the crayfish neuromuscular junction. Neurosci Lett 90:152–158

1112. Wong K, Redman S (1980) The recovery of a random variable from a noisy record with application to the study of fluctuations in synaptic potentials. J Neurosci Meth 2:389–409

1113. Wong RKS, Watkins DJ (1982) Cellular factors influencing GABA response in hippocampal pyramidal cells. J Neurophysiol 48: 938–951

1114. Wong RKS, Miles R, Traub RD (1984) Local circuit interactions in synchronization of cortical neurones. J Exp Biol 112:169–178

1115. Woodson PBJ, Schlapfer WT, Tremblay JP, Barondes SH (1976a) Resting and stimulated values of model parameters governing transmitter release at a synapse in *Aplysia californica*. Brain Res 109:21–40

1116. Woodson PBJ, Schlapfer WT, Tremblay JP, Barondes SH (1976b) Synaptic depression at a synapse in *Aplysia californica*: analysis in terms of a material flow model of neurotransmitter. Brain Res 109:41–59

1117. Woodson PBJ, Schlapfer WT, Barondes SH (1978) Amplitude and rate of decay of post-tetanic potentiation are controlled by different mechanisms. Brain Res 157:33–46

1118. Woody CD (1982) Memory, Learning, and Higher Function. Springer, Berlin Heidelberg New York

1119. Woody CD (1986) Understanding of the cellular basic of memory and learning. Ann Rev Psychol 37:433–493

1120. Woody CD, Black-Cleworth P (1973) Differences in excitability of cortical neurones as a function of motor projection in conditioned cats. J Neurophysiol 36:1104–1116
1121. Woody CD, Schwartz DE, Gruen E (1978) Effects of acetylcholine and cyclic GMP on input resistance of cortical neurons in awake cat. Brain Res 158:373–395
1122. Woody CD, Alkon DL, McGaugh JL, eds (1988) Cellular Mechanisms of Conditioning and Behavioural Plasticity. Plenum, New York
1123. Xie Z, Morishita W, Kam T, Maretic H, Sastry BR (1991) Studies on substances that induce long-term potentiation in guinea-pig hippocampal slices. Neuroscience 43:11–20
1124. Yaglom AM (1981) Korrelyatsionnaya Teoriya Statsionarnykh Sluchainykh Funktsii (The Correlation Theory of Stationary Random Functions). Gidrometeoizdat, Leningrad
1125. Yamamoto C (1972) Activation of hippocampal neurons by mossy fiber stimulation in thin brain section in vitro. Exp Brain Res 14:423–435
1126. Yamamoto C (1982) Quantal analysis of excitatory postsynaptic potentials induced in hippocampal neurons by activation of granule cells. Exp Brain Res 46:170–176
1127. Yamamoto C, Chujo T (1978) Long-term potentiation in thin hippocampal section studied by intracellular and extracellular recordings. Exp Neurol 58:242–250
1128. Yamamoto C, Matsumoto K, Takagi M (1980) Potentiation of excitatory postsynaptic potentials during and after repetitive stimulation in thin hippocampal sections. Exp Brain Res 38:469–477
1129. Yamamoto C, Higashima M, Sawada S (1987) Quantal analysis of potentiating action of phorbol ester on synaptic transmission in the hippocampus. Neurosci Res 5:28–38
1130. Yang XD, Korn H, Faber DS (1990) Long-term potentiation of electrical coupling at mixed synapses. Nature 348:542–545
1131. Yip JW, Dennis MJ (1976) Suppression of transmission at foreign synapses in adult new muscle involves reduction in quantal content. Nature 260:350–352
1132. Zalutsky RA, Nicoll RA (1990) Comparison of two forms of long-term potentiation in single hippocampal neurons. Science 248:1619–1624
1133. Zefirov AL (1981) Analysis of frequency facilitation in neuromuscular synapses of the frog sartorius muscle. Bull Exp Biol Med 92:389–392
1134. Zefirov AL (1983) Secretion of the transmitter from proximal and distal sites of the neural ending at the frog sartorius muscle. Neurophysiology (Kiev) 15:362–369
1135. Zefirov AL (1985) Location and functioning of transmitter release sites in frog neuromuscular junction. Neurophysiology (Kiev) 17:152–160
1136. Zefirov AL, Khalilov (1985) Peculiarities of electrical activity in various parts of the neural terminal in the frog. Bull Exp Biol Med 99:7–10
1137. Zefirov AL, Stolov EL (1982) A model of the transmitter secretion in the neuromuscular synapse based on the nonuniformity of the release probability of acetylcholine quantum. Neurophysiology (Kiev) 14:233–240
1138. Zefirov AL, Stolov EL (1984) Analysis of factors limiting the application of the binomial statistics to the study of transmitter release in neuromuscular synapse. Neurophysiology (Kiev) 16:182–189
1139. Zefirov AL, Kashapova LA, Moshkov DA, Bezgina EN, Khalilov IA (1986) Electrophysiological and ultrastructural study of the topography of active zones in motor nerve terminal of the frog. Doklady Biol Sci 290:1277–1280
1140. Zemskova SN, Eder K, Giniatullin PA, Zefirov AL, Khazipov PN (1991) Early postdenervation changes on properies of postsynaptic membrane of amphibian muscle fibers (in Russian). Fiziol Zh SSSR 77:57–65
1141. Zimmermann H (1979) Vesicle recycling and transmitter release. Neuroscience 4:1773–1804

1142. Zola-Morgan S, Squire LR, Amaral DG (1986) Human amnesia and the medial temporal region: memory impairment following a bilateral lesion limited to field CA1 of the hippocampus. J Neurosci 6:2950–2967
1143. Zorychta E, Esplin DW, Capek R (1975) Action of halothane on transmitter release in the spinal monosynaptic pathway. Fed Proc Amer Soc Exp Biol 34:2999
1144. Zucker RS (1972) Crayfish escape behavior and central synapses. II Physiological mechanisms underlying behavioral habituation. J Neurophysiol 35:621–637
1145. Zucker RS (1973) Changes in the statistics of transmitter release during facilitation. J Physiol (Lond) 229:787–810
1146. Zucker R, Bruner J (1977) Long-lasting depression and the depletion hypothesis at crayfish neuromuscular junctions. J Comp Physiol A121:223–240

Subject Index

MIX
Papier aus verantwortungsvollen Quellen
Paper from responsible sources
FSC® C105338

If you have any concerns about our products,
you can contact us on
ProductSafety@springernature.com

In case Publisher is established outside the EU,
the EU authorized representative is:
Springer Nature Customer Service Center GmbH
Europaplatz 3, 69115 Heidelberg, Germany

Printed by Libri Plureos GmbH
in Hamburg, Germany